Tropical Forest Insect Pests

Forest entomology is rich in theory, but much of this is based on observations of temperate forest insects. This comprehensive volume, by a leading researcher in tropical forest entomology, aims to promote a more global theoretical understanding of pest population dynamics and the causes of forest insect outbreaks.

Covering pests of both natural forests and plantations, the book examines the diversity of tropical forest insects; their ecological functions; the concept of pests and the incidence of pests in natural forests, plantations and stored timber. It explores the circumstances under which insect populations increase and acquire pest status. General issues on which foresters and forest entomologists hold strong traditional views, such as the severity of pest incidence in plantations vs. natural forests, in plantations of exotics vs. indigenous tree species and in monocultures vs. mixed plantations are discussed. The final chapter looks in detail at specific insect pest problems of the common plantation tree species across the tropics, and provides recommendations for control.

Containing a wealth of information about tropical forest insects, this book will be valuable for graduate and postgraduate students of forestry, research scientists interested in tropical forest entomology and forest plantation managers in the tropics.

Dr K. S. S. NAIR obtained his PhD in Zoology, specializing in Entomology, from the M. S. University of Baroda in 1964. From 1976 to 1994 he headed the Entomology Division of the Kerala Forest Research Institute in India, and carried out pioneering research on the management of tropical forest insect pests. He was later made Director of the Institute and guided research on various aspects of tropical forestry. Since 1999 he has served as an Editor of the journal Entomon and President of the Association for Advancement of Entomology. Earlier, he also served as Chairman of the IUFRO (International Union of Forestry Research Organizations) Working Party on 'Protection of Forest in the Tropics', for nine years. Dr Nair is now retired and lives in Kerala, India.

Tropical Forest Insect Pests

Ecology, Impact, and Management

K. S. S. Nair

Formerly Head,
Division of Entomology and Director
Kerala Forest Research Institute
Peechi, Kerala, India

CAMBRIDGE
UNIVERSITY PRESS

CAMBRIDGE UNIVERSITY PRESS
Cambridge, New York, Melbourne, Madrid, Cape Town, Singapore, São Paulo

Cambridge University Press
The Edinburgh Building, Cambridge CB2 8RU, UK

Published in the United States of America by Cambridge University Press, New York

www.cambridge.org
Information on this title: www.cambridge.org/

First published 2007

Printed in the United Kingdom at the University Press, Cambridge

A catalogue record for this publication is available from the British Library

Library of Congress Cataloguing in Publication data

Nair K. S. S.
Tropical forest inspect pests: ecology, impact, and management / K. S. S. Nair.
p. cm.
Includes bibliographical references.
ISBN-13: 978-0-521-87332-1
ISBN-10: 0-521-87332-0
1. Forest insects--Tropics. 2. Forests insects--control--Tropics. I. Title
SB764.T73N35 2007
634.9′670913--dc22

2006100254

ISBN-13 978-0-521-87332-1 hardback

Dedication

Dedicated to:
Late Professor J. C. George, who introduced me to research
and
Dr P. M. Ganapathy, who introduced me to forestry

Contents

Illustrations

Preface

This book has grown out of my feeling that tropical forest insects have not received the research attention they deserve. Most books on forest entomology deal with only temperate forest insects and those few that deal with tropical forest insects cover only small regions of the tropics and mostly contain descriptions of pest biology. An exception is a recent book by M.R. Speight and F.R. Wylie (2001) which covers the entire tropics and lays stress on pest management, although their coverage of the subject is very general. Other books on tropical forest entomology covering parts of the tropics are mentioned in the introduction to Chapter 2: particular mention must be made of C. F. C. Beeson's (1941) excellent treatise on the ecology and control of forest insects of India and the neighbouring countries. This book, published some 65 years ago, contains much information that is valid and relevant even today, although it is not accessible to many. Extensive new knowledge has now accumulated on tropical forest insects across the world, but it lies scattered in innumerable journal articles and reports. I have made an attempt in this book to bring this knowledge together and present it in an ecological framework. Knowledge is seldom created by one individual and I have used the knowledge accumulated over time by the dedicated work of innumerable researchers. What is new here is a new framework on which the accumulated knowledge is organized to convey some central ideas relevant to the management of tropical forest insect pests. Facts or observations make sense only when arranged logically and interpreted. My attempt has been to provide an overview of tropical forest insect pests and discuss the basic principles of their ecology in the forest environment, using information about commonly encountered insects across the tropics.

Forest entomology is rich in theory. Much of this is based on observations on temperate forest insects. These theories, particularly, those on population dynamics, have not been static. For example, new theoretical alternatives to the conventional equilibrium viewpoint of population regulation have emerged in recent years. It is an open question whether the study of tropical forest insects might lead to modification of some of the existing theories, or reinforce them. Tremendous opportunities exist for using long-term observational and experimental data from tropical forests to test theories on insect population regulation. This is because the warm temperatures of the tropics permit year-round growth and multiplication of insects. While many temperate forest insects pass through only one generation per year, many tropical forest insects pass through 12–14 generations in the same period. Therefore testing theories should be easier in the tropical forests. Wider dissemination of knowledge about the tropical forest insects and the research opportunities they offer will promote collaborative work among scientists from developed and developing countries, for the benefit of both and the science of entomology in general. This thought has been one of my main motivations for embarking on this work.

The book is organized into 10 chapters. Chapter 1 gives an overview of the broad features of the tropical forests and their management. This is followed by an overview of tropical forest insects in which their structural and functional diversity and the concept of pests are discussed. Chapter 3 then discusses the several ecological functions the insects perform in the forest ecosystem, and how they influence plant succession. Against the background of these three chapters, the next three describe pest incidence in natural forests (Chapter 4), plantations (Chapter 5) and stored timber (Chapter 6). Characteristics of pest incidence in the three situations are described with examples (except for plantations, where the details are reserved for the last chapter) and generalizations drawn. Pest problems arise when insect numbers increase beyond a certain limit. Therefore, Chapter 7 examines the circumstances under which insect populations increase and how their numbers are regulated in nature.

In Chapter 8, some general issues on which foresters and forest entomologists hold strong traditional views are discussed critically in the light of available evidence. These include the severity of pest incidence in plantations vs. natural forests, in plantations of exotics vs. indigenous tree species and in monocultures vs. mixed plantations. With this background, Chapter 9 examines the pest management options, current practices and constraints in the tropical forestry setting, and suggests guidelines for practice. The last chapter, which occupies nearly half of the book, is devoted to detailed case studies of pest problems of the most common plantation tree species across the tropics. For each of the selected tree species, a tree profile is given which is followed by an overview of pests and detailed pest profiles of the major pests, including control options and knowledge gaps. This chapter contains the core of the data on which the generalizations made in the other chapters rest. But for the bulk, the information in this chapter should have been incorporated into Chapter 5. But placing it as a separate chapter at the end of the book will facilitate easy reference by practising foresters, planters and researchers who may want specific information on pest problems of particular tree species or details regarding specific pests.

Some observations on the general features of the book seem desirable here. Several changes have occurred recently in the scientific nomenclature of tree species, and the plant families in which they are placed. For example, the tree which was known as *Paraserianthes falcataria* until recently is now *Falcataria moluccana* and the teak tree which has been traditionally placed in the family Verbenaceae is now in the family Lamiaceae. Although these changes are not necessarily accepted by all, some standard is necessary. I have used the Forestry Compendium (2005, CD version) published by the Commonwealth Agricultural Bureau International as the standard for this purpose. Synonyms are given, both for plants and insects, when they are common in recent literature. On countries of occurrence of pests, only known information can be given; updating is necessary in many cases.

I have used the example of the teak defoliator *Hyblaea puera* at several places in the book, in several contexts, to illustrate some points. Also, the pest profile on this species is the longest. This is partly due to the knowledge available and partly to my personal familiarity with the insect. I hope the reader will bear with me for this indulgence.

This book is primarily intended for graduate and postgraduate students in forestry, and research students and research scientists interested in tropical forest entomology. Since its major focus is the researcher, I have included references to published scientific papers to substantiate the statements, at the cost of increasing the work's bulk, although many text books omit these while summarising the knowledge. Unfortunately, published literature is generally taken as truth, which need not always be the case. By including the references, I wish to encourage researchers to be critical and read the original article wherever possible, to understand the conditions under which the reported results were obtained. To stimulate further research, I have included comments on knowledge gaps under each pest profile.

A large part of the knowledge assembled in this book, from the field as well as from literature, was gathered during my career as a research scientist at the Kerala Forest Research Institute, Peechi, India, and I am indebted to the Institute, particularly to its former director, Dr P. M. Ganapathy, for creating an excellent work environment. Gathering of information was also facilitated by a short research assignment at the Center for International Forestry Research, Bogor, Indonesia, and from my association with the International Union of

Forestry Research Organizations (IUFRO) Working Group on 'Protection of Forest in the Tropics', founded by Dr Heinrich Schmutzenhofer. I am deeply indebted to my former entomologist colleagues at the Kerala Forest Research Institute – Dr R. V. Varma, Dr George Mathew, Dr V. V. Sudheendrakumar and Dr T. V. Sajeev – for help in various forms, including supply of photographs or specimens for photographing, reading and making suggestions on parts of the manuscript, providing literature and, above all, encouraging me to undertake this work. The draft of the book was prepared at the Department of Zoology, University of Kerala, Trivandrum, India, where Professor Oommen V. Oommen, Professor D. Muraleedharan and Dr Mariamma Jacob extended various kinds of help and made it pleasant to work. I thank Professor Alan Berryman, Dr Ronald F. Billings, Professor T. N. Ananthakrishnan, Professor A. Mohandas, Professor T. C. Narendran and Dr P. T. Cherian, who read parts of the manuscript and made helpful suggestions. Thanks are also due to Mr Sajan Bhaskaran who made the diagrams and Mr A. M. Shanmugam who processed some of the illustrations. A few of the illustrations were reproduced from other publications with the permission of the publishers, for which I am thankful to them; the sources are acknowledged in the respective legend. Some photographs used in the book were kindly provided by colleagues who are also acknowledged in the respective legend; others were taken by me at various places and times over the years, except a few taken by Dr T. V. Sajeev. This work would not have been possible but for the unstinted support rendered by my wife, Mrs Sathi Nair, in many different ways, including the long, lonely hours spent by her while I was engrossed in the work, particularly at the final stages of preparation of the document.

Last, but most important, the writing of this book was catalysed and supported by the Department of Science and Technology, Government of India, under its Utilisation of Scientific Expertise of Retired Scientists Scheme.

K. S. S. Nair
June 2006

Foreword

This book forms a comprehensive and thoroughly up-to-date text on tropical forest entomology written by an author who has spent his entire career working and living in the tropics. It is both a broad treatment of the principles and practice of tropical forest entomology, and a detailed and penetrating exploration of specific insect pests and the methods used to manage them. What is most significant about this work is its organization of an enormous body of information on tropical insect pests within a general theoretical framework. This is particularly important to students of forest protection, who need to understand the theory of population dynamics and pest outbreaks before they can intelligently manage insect pests.

Dr K. S. S. Nair is eminently qualified to write such a book. He has served as head of the Entomology Division of the Kerala Forest Research Institute in India for some 18 years, and as its director for a further five years, and has also worked at the Centre for International Forestry Research in Indonesia. He has been an active member of the International Union of Forestry Research Organizations (IUFRO) Subject Group 'Entomology' for many years, and has served as chairman of the Working Party on 'Protection of Forest in the Tropics' for eight years, and as deputy coordinator of the subject group 'Forest Health' for nine years. This has given him a broad experience in international forest entomology, both in tropical and temperate forests.

This book will be invaluable to teachers, researchers and forest protection specialists in the tropics. I expect it to become the major textbook in tropical forest entomology as well as an important reference for those involved in research and management of tropical forest pests. It should also bring tropical forest entomology to the attention of a broader audience and, as the author hopes, stimulate collaborative research between scientists in the developed and developing countries. Forest entomology evolved as a science in the Northern Hemisphere. Nair's book will help to correct this bias and thereby lead to a more global theoretical understanding of pest population dynamics and the causes of forest pest outbreaks.

On a personal note, I remember with pleasure my visit to Kerala in 1986 and, in particular, my walks in the teak plantations with K. S. S. where we contemplated the ways of that mysterious teak defoliator, *Hyblaea puera*. We once stumbled upon an aggregation of moths resting in the undergrowth of a natural forest. When the small shrubs were disturbed they emitted clouds of moths identified as *Hyblaea* by their orange wing-flashes. This discovery helped us to understand the sudden appearance of concentrated, single-aged populations of larvae that completely defoliate stands of teak trees, and made us think of this insect more like a locust than a moth.

Alan Berryman
Emeritus Professor,
Department of Entomology and Natural Resource Sciences,
Washington State University,
Pullman, Washington, USA
June 20, 2006

1

Forestry in the tropics

1.1 Introduction

Tropical forests have always attracted the world's attention because of their magnificence and potential for economic exploitation. For centuries, they catered to the people's livelihood needs for timber, fruits, firewood, medicinal plants etc., and also, indirectly, animal meat. The native people lived in harmony with the forest as their populations were small and their demands did not exceed the forest's capacity to regenerate. The situation changed drastically in the colonial era between the mid seventeenth and mid twentieth centuries. During this period, large areas of tropical forests were cleared for human settlement and large-scale cultivation of agricultural and estate crops like sugar cane, tea, coffee, rubber and wattle. Forests were also logged for selective extraction of valuable timbers such as teak and rosewood in Asia, mahogany in Latin America and khaya in Africa, mainly for export. By the mid eighteenth century, forest plantation technology had developed and the natural forests were increasingly replaced by plantations. After the Second World War, forest plantation programmes received a further boost in the newly independent nations due to international exchange of information and availability of international development funds (Evans, 1992). Exotic, fast-growing eucalypts and pines were raised in the tropics on a large scale during this period. As industrialization progressed, more extensive plantations were established, mostly with exotic fast-growing species, and on land cleared of natural forests. Most of these were intended to produce pulpwood for paper, rayon and fibreboard. As a result, vast stretches of natural tropical forests across the continents were destroyed or degraded.

While tropical deforestation was thus progressing steadily, the environmental value of tropical forests was also being slowly recognized worldwide, particularly after the 1960s. The role of natural forests in maintaining the climate, soil, hydrological regime, biodiversity, the global carbon balance and the overall security of the local people's livelihood was recognized. Campaigns against the indiscriminate destruction of tropical forests gathered momentum. A large number of local and international voluntary organizations were established to push the cause of conservation, with particular emphasis on tropical forests. Although many of them had a negative agenda, opposing all sorts of developmental activities, their dramatic and emotional campaigns helped to create wide public awareness of the ecological importance of tropical forests. As a result, national governments and international bodies such as UN agencies took initiative in conservation action. For example, roughly 10% of the world's tropical forests are now set aside as national parks or undisturbed reserves. In some places like the hilly State of Kerala in India, for instance, as much as 25% of about one million ha of forest has been designated as wildlife sanctuaries and national parks. In spite of this awareness, deforestation in the tropics continues, albeit at a slower pace, driven by the profit motive of pulpwood industries and the gullibility of the governments of economically stressed tropical countries. As Whitmore (1998) observed, 'logging proceeds as fast as ever and moves on to fresh countries'. The tropical forests of South and Southeast Asia have been heavily depleted and the timber lobby is now focussing on Latin America.

Although the progress of deforestation was concomitant with the growth of the human population, and some of it was essential to ensure civilization, recent decades have witnessed an unprecedented destruction of tropical forests with the growth of the pulp and paper industry. Can we continue to destroy the tropical forests at the current rate of 17 million ha annually, and degrade much of the remaining area, without endangering our own future survival? Can we manage the remaining tropical forest, or at least a reasonable chunk of it, in a sustainable manner so that we will continue to be sustained by it? What is unique about tropical forests? It is beyond the scope of this book to answer all these questions, on which there is a vast literature. For details on the state of the world's forests the reader is referred to the periodic reports of the Food and Agriculture Organization of the United Nations (FAO) (2005), which are updated every two years. Whitmore (1998) gives a balanced account of the tropical rain forests and discusses the key tropical forestry issues. What is attempted here is a brief overview of the broad features of tropical forests to facilitate an appreciation of the role and importance of the forest insects.

1.2 The tropics

Although the tropics can be easily defined as the geographical area lying between the tropic of Cancer and the tropic of Capricorn (latitudes 23° 27' north and south, respectively, of the equator), it is not possible to discuss tropical forestry exclusively within these geographical limits. For one thing, the distribution of many tropical forest tree species does not coincide with the limits of the tropics; it often extends beyond. For example, the natural distribution of *Eucalyptus tereticornis* extends from 9° S to 38° S, and that of the dipterocarp *Shorea robusta*, from 18° N to 32° N, covering both tropical and subtropical zones. Teak (*Tectona grandis*) has a natural distribution mostly confined to the tropics (25° N to 9° N), but is also planted widely in the subtropics (e.g. Bhutan, Japan, Korea, Nepal, Pakistan, Turkey). Even the tropical rain forest, the most characteristic forest formation of the tropics, extends, for example, into southern China at 26° N, with ill-defined change into subtropical rain forest (Whitmore, 1998). Secondly, most information related to forestry is available according to country, and countries known as tropical countries do not fall neatly within the tropics either. According to the FAO definition, if more than 50% of the area of a country falls within the tropics, it is designated as a tropical country. Thus India, situated between latitudes 8° 4' N and 37° 6' N, is a tropical country but has a substantial area outside the tropics. And a non-tropical country such as China or Taiwan has areas that fall within the tropics. Thirdly, even within the tropics, temperate conditions are obtained on high mountains. For example, Honduras in Central America lies between latitudes 13° N and 16° N and the climate is tropical, but most forests lie in the cooler highlands (plateaus) where the mean annual temperature is about 21.1 °C. These forests are dominated by oak and pine, while the coastal regions are warmer, with a mean annual temperature of 26.7 °C (Simon and Schuster, 1999). Because of these overlaps, in general, the term 'tropics' is loosely used. The regions lying not only between but also *near* the tropics of Cancer and Capricorn are usually included under the tropics. In this book, we will use the term in a similar broad sense. This will allow us to deal with insects associated with predominantly tropical trees even when these trees' natural or planted distribution extends into the subtropical zone. In fact, strict distinction into tropics based on the imaginary latitudinal lines is artificial, because the tropics merge imperceptibly into the subtropics, often termed 'warm temperate'.

What distinguishes the tropics more easily from other parts of the world is the consistently warm atmospheric temperature, with no drastic difference between seasons and all months without frost. Climatically, the tropical zone is characterised by annual and monthly average temperatures above 20 °C and a difference of not more than 5 °C between the mean monthly temperatures of the warmest and coolest months. This permits biological activity almost throughout the year except where seasonal drought limits the activity.

The tropics encompass many continents – parts of Asia, Australia, Africa, North America and South America, and several islands in the Pacific, Atlantic and Indian oceans. Conventionally, the tropical countries are grouped under three major regions, that is Asia-Pacific, Africa and Latin America. The countries that fall within the tropics are listed in Table 1.1. Together, these tropical countries cover a substantial portion of the earth's land surface, nearly 37%, comprising about 4800 million ha.

Table 1.1. *List of tropical countries/areas*[a]

Africa	Asia-Pacific	Latin America
Angola	Bangladesh	***Central America***
Benin	Brunei Darussalam	Belize
Botswana	Cambodia	Costa Rica
Burkina Faso	Fiji	El Salvador
Burundi	India	Guatemala
Cameroon	Indonesia	Honduras
Central African Rep.	Lao People's Dem. Rep.	Mexico
Chad	Malaysia	Nicaragua
Comoros	Myanmar	Panama
Congo	New Caledonia	
Côte d'Ivoire	Papua New Guinea	***The Caribbean***
Dem. Rep. Congo (Zaire)	Philippines	Antigua and Barbuda
Equatorial Guinea	Samoa	Bahamas
Ethiopia	Singapore	Barbados
Gabon	Solomon Islands	Cuba
Gambia	Sri Lanka	Dominica
Ghana	Thailand	Dominican Republic
Guinea	Vanuatu	Grenada
Guinea-Bissau	Vietnam	Guadeloupe
Kenya		Haiti
Liberia		Jamaica
Madagascar		Martinique
Malawi		Puerto Rico
Mali		St. Kitts and Nevis
Mauritania		St. Lucia
Mauritius		St. Vincent and the Grenadines
Mozambique		Trinidad and Tobago
Namibia		

Table 1.1. (cont.)

Africa	Asia-Pacific	Latin America
Niger		***South America***
Nigeria		Bolivia
Réunion		Brazil
Rwanda		Colombia
Senegal		Ecuador
Seychelles		French Guinea
Sierra Leone		Guyana
Somalia		Paraguay
Sudan		Peru
Togo		Suriname
Uganda		Venezuela
United Rep. Tanzania		
Zambia		
Zimbabwe		

[a]Countries more than 50% of whose area falls between the tropic of Cancer and tropic of Capricorn. It must be noted that smaller parts of other countries also fall within the tropics.

1.3 The tropical forests

As per recent estimates (FAO, 2001b), about 47% of the world's total forests (1818 million ha out of 3870 million ha) lie in the tropics and 8% (323 million ha), in the subtropics, together making up 55% of the total. Of the tropical forests, the largest portion is in Latin America (52%), centred on the Amazon river basin; followed by Africa (28%), centred on the Congo river basin; and the rest in the Asia-Pacific (19%), where it is more scattered (Fig. 1.1).

1.3.1 *Characteristics of tropical forests*

In general, tropical forests are characterised by high species diversity, in comparison to temperate and boreal forests. The composition varies considerably across the tropics, mainly depending on the temperature and moisture regimes, the soil and the geological history. The richest in species are those in Latin America, followed by those in Asia-Pacific and Africa. Between these three regions, there is very little similarity in the tree species present, although there are some common genera and similarities at family level. Some plant families are unique to certain regions. For example, Dipterocarpaceae, an important family of timber-yielding trees, is characteristic of Asia-Pacific where they are particularly abundant in Southeast Asia (Indonesia, Malaysia, Philippines). Similarly, Australian forests (excluding rain forests) are dominated by the tree

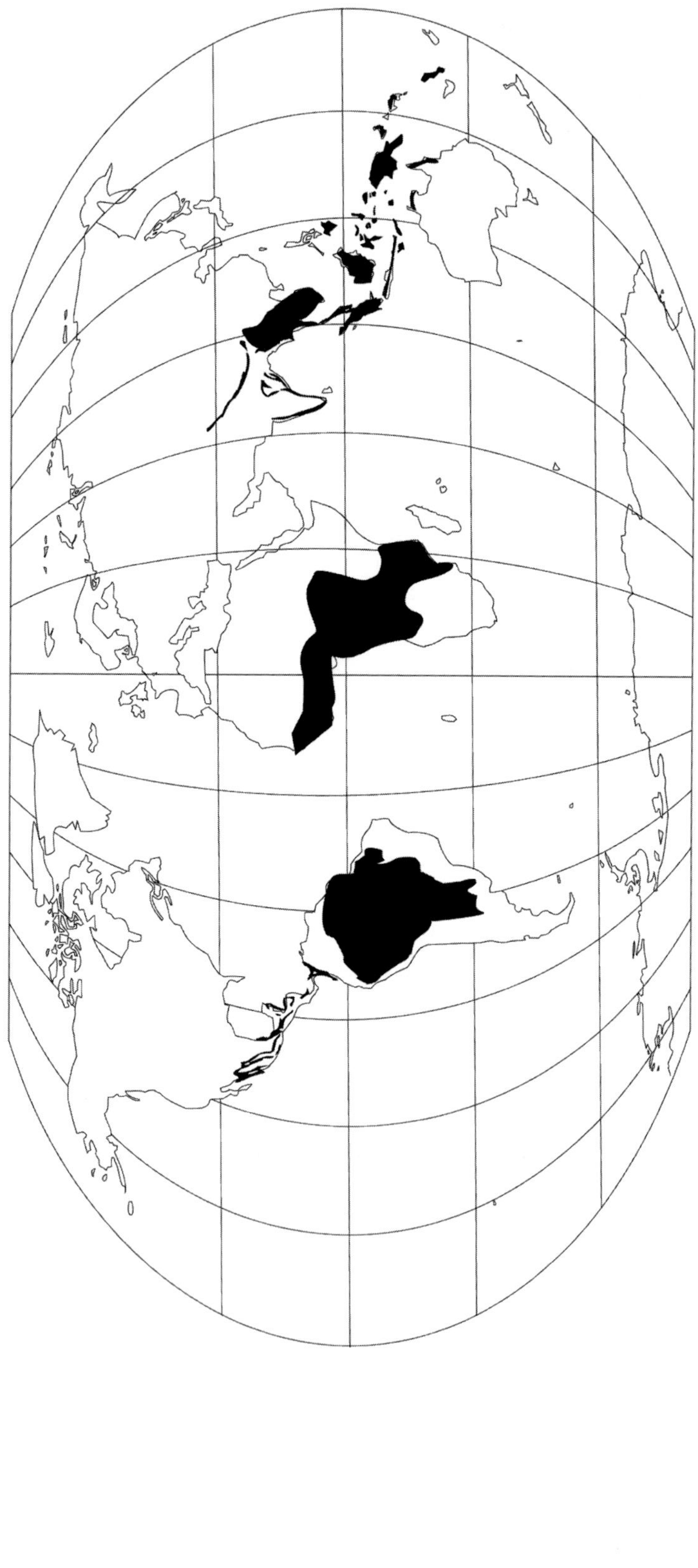

Fig. 1.1 Distribution of the world's tropical forests. The figure gives only a rough indication, as the scale does not permit depiction of small forest areas and the density of the forest cover is not represented. Based on FAO (2001b).

genus *Eucalyptus*. Africa is generally poor in flora. For instance, there are only four species of bamboos in mainland Africa (Whitmore, 1998), compared to 87 species in India alone within Asia-Pacific and about 1250 species worldwide. Africa is also characterised by the presence of savannas, plant communities dominated by grasses which may also contain scattered populations of trees that do not form a closed canopy. There are also differences within each region. For instance, within the Asia-Pacific, the natural distribution of teak is confined to two disconnected patches, one in peninsular India and the other covering most of Myanmar, northern Thailand and a small part of northwest Laos.

In spite of these differences between and within the tropical regions, in comparison to temperate forests, there are some broad features that can be considered as characteristic of tropical forests. These include high species diversity, year-round growth, existence of crown tiers, presence of lianas and understorey palms, development of buttresses on tree trunks and cauliflory (trunk-borne flowers). The major characteristic traits are discussed briefly below.

Species diversity

The diversity of life forms present in tropical forests, both of plants and animals, is staggering, and has not yet been fully scientifically documented. The number of species in a small spatial unit (generally one hectare or less) is usually referred to as 'alpha' diversity. It represents diversity within the community or local diversity, as compared to diversity among different communities on a larger spatial scale, referred to as 'gamma' diversity. The alpha diversity for tree species in tropical forests, particularly the tropical rain forests, is high. Typically, between 120 and 180 tree species (stems 10 cm or above in diameter) are present per ha in most tropical rain forest sites in the Far East (Whitmore, 1998). On the higher side, 307 tree species per ha were recorded at Cuyabeno in the western Amazon in Ecuador, while on the lower side, in Nigeria there may be only 23 per ha. Species numbers rise with increasing plot area. For example, 830 tree species were recorded in a 50-ha plot at Pasoh, Malaysia. Such high species diversity is in striking contrast to what is observed at higher latitudes.

In general, tree species diversity falls with increasing latitude (Fig. 1.2). Whitmore (1998) notes that the whole of Europe, north of the Alps and west of Russia has only 50 indigenous tree species and Eastern North America has only 171. Similarly Finland, a country with more than two-thirds of its land area under forest cover, lying between latitudes 60°N and 70°N in the boreal forest zone, has only 23 natural tree species in about 20 million ha of forest area, with Scots pine (*Pinus sylvestris*), Norway spruce (*Picea abies*) and birch (*Betula pendula* and *B. pubescens*) accounting for 97% of the forest's growing stock (Hakkila, 1995).

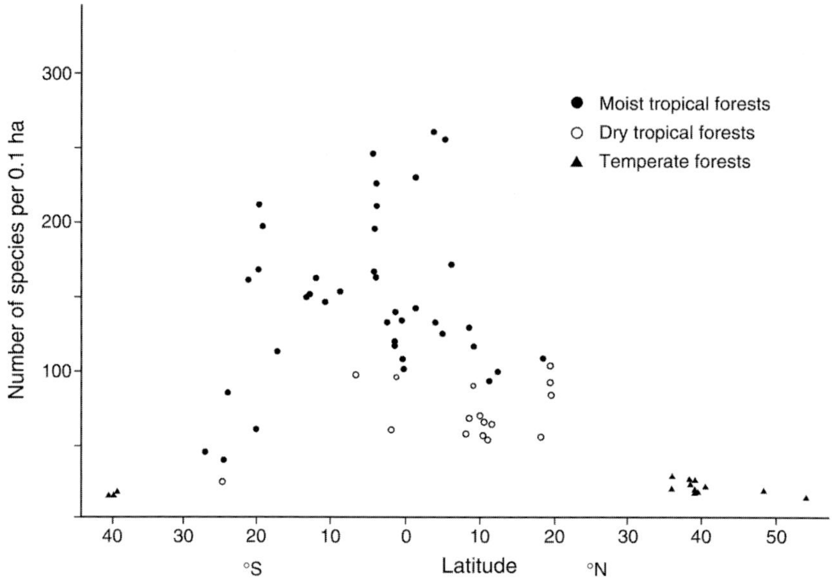

Fig. 1.2 Relationship between number of tree species and latitude. The number of tree species per standardized 0.1 ha plots at various latitudes is shown. Note that the species richness falls with increasing latitude on both sides of the equator. Reproduced, with slight modification, from *Annals of the Missouri Botanical Garden* (Gentry, 1988).

By comparison, the Kerala State in India, lying between latitudes 8°N and 13°N in the tropical zone, has 740 native tree species (Sasidharan, 1997) in about one million ha of forest. Gentry (1988) noted that in standardised 0.1 ha sample plots, temperate forests generally have 15–25 species, tropical dry forests 50–60 species and moist and wet tropical forests an average of about 150 species.

Some plant families like Myristicaceae are distributed only in the humid tropical climates while some others like Annonaceae, Musaceae and Ebenaceae are mostly concentrated there, with a few temperate outliers (Whitmore, 1998). The high species diversity of tropical forests is attributable mainly to environmental stability and possibly higher levels of speciation due to year-long biological activity. At higher latitudes, trees had to face great climatic variations during recurrent ice ages in the past and only a limited number of species were able to survive under such harsh conditions or recolonize from warmer areas.

Because of high species diversity, the number of individual stems of a species present per unit area (i.e. species density) is usually low. Often, the most abundant species do not make up more than 2.5% of all stems (He, Legendre and LaFrankie, cited by Kellman and Tackaberry, 1997) and many species are present at a density of less than one tree per ha. For instance, a density of one mature tree per ha represents the upper limit of density for mahogany (*Swietenia macrophylla*) in Brazil although its density may vary widely between regions. Thus, one to two mature trees of *S. macrophylla* per ha have been recorded in Mexico, four to eight trees per ha in Venezuela and 20–60 in Bolivia (Mayhew and Newton, 1998). There are exceptions to the general trend of low species density of tropical forests. As we proceed along the moisture gradient from wet evergreen to dry deciduous tropical forests and along the temperature gradient from lower to higher latitudes, some species become more abundant and in some cases lead to monoculture-like stands, as in higher latitudes. For example, teak (*Tectona grandis*) may constitute 10% to nearly 100% of the tree species present in the moist to dry deciduous forests in different parts of India. Similarly, sal (*Shorea robusta*) often occurs in high density stands in central and northern India. Many other dipterocarp species also occur in high densities in lowland evergreen forests of Southeast Asia. Monocultures tend to develop when competing species are eliminated mainly by climatic factors. For example, the northern limit of natural teak in India is 25° N latitude, beyond which sal takes over, because teak seedlings, unlike sal, cannot survive frost. In Finland, pure stands of spruce develop in areas prone to harsh winters. In winters with heavy snowfall, the load of ice and snow on trees can be as much as 100–150 kg/m of stem. Spruce withstands the load, but pine and birch are easily broken (Hannelius and Kuusela, 1995). Snow thus promotes the development of pure spruce stands. Tree species diversity and density have implications for pest outbreaks, as will be discussed later.

Forest structure

In tropical forests, tree growth is luxuriant and the stand is usually dense. The stem density (trees 10 cm or above in diameter) has been estimated at 497.4 ± 135.0 per ha for tropical lowland evergreen forests (Meave and Kellman, cited by Kellman and Tackaberry, 1997). Woody lianas are common and a few monocots such as canes and reed bamboos are sometimes present. Trees often harbour ferns, aroids, orchids, mosses and lichens, their presence and density varying with the moisture regime. At least three crown layers can often be distinguished. The chief, middle layer, may be between 15 and 35 m above ground, depending on the forest subtype. The top layer is formed by dominant species whose crowns may reach up to 40–45 m. Usually these trees are buttressed at the base, have unbranched, cylindrical boles and possess an umbrella-shaped top crown. The bottom layer consists of shade-tolerant trees, less than 15 m in height. A 'profile diagram' is generally used to depict the vertical layering of the trees in tropical forests (Fig. 1.3). Stratification of

Fig. 1.3 Profile diagram of a tropical evergreen forest at Pothumala in Nelliampathy Forest Range, Kerala, India. Courtesy: U. M. Chandrashekara, Kerala Forest Research Institute.

AcM, *Actinodaphne malabarica*; AM, *Antidesma manasu*; CM, *Cinnamomum malabatrum*; CE, *Cullenia exarillata*; DL, *Dimocarpus longan*; DM, *Dysoxylum malabaricum*; DW, *Drypetes wightii*; GA, *Garcinia morella*; MN, *Mesua nagassarium*; MP, *Meiogyne pannosa*; PC, *Polyalthia coffeoides*; PE, *Palaquium ellipticum* tropical forest canopies into distinct layers is an abstraction and simplification of a complex structure that is in a dynamic state due to growth of the forest stand.

Growth dynamics

In contrast to the temperate climates where winter imposes an annual break in the growth of trees, in the tropics, growth is possible throughout the year, provided moisture is not limiting. In tropical forests, seasonality in growth is largely imposed by moisture availability. In general, the primary productivity of tropical forests is higher than in temperate forests due to the longer growth period, although part of this productivity is lost due to greater respiratory loss in the higher temperatures.

Although a climax tropical forest may appear uniform, on a larger spatial scale it is a mosaic with smaller parts in a continual flux. As Whitmore (1984) wrote, trees are mortal and eventually die of old age. This creates gaps in the forest canopy, initiating a forest growth cycle. Gaps of various sizes can also be created by other causes as when trees are killed by lightning strikes, fire, pest outbreaks, blown over by wind or swept aside by landslides. When the gap is small, pre-existing seedlings and saplings of shade-tolerant species (climax species) grow up to fill the gap and eventually become mature trees. When the gap is large, seeds of fast-growing, light-demanding species (pioneer species) germinate and colonize the gap, and as the short-lived pioneer species mature and decline, climax species take over. Three arbitrary phases, gap phase, building phase and mature phase, have been recognized in this growth cycle, and any stretch of forest is a mosaic of small and large patches in each of these three phases (Fig. 1.4). In drier tropical forests, fire often initiates the process of regeneration over larger areas and the stand dynamics may differ in detail from the gap phase dynamics.

1.3.2 *Types of tropical forests*

The structure and floristic composition of tropical forests varies from place to place according to latitude, altitude, amount and pattern of rainfall, nature of the underlying soil, past geologic and climatic history etc., and in transitional zones one kind of forest merges imperceptibly into another. Therefore classification of tropical forests into neat subtypes has been difficult and various schemes of classification have come into vogue in different parts of the world, causing some confusion. Forest stands that differ in overall vegetation structure (height of trees, crown tiers, presence of climbers and epiphytes), physiognomy (individual tree characteristics such as crown shape, presence of buttress, leaf shedding habit etc.) and floristic composition are often called **forest formations**. Examples of forest formations are lowland evergreen rain

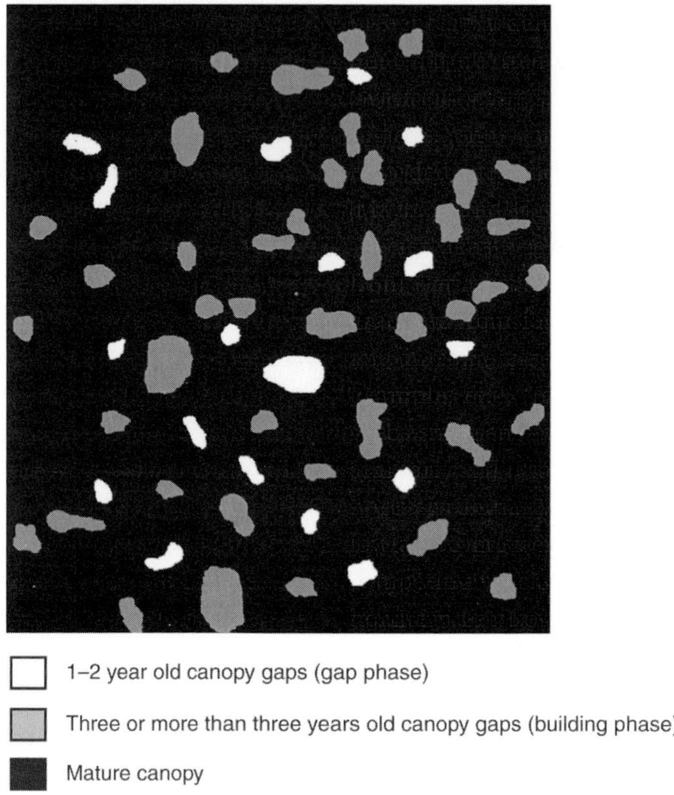

Fig. 1.4 Gap-phase dynamics. Map of the distribution of three phases of canopy development on a 10 ha block of tropical evergreen forest at Pothumala in Nelliampathy Forest Range, Kerala, India. The gap size varied from 86 – 665 m^2. A natural forest is a mosaic of patches in these three phases of development. Courtesy: U.M. Chandrashekara, Kerala Forest Research Institute.

forest, upper and lower montane rain forest, heath forest, mangrove forest, peat swamp forest, semi-evergreen rain forest, moist deciduous forest etc. For a detailed description of the various tropical forest formations the reader is referred to Whitmore (1984). Only the broad categories of tropical forests are discussed here.

As per FAO global ecological zoning, the tropical forests can be grouped into four broad categories – tropical rain forest, tropical moist deciduous forest, tropical dry forest and tropical mountain forest. Their extent and distribution in the three major tropical regions is shown in Table 1.2.

Of the total tropical forests, tropical rain forest constitutes the largest proportion (60%), followed by tropical moist deciduous forest (23%), tropical dry deciduous forest (11%) and tropical mountain forest (9%) (Table 1.2).

Table 1.2. *The major types of tropical forests and their distribution*

	Area in million ha			
Forest type	Total[a]	Africa	Asia-Pacific	Latin America
Tropical rain forest	1083 (60%)[b]	260	184	639
Topical moist deciduous forest	426 (23%)	170	86	170
Tropical dry forest	193 (11%)	75	44	76
Tropical mountain forest	155 (9%)	17	45	94
Total tropical forest	1818 (100%)	509 (28%)	345 (19%)	945 (52%)
Total subtropical forest	323	7	170	146

[a]The totals may not tally exactly as the individual values were calculated from rounded percentages from the source.
[b]Percentage of total tropical forests.
Source of data: FAO (2001b)

Tropical rain forest

The rain forest is unique tropical forest, well known for its extreme richness in species of plants and animals. Although it occupies only about 8% of the world's land area, it is believed to harbour about half the world's species (Whitmore, 1998). It occurs in all three tropical regions, with the largest proportion in Latin America, constituting about 59% of the global total. Brazil has more rain forest than any other nation. The second largest area (24%) of tropical rain forest is in Asia-Pacific, where Indonesia tops the list, with substantial areas also in Malaysia, Thailand, Myanmar, Cambodia, Laos and Vietnam. Africa has the smallest extent of tropical rain forest; it is centred on the Congo basin with extension into western Africa as a coastal strip and small patches on the eastern coast of Madagascar.

Rain forest develops in places where every month is wet, with 100 mm or more rainfall (Whitmore, 1998). Different climate, soil characteristics, soil water conditions and altitude have led to various rain forest formations. The major tropical rain forest formations are lowland evergreen rain forest, lower or upper montane rain forest, heath forest, peat swamp forest, freshwater swamp forest and semi-evergreen forest.

Tropical **lowland evergreen forest** is the most magnificent tropical forest formation, with lofty trees reaching 45 m or more in height at the top tier, over a main stratum between 20–35 m, and with smaller shade-tolerant trees below that. As the name indicates, it occurs at low elevations. In Indonesia, Malaysia and the Philippines, species of Dipterocarpaceae, many of which are of high commercial value, dominate the lowland rain forests. The **semi-evergreen rain forest** contains an intimate mixture of deciduous trees in the top canopy and the tree heights are smaller. It develops in places where there is a regular annual dry period of at least a few weeks. **Heath forest** is the name given to stunted forests that develop on sites of extreme infertility, with limited flora showing high specialization such as leathery leaves. Most often they occur on podzolic soil originating from siliceous sand that degrades quickly when the trees are removed. A variant of heath forest that occurs on peat swamps that are very rich in organic matter but highly deficient in mineral nutrients, formed as a result of incomplete decomposition of plant remains under wet anaerobic conditions, is known as **peat swamp forest. Fresh water swamp forests** develop where the underlying land is subject to inundation by fresh water, as on the coasts of the Amazon and its main tributaries, while **mangrove forests** develop where the underlying land is inundated by salt water, such as shorelines not subject to severe wave action and saline river mouths where large quantities of sediments are deposited. Each of these specialized forest types has its characteristic tree flora, generally species poor, capable of withstanding the harsh conditions.

The rain forests contain a large number of commercially valuable broad-leaved tree species, yielding timber of various qualities useful for a variety of purposes. Because of the large number of species it is not feasible to enumerate them here but the important ones will be referred to at appropriate places in connection with their pests.

Tropical moist deciduous forest

Tropical moist deciduous forest, sometimes called monsoon forest or seasonal forest, develops in places where several dry months regularly occur in a year. In general, these forests are less tall than rain forests and have a lower biomass. The dominant trees in these forests are mostly deciduous, with evergreen trees occupying the lower level. Characteristically bamboos are present.

Tropical moist deciduous forest occurs extensively in the Asia-Pacific countries of India, Myanmar, Thailand and Indonesia (Java) and in Africa and South America. Many valuable commercial species like teak (*Tectona grandis*), rosewood (*Dalbergia latifolia*), mahogany (*Swietenia macrophylla*), *Terminalia*, *Pterocarpus*, *Lagerstoemia* and *Albizzia* which yield construction wood as well as many species like *Ailanthus triphysa* and *Bombax ceiba* which yield plywood and matchwood timber occur in this forest type.

Tropical dry forest

Tropical dry forest, also called dry deciduous forest, occurs in areas receiving less rainfall. It is less species-rich than moist deciduous forest but has many valuable species like sandal (*Santalum album*), *Terminalia* spp., bamboos etc. Many species like teak and mahogany also occur in dry forest, forming a continuum with the moist deciduous forest.

Tropical mountain forest

Tropical forests that lie on mountains above a level of approximately 1000 m altitude constitute tropical mountain forests. On these mountains, which experience lower temperatures and other altered atmospheric conditions such as barometric pressure, solar radiation, moisture, wind and oxygen concentration, the forests take different forms that vary according to the altitude. Mountain rain forests have been further categorized into lower montane, upper montane and subalpine forests. There is a progressive diminution in the height and biomass of trees as the elevation increases. The leaf sizes decrease. There is an increase in the volume of epiphytes, particularly at medium heights as the cloud zone is reached. Floristic composition also changes drastically, with the flora becoming poorer as the height increases. Many plant families common to lowland tropical forests such as Anacardiaceae, Burseraceae, Capparidaceae, Combretaceae, Connaraceae, Dilleniaceae, Dipterocarpaceae, Flacourtiaceae, Marantaceae, Myristicaceae and Rhizophoraceae are replaced by families common to temperate regions such as Aceraceae, Araucariaceae, Clethraceae, Cunoniaceae, Ericaceae, Fagaceae, Lauraceae, Myrtaceae, Pentaphylacaceae, Podocarpaceae, Symplocaceae and Theaceae (Whitmore, 1984).

In drier forests, fire occurs often, resulting in the replacement of trees by tall grasses.

1.4 Management of tropical forests

Traditionally, the tropical forests have served as the source of wood and other building materials, and a variety of non-wood products such as fruits and tubers, dyes, resins, gums and medicines. With the progress of industrialization, exploitation of forests for wood increased, both for domestic use and national and international trade. Selective logging, that is, cutting and removal of selected species, was the initial practice. Among the large variety of tree species present in a tropical forest, only some had properties suited to meet constructional requirements and these were selected and removed. For example, in India, teak (*Tectona grandis*) was selectively logged from natural forests to meet the demands of the British shipbuilding industry. By the middle of eighteenth century, the once plentiful teak was becoming scarce on the Malabar coast (Western Ghats), as natural regeneration was not able to cope with the heavy extraction of logs. Eventually this led to the establishment of teak plantations during the 1840s, and the extent of teak plantations has grown steadily since then. Initially plantations were raised after the clear-felling of natural forests, but as a result of increasing environmental awareness since the 1960s this practice is now discouraged in most tropical countries. In most countries, large forest areas are now set aside for conservation purposes. They range from national parks, wildlife reserves, biosphere reserves etc., where limited management actions are permitted, to simple nature reserves where no management is carried out. The remaining forests are managed for timber production. This involves selective logging of desired tree species, although the logging intervals and methods used to promote the growth and regeneration of the residual trees vary. These silvicultural systems are summarised below.

Polycyclic selective logging

This involves removal of selected trees in a series of felling cycles. Usually, a large forest area is divided into smaller units called working 'coupes' and the well-grown trees of the desired species are removed from the coupes in a cyclic operation. That is, the logging is repeated in the same coupe after a fixed interval of, say, 10 to 30 years, by which time new trees will have reached maturity. For example, in the evergreen forests of the Western Ghats in Kerala, India, the felling cycle is 30 years and only 8 to 12 trees of girth above 180 cm at breast height can be felled. Damage to juvenile trees is kept to a minimum and when necessary obstructive climbers are cut down to remove crown competition.

Polycyclic selective logging is carried out in the moist deciduous forests of India and Myanmar for teak (*Tectona grandis*) and rosewood (*Dalbergia latifolia*); in dry to wet tropical forests of Latin America for mahogany (*Swietenia macrophylla*) and in the rain forests of west Africa for African mahoganies (*Khaya* spp.), among others. In Bolivia, where 20–60 mature mahogany trees may occur per ha, a 25-year cutting cycle is prescribed for mahogany and 10% of commercial-sized trees must be left behind as seed trees (Mayhew and Newton, 1998).

Monocyclic selective logging

In this system, practised in natural forests with a high density of commercially valuable tree species as in the dipterocarp forests of Malaysia, all marketable trees are harvested in a single operation at an interval that approximates to the rotation age of the trees. This is also known as the Malayan Uniform System. When all the marketable trees are removed, the canopy is opened up and a more or less uniform crop of seedlings/saplings is left behind, which is expected to grow to maturity by the next cutting cycle (70 years for dipterocarp forests of Malaysia). Less valuable trees are poisoned to 'release' the valuable dipterocarp seedlings/saplings that are in an advanced stage of regeneration. Climber-cutting and canopy opening are carried out at 10-year intervals.

A number of variants of selective logging exist in different tropical countries, with modifications made to suit local demands and stand characteristics. In the forests of West Africa, canopy opening is carried out several years prior to the logging operation to encourage advance regeneration, and several weeding and thinning operations are done following logging. In some places, an inventory is made prior to logging to ascertain the size–class distribution of trees, based on which the girth limits for felling are determined.

Clear-cutting in strips

A common practice in the Amazon region is to carry out clear-cutting in strips, 30–40 m wide, and allow natural regeneration to take place. The cutting cycle is 30–40 years and silvicultural treatments are carried out intermittently.

Multiple-use management

Increasing environmental and biodiversity considerations have led, in recent years, to attempts to manage tropical forests with minimal disturbance, for multiple benefits. It has been increasingly recognized that forests produce not only wood but also many other goods and services, such as a variety of non-wood forest products (see section 1.6 below), and environmental benefits such as soil protection and regulation of the hydrological regime. Multiple-use management of forests envisages management for two or more of these benefits simultaneously. The concept of multiple-use management has received wide acceptance but practical implementation is still elusive. Attempts range from 'reduced-impact logging' of natural forests by using aerial lifting of logs to reduce the damage to standing crop caused by large falling trees, to management of forest with people's participation at local level, including sharing of benefits. Some simple systems like cultivation of the shade-loving cardamom as an undergrowth in natural forest, with necessary shade regulation by lopping of trees, is already in practice in the evergreen forests of the Western Ghats in India. Here there is economic gain from cardamom cultivation, the soil and water regimes are reasonably protected and timber production is also ensured. However, while established trees continue to grow, natural regeneration of trees is hindered, affecting the long-term sustainability of the forest. Such trade-offs seem inevitable in multiple-use forest management. Suitable models of multiple-use management are still being developed.

1.4.1 *Problems of natural forest management in the tropics*

In spite of the best intentions at policy level, management of natural forests in the tropics is beset with practical problems in implementation. Apart from the minor floristic alterations over the years that may be caused by selective removal of some tree species, incidental damage caused by logging operations can be substantial. Damage is caused by building an often extensive road network in the hilly forest terrain to transport logs and by the unintentional breakage of residual trees caused by the falling crowns of large trees when felled. In most tropical countries, the natural forests are government owned but the felling operations are carried out by private contractors. Inadequate supervision and/or collusion between unscrupulous government officials and contractors often results in excessive over-cutting of trees which destroys the capacity of the forest to regenerate adequately, leading to degeneration and formation of secondary forests.

Indonesia provides a typical example where, in the past, rules and regulations have been grossly violated with the patronage of corrupt politicians (Cossalter and Nair, 2000). Management of the vast Indonesian tropical forests has traditionally been vested with forest concessionaires. In the forest-rich outer islands of Kalimantan (Indonesian Borneo) and Sumatra, concession right was granted to private and state-owned enterprises to exploit and manage natural forests. In 1979, the management of about 50 million ha of forest was entrusted to the concessionaires. Although the concessionaires were required to rehabilitate the logged areas, the net result was large-scale clearing of natural forests and degradation of vast areas into unproductive grasslands. In 1987, such grasslands were estimated to occupy about 30 million ha. In another government programme initiated in 1980, additional areas were allotted to domestic and foreign concessionaires in order to establish industrial forest plantations, with liberal assistance such as capital in the form of government equity, interest-free loans etc. Although some plantations were established, many forest concessionaires catered only to their own short-term financial interests rather than the long-term ecological, economic and social prosperity of the country. In a review after 20 years of the 35-year concession right, the rights of over 60% of the 359 concession holders were revoked for not conforming to the prescribed rules and regulations (Cossalter and Nair, 2000), which is indicative of the level of mismanagement. Substantial damage to forests had already been caused by that time. The Indonesian scenario may be an extreme case, but it is indicative of the problems in many of the economically stressed tropical countries.

1.5 Plantation forestry

For timber production, plantations offer many advantages over natural forests. In mixed-species tropical forests, the low density of usable trees limits timber productivity. In plantations, the species composition can be changed according to requirements. There is also large inter-tree variation in growth rate in a natural forest due to genetic variability, soil properties, exposure to sunlight, competition among trees etc. These factors can be controlled to our advantage in a plantation, by genetic selection, fertilization, suitable spacing etc., thereby enhancing the growth rate. Whitmore (1998) observed that mixed-species tropical rain forests will yield 3.6–12 tons above-ground dry weight of bole timber/ha per year. In comparison, plantations of broad-leaved trees on moderately good soil will yield 6–17 tons and conifer plantations, as well as some eucalypts, 12–35 tons of timber/ha per year. Suitable spacing in plantations also facilitates silvicultural operations, including mechanical thinning and harvesting.

Due to these advantages, over the past few decades there has been a steady expansion of forest plantations across the tropics. Plantation expansion has also been driven by increasing demand for various wood products and the dwindling supply of wood from natural forests, as environmental concerns imposed increasing restrictions on the cutting of natural forests.

The beginnings

Trees have been planted and protected in the tropics since ancient times, for both religious and aesthetic reasons and to provide shade along the roadsides. However, planting trees over large contiguous areas to create plantations was largely initiated in the mid-nineteenth century by the colonial governments. Evans (1992) and Bass (1992) give detailed accounts of the historical development of plantation forestry in the tropics. The Dutch in Indonesia tried planting teak on a small scale around 1650. In India, teak plantations were first raised in the 1840s, mainly to cater to the needs of the British navy for shipbuilding. The success of teak plantations in India led to their subsequent extension throughout the tropics – Bangladesh, Myanmar, Indonesia, Philippines, Sri Lanka and Vietnam in Asia; Tanzania in Africa; and Trinidad in tropical America. During this period, plantations of eucalypts were also raised in India, Brazil, Peru and Tanzania. Wattle (*Acacia mearnsii*) was planted in India and Tanzania to produce tannin from its bark.

The expansion

Much of the expansion in forest plantations took place during 1900–2000. While plantations of teak in India and of eucalypts in Brazil,

Ethiopia etc. had been built up steadily during the first half of the century, since the 1960s there has been a boom in the establishment of tropical forest plantations. As noted earlier, this was triggered by international exchange of information and the availability of international development funds following the Second World War. In addition, some transnational private entrepreneurs, supported by liberal incentives from local governments, took an interest in plantation development in some tropical countries. The ambitious Jari venture in Brazil, initiated in 1968 by an American entrepreneur to convert 400 000 ha of Amazon forest to plantations of pines, eucalypts and *Gmelina arborea* to produce pulpwood, is well known. Due to several problems (labour management, social strife, poor performance of *Gmelina* etc.) only about 120 000 ha were actually planted up by 1980, but eventual transfer of ownership to a Brazilian consortium and the choice of more suitable tree species such as *Pinus caribaea* var. *hondurensis* and *Eucalyptus deglupta* made it one of the most successful commercial plantations in the tropics (Bass, 1992). The later Aracruz plantation of eucalypts in Brazil is also well known for the high growth rates achieved through genetic selection and vegetative propagation. Other notable examples of plantations established between 1960 and 1980 are about 2.8 million ha, mainly of teak, eucalypts and other species in India; 28 000 ha of *Pinus caribaea* in Fiji; and about 15 000 ha of eucalypts in Congo (Evans, 1992).

Much of Indonesia's industrial forest plantations of eucalypts and acacias were established in the 1980s. Another significant development in the late twentieth century was the emergence of a large number of voluntary agencies worldwide who promoted tree planting in degraded and waste lands, for environmental protection. Agroforestry, that is, cultivation of forest trees along with agricultural crops, also became popular during this period.

Present status

Evans (1992) estimated that tropical and subtropical forest plantations covered about 6.7 million ha in 1965, about 21 million ha in 1980 and 43 million ha in 1990. A recent study by the FAO (Pandey, 1997) puts this figure at 55.4 million ha for 1995, representing nearly 45% of the global area of forest plantations. In 1995, the annual rate of plantation establishment in the tropics was estimated at 1.7 million ha. As noted above, there has been a rapid increase in the tropical forest plantation area since the 1960s (Fig. 1.5). Much of this expansion took place in Asia-Pacific, particularly in Indonesia, India and Malaysia. In 1990, the distribution of plantation area by region was Asia-Pacific (including southern China, Australia and the Pacific Islands) – 29.7 million ha (71%); Latin America – 8.2 million ha (20%) and Africa – 3.8 million ha (9%). Several countries in Asia-Pacific (Bangladesh, Myanmar,

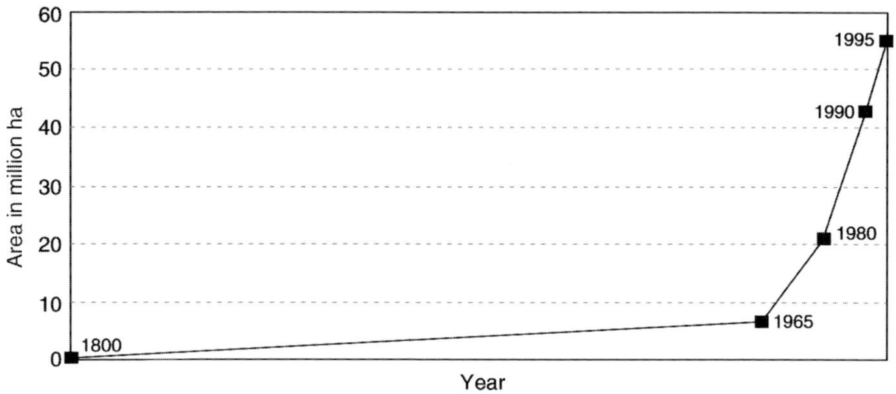

Fig. 1.5 Growth of tropical forest plantations.

India, Indonesia, Philippines, Sri Lanka, Thailand, Vietnam), Latin America (Cuba, Mexico, Brazil, Colombia, Peru, Venezuela) and Africa (Angola, Ethiopia, Kenya, Madagascar, Nigeria, Sudan, Zimbabwe) had more than 100 000 ha of forest plantations by 1990 (Evans, 1992). Thus plantation forestry in the tropics, under both commercial ventures and social forestry programmes, is poised for great expansion in the coming years.

1.5.1 *Tree species planted*

Over a hundred tree species are raised in plantations in the tropics, but a few dominate. The choice of species is determined by the purpose for which the plantations are raised and the nature of the site. The different categories of plantations include the following.

Plantations for constructional timber

Many species traditionally obtained from native forests and found valuable for use as constructional timber have been raised in plantations in the respective countries. An added criterion is comparatively faster growth rate. For example, rosewood (*Dalbergia latifolia*) may take about 300 years to yield a good commercial log while teak (*Tectona grandis*) can produce the same volume of wood in 60–80 years. Teak is one of the most valuable constructional timbers grown in plantations in the tropics. Its success in native plantations in India and exotic plantations in Indonesia has led to its cultivation in many tropical countries across the world; CABI (2005) lists about 50 countries (including subtropical ones) where it is grown. The global area under teak plantations in the year 2000 was estimated at 5.7 million ha, most of it in the Asian tropics (92%), followed by tropical Africa (4.5%) and Latin America (3%) (Ball *et al.*, 2000; FAO, 2001a). Some of the important plantation species for constructional timber and the major countries in which they are grown are given in Table 1.3. Mahogany (*Swietenia macrophylla*) from Central and South America and *Khaya ivorensis* from West Africa (known as African mahogany) need special mention as they are well-known, all-round timber species of the tropics. Together with teak, they share common properties of medium density, medium strength, good workability, medium to high durability, good dimensional stability, fine finish and a red–gold lustrous brown colour (Appanah and Weinland, 1993).

In addition to the species listed in Table 1.3, there are many that are locally important, such as *Manglietia conifera* (Magnoliaceae) in Vietnam, *Hopea odorata* (Dipterocarpaceae) in Bangladesh, *Pterocarpus macrocarpus* (Fabaceae) in Thailand, *Peronema canescens* (Verbenaceae) in Indonesia and several species of dipterocarps in Malaysia. Plantations of these species are on an experimental scale and little information is available about areas planted in the different countries. Cossalter and Nair (2000) reported 4963 ha under *Peronema canescens* and 4456 ha under *Octomeles sumatrana* in Indonesia as of 1999, based on information supplied by plantation companies. Appanah and Weinland (1993) give a long list of species under plantation trial in Peninsular Malaysia.

Plantations for pulpwood

Australian eucalypts are the most widely planted for pulpwood. Species commonly planted include *E. camaldulensis*, *E. globulus*, *E. grandis*, *E. robusta*, *E. saligna* and *E. tereticornis*, all of Australian origin, and *E. deglupta* and *E. urophylla* of Southeast Asian origin. The global area under tropical and subtropical eucalypt plantations in 1995 was 10 million ha (Brown and Ball, 2000). India had 3.1 million ha and Brazil 2.7 million ha (Pandey, 1997), but the exact total figure for the tropics is not available. Tropical pines, especially *Pinus caribaea* and *P. merkusii*, have also been planted extensively in the past, but they have not done well in some places due to the absence of mycorrhiza. Moreover, most pine plantations are in the subtropical or montane zones of the tropical countries. In recent years, *Acacia mangium* has taken over second place, with plantations in Bangladesh, China, India, Indonesia, Laos, Malaysia, Philippines, Sri Lanka, Thailand and Vietnam, accounting for about 600 000 ha of which 500 000 ha are in Indonesia (Turnbull *et al.*, 1998). Its fast growth rate and ability to compete with grasses in degraded secondary forests, combined with the suitability of its pulp for medium density fibreboard, has made it a preferred species for industrial plantations. *Gmelina arborea* which produces good quality timber has also been planted extensively as a short rotation crop for pulpwood in Brazil, West Africa and Indonesia. With recent improvements in pulping technology, plantations of many other fast-growing tree species such as other species of

Table 1.3. *Major plantation tree species for constructional timber*

Species	Common name	Major countries where grown
Tectona grandis (Lamiaceae)	teak	India, Indonesia, Myanmar, Thailand, Bangladesh, Sri Lanka, Nigeria, Côte d'Ivoire, Ghana, Trinidad, Brazil
Swietenia macrophylla (Meliaceae)	mahogany	Indonesia, Fiji
Khaya ivorensis (Meliaceae)	African mahogany	Nigeria, Côte d'Ivoire, Costa Rica, Trinidad and Tobago, Malaysia
Gmelina arborea (Lamiaceae)		India, Indonesia, Thailand, Malaysia, the Philippines, Nigeria
Agathis spp. (Araucariaceae)	damar	Indonesia, Malaysia
Araucaria cunninghami (Araucariaceae)	hoop pine	Papua New Guinea, Kenya
Dalbergia spp. (*latifolia, sissoo, cochinchinensis*) (Fabaceae)	rosewood, shisham, Siamese rosewood	India, Indonesia, Thailand, Vietnam, the Philippines, Sri Lanka, Kenya, Nigeria, Ghana, Sudan
Shorea spp. (*robusta, leprosula, parviflora* etc.) (Dipterocarpaceae)	sal, red meranti	India, Malaysia
Terminalia ivorensis (Combretaceae)		Many West African countries, many South American countries, Fiji, Solomon Islands
Triplochiton scleroxylon (Sterculiaceae)		Cameroon, Ghana, Nigeria, Solomon Islands
Eusideroxylon zwageri (Lauraceae)	ironwood	Indonesia, Malaysia
Koompassia spp. (Fabaceae)		Malaysia, Indonesia
Xylia xylocarpa (Fabaceae)	irul	India, Myanmar, Laos, Malaysia, Thailand, Vietnam, Nigeria, Uganda

Acacia, Eucalyptus deglupta, Falcataria moluccana, Maesopsis eminii and *Neolamarckia cadamba* are now being grown for pulpwood.

Plantations for plywood, matchwood, packing cases and light construction

Plantations of several species which produce light timber have been raised in different countries to provide wood for manufacture of plywood, matchsticks, veneers, packing cases and light construction. These include *Alstonia scholaris* (India, Indonesia, Laos, the Philippines, Sri Lanka), *Bombax malabaricus* (India, Bangladesh, Sri Lanka), *Falcataria moluccana* (Bangladesh, India, Indonesia, Malaysia, Philippines, Sri Lanka, Cameroon, Cote d'Ivoire, Malawi, Nigeria, Mexico, Hawaii, Samoa), *Neolamarckia cadamba* (India, Indonesia, Laos, Malaysia, Philippines, Puerto Rico, Suriname, Fiji, Solomon Islands), *Populus deltoides* (India, Bangladesh, Kenya, Uganda, Zambia, Zimbabwe, Fiji) and *Maesopsis eminii* (India, Indonesia, Malaysia, Philippines, several countries in Africa, Fiji).

Plantations of multipurpose tree species

Widespread destruction in the tropics of natural forests which once catered to the needs of the local population for firewood, building materials, fodder, shade etc. has necessitated tree planting on a large scale to meet social needs. This has often been promoted by government-sponsored 'social forestry' schemes supported by leading agencies like the World Bank and voluntary non-governmental organizations (NGOs). Some tree species, both native and exotic, have been grown under such programmes for a variety of purposes. These trees are often called 'Multipurpose Tree Species' (MPTS). The best-known example is *Leucaena leucocephala*, native to tropical America, now cultivated very widely across the tropics for fodder, green manure, fuel, erosion control, nitrogen fixation etc. Alley cropping of leucaena, with one row of leucaena between 4–6 rows of food crops in agricultural fields, is common in many countries. Other popular MPTS include *Calliandra callothyrsus* (ornamental, fuel, nitrogen fixing, green manure, erosion control), *Casuarina equisetifolia* (poles, erosion control on coastlines, windbreak), *Acacia auriculiformis* (ornamental, shade, nitrogen fixing, afforestation of impoverished sites, erosion control, fuel) and *Azadirachta indica* (shade, wood, fuel, medicinal/insect repellent, arid zone afforestation). Locally important tree species which produce edible fruits along with timber, such as jack (*Artocarpus heterophyllus*), *Tamarindus indica* and mango (*Mangifera indica*), are often planted.

Plantations for special products

Extensive plantations of some species have been raised in the tropics for tapping or harvesting special products. These include *Hevea brasiliensis* to produce latex for rubber (several countries in Asia-Pacific and Africa), *Pinus merkusii* for resin (Indonesia), *Melaleuca cajuputi* (syn. *M. melanoxylon*) for cajuput oil (Indonesia), *Acacia mearnsii* (black wattle) for tannin (a number of countries in eastern and southern Africa, Brazil, India, Indonesia) and *Diospyros melanoxylon* for harvesting leaves for use as wrappers for a kind of cigar (India).

Planting for environmental protection

There has been a significant increase in tree planting for environmental protection over the past few decades, due to increasing recognition of the role of trees in stabilizing soil, water conservation, regulation of climate, carbon dioxide fixation etc. A large variety of tree species have been used for this purpose, depending on the specific objective and the locality.

In degraded forests devoid of adequate natural regeneration, enrichment planting is usually carried out with seedlings or saplings of indigenous trees commonly present in the locality.

Species commonly used for afforestation of wasteland or degraded land include the nitrogen-fixing tree species, *Leucaena leucocephala*, *Calliandra callothyrsus*, *Acacia auriculiformis* and *Falcataria moluccana*. Bamboo species are often used to afforest dry degraded lands. For afforestation of arid zones, *Tamarindus indica* has been used in northern Kenya, *Erythrina senegalensis* in Sahel, *Azadirachta indica* in West Africa, *Prosopis cineraria* in the Middle East and Pakistan (Evans, 1992) and *Acacia nilotica* in India. *Casuarina equisetifolia* is planted to stabilize sand dunes.

Some species known to absorb toxic gases from the atmosphere and neutralise them through physiological processes have been planted near the source of the pollutants. For example, *Ficus* is capable of absorbing fluorides, and mango can absorb chlorides (Nair *et al.*, 1999). Similarly, *Ailanthus excelsa* is tolerant to sulphur dioxide, fluorides and chlorides. Some species are suitable for rehabilitation of degraded sites such as mined lands, for example, *Eucalyptus tereticornis* for copper, *Pinus caribaea* var. *hondurensis* for iron/nickel, acacias and eucalypts for dolomite and bauxite, and eucalypts for tin (Evans, 1992).

1.5.2 Native versus exotic tree species in plantations

Exotic tree species have dominated plantation forestry in the tropics for various reasons, ever since plantation forestry began. Teak, the highly prized all-purpose timber tree which is native to some parts of Asia, has been planted extensively across Asia, Africa and the Caribbean since the nineteenth century, largely on the initiative of the European colonisers. It is currently expanding its reach into Brazil and other Latin American countries, as a fairly fast-growing tree when under high-input management, and is promoted by multinational plantation enterprises. Several species of eucalypts originating from Australia have been planted very extensively across the tropics as a staple pulpwood source. The global area under eucalypt plantations, including subtropical plantations, is estimated at 10 million ha. Some species of pines, notably *Pinus caribaea*, have been widely planted outside their natural range in many countries in the Asia-Pacific, Africa and Latin America in the past, although the area under pines is now declining due to their poor performance. The black wattle *Acacia mearnsii*, of Australian origin, has been planted very extensively in all three tropical regions, for the production of tannin. The rubber tree *Hevea brasiliensis*, which is indigenous to South America, is another tree species planted very widely in the tropics, originally for tapping latex but now also for timber. *Leucaena leucocephala*, a multipurpose tree of Central American origin, has been planted widely as an agroforestry crop. Plantations of each of the above species occupy several millions of hectares of forest land in the tropics. Another traditional exotic forest plantation tree is mahogany, *Swietenia macrophylla*, of tropical American origin, which occupied about 1.5 million ha in 1995 in Asia-Pacific and Africa. The list of exotic plantation species continues to grow. Those that have been planted on a large scale in recent years include *Acacia mangium, A. auriculiformis, Gmelina arborea* and *Falcataria moluccana*.

It is evident that a substantial proportion of tropical forest plantations consists of a few exotic species although some indigenous species have also been planted extensively. Examples of the latter include teak in India and Thailand; *Pinus merkusii* in Indonesia; *Eucalyptus deglupta* in the Philippines and Papua New Guinea; *Araucaria* spp. in Papua New Guinea; *Terminalia ivorensis, Nauclea diderrichii* and *Triplochiton scleroxylon* in western Africa; and *Cordia alliodora, Swietenia macrophylla* and *Cupressus lusitanica* in Central America (Evans, 1992). In addition, small-scale trial plantations of dozens of indigenous species have been established in several tropical countries. Although many of these indigenous species may be potentially useful, the choice of species for large-scale plantations has been driven by market demand and financial support for planting. Many of the recent plantings have been for industrial purposes, mainly to ensure a steady supply of uniform pulpwood material for the manufacture of paper and fibreboard, an industry dominated by a few multinational companies. The choice fell on a few exotic species for which good pulp quality has been proven and the pulping technology standardized by the industrially developed countries. Financial aid in the form of liberal loans from international or bilateral agencies was made available to the poorer developing countries in the tropics in the name of promotion of industrialization, and the donors often influenced the choice of species. Thus large-scale industrial monocultures of a few exotic species soon overtook small-scale planting efforts with diverse indigenous species.

1.6 Non-timber forest products

The value of tropical forest lies not only in its timber, although, unfortunately, organized forest management in all the tropical countries has been largely directed towards commercial timber production. Unlike temperate forest, species-rich tropical forest yields a large variety of non-timber forest products (NTFPs), the value of which may surpass that of timber. It is estimated that NTFPs, excluding firewood and small wood, account for about 40% of the forest revenue and are the source of over 55% of the total employment in the forestry sector in India (Gupta and Guleria, 1982). They also contribute to the sustenance and wellbeing of the predominantly rural populations of the tropics. It is beyond the scope of this book to discuss NTFPs in detail, but as a typical case study we will examine those of Indian forests (Box 1.1). Since the products are too numerous to list, a broad indication is given by classifying them into major kinds. The annual production of some of the items and the employment generated by their collection is shown in Table 1.4.

The example of India shows that non-timber forest products play an important role in the tropical developing countries by generating rural

Table 1.4. *Production and employment in collection of some non-timber forest products in India*

Item	Production (in tons)	Employment (in man-years)[a]
Grasses	350 000	1 200 000
Fibres and flosses	5 500	14 400
Bamboo	1 932 000	48 300
Canes	14 000	700
Essential oils	1 698	27 220
Oil seeds	342 700	109 037
Tans and dyes (bark and myrobalans)	187 400	21 170
Gums and resins	91 200	87 000
Lac and lac products	22 000	7 300
Cinchona	1 420	23 635
Tendu leaves	210 000	74 900

[a]Employment relates to collection only, not to processing and marketing.
Source: Gupta and Guleria (1982)

Box 1.1 Indian Forest NTFPs

Fuel: The most important non-timber forest product of India is fuelwood. Wood continues to be the major source of house-hold energy for the large rural population of India, as for many other developing countries. Forests contribute a substantial portion of this requirement, part of it through illicit removal.

Small wood: A large quantity of small-diameter poles are used for making agricultural implements, tool handles and sheds, and as fence posts, banana props and scaffolding. These are used mostly by people living in or near forests.

Fodder and green manure: Fodder in the form of *in situ* grazing, as well as cut grasses and leaves, is another important product of forests. About 90 million domestic animals are estimated to graze in the forests of India. Forests are also an important source of green manure for agriculture.

Bamboos and canes: India has dozens of species of bamboos that are used for a variety of purposes by the rural and urban populations, and for producing pulp for paper and rayon. It is estimated that bamboos grow over 10 million ha of forest in India. Many of the more than 50 species of canes occurring in India are commercially important, and are used for making high quality furniture, baskets and handicrafts.

Edible fruits, flowers and seeds: The forests of India produce a large variety of edible fruits, flowers and seeds. In addition to their use for human consumption, they sustain a wide variety of animal life in the forests.

Fibres and flosses: The bark of several forest tree species yields fibres that are used for a variety of purposes. Floss from *Bombax ceiba*, used in pillows and mattresses, is an important item of trade.

Essential oils: A large variety of essential oils produced by the flowers, fruits, leaves, bark etc., of various trees find uses in the manufacture of soaps, sprays and deodorants, as medicines or pharmaceuticals and for other uses. Sandalwood oil is an example. There is a multitude of such potential products still unexploited.

Oil seeds: About 100 tree species occurring in natural forests have been identified for production of oil from seeds, of which about 25 are regularly used. *Madhuca longifolia* (mahua), *Azadirachta indica* (neem), *Shorea robusta* (sal) and *Pongamia pinnata* are some of the more important ones. The full potential of oilseeds of forest origin has not been explored and utilized.

Tans and dyes: Several forest tree species produce tannins and dyes which are of high commercial value. They are used for tanning of leather, manufacture of ink, as mordants, astringents etc. The tannins from *Acacia* spp., myrobalans from *Terminalia chebula* and dyes from various trees are widely used.

Gums and resins: Gum Arabic from *Acacia* spp. and gum Karaya from *Sterculia urens* are important commercial products. The resin of *Pinus roxburghii* is tapped to manufacture turpentine. Gum of *Ailanthus triphysa* is used for manufacture of incense sticks ('agarbathis').

Drugs: Thousands of chemical compounds with various potential uses are produced by forest trees and plants. The indigenous system of medicine known as 'Ayurveda' which caters to the health needs of a large proportion of the population, is dependent on raw materials obtained from the forest. The small state of Kerala with about a million ha of forest alone uses at least 350 forest plants or plant products as ingredients of Ayurvedic drugs. In spite of the vast ancient and modern literature on medicinal plants and their uses, their potential has not yet been fully utilized.

Spices: India is the legendary home for various spices used across the world. Most of the spices come from the forests.

Heartwood extractives: Several valuable products are extracted from the heartwood of certain trees. Examples are 'katha' and 'kutch' from *Acacia catechu*, which are used as masticatory, and teak heartwood oil used against hoof disease in cattle. Another example is agarwood, a highly prized resinous product used in perfumery, which is produced in the heartwood of *Aquilaria* sp., possibly when infected by a disease of unknown etiology.

Tassar silk and lac: Tassar silk is produced from the cocoons of tassar silk moth collected from natural forests. Lac, which is used for the manufacture of shellac and for many other purposes, is produced by the lac insect infesting some tree species and is both collected from the wild and cultivated.

Tendu leaves: Leaves of the tree *Diospyros melanoxylon* are used to wrap tobacco to produce a kind of cigar, known as 'bidi'. It is a major small-scale industry in India.

income and employment. This is more important than their commercial value which nevertheless is not small.

1.7 Contemporary issues in tropical forestry

Deforestation

As pointed out in the Introduction, campaigns against tropical deforestation have gathered momentum since the 1960s due to increasing recognition of the environmental benefits conferred by tropical forests. The need to halt wanton destruction of tropical forest is now well recognized and the governments of most tropical countries have taken policy decisions to this end. Because of expanding populations in tropical countries, pressures for diversion of forest land for urbanization, agriculture etc. will continue to grow and deforestation is likely to continue although at a reduced pace. In the past, human civilization and progress entailed clearance of some natural forest, but unfortunately we cannot scientifically determine how much of the tropical forest needs to be retained. The commonly advocated dictum of conserving 33 % of the geographical area of each country as forest fails to convince governments, and environmental enthusiasts often oppose all developmental activities in the name of conservation, creating a negative reaction from the decision makers.

Degradation of the quality of the natural forest leading to slow destruction is another factor to be reckoned with. Even where mature forests are retained, the undergrowth is often cleared, particularly in areas close to human habitation, endangering the natural regeneration process.

While the interest shown by the general public, voluntary organizations and governmental agencies for conservation of tropical forests is welcome, scientists need to concentrate their effort to determine how much of the tropical forests need to be conserved and where.

Sustainable management

Obviously, all the existing tropical forests cannot be maintained without any human intervention. After setting aside part of the natural forest for total conservation (without any human interference), and part for conservation with limited intervention (for eco-tourism, parks etc.), a reasonable part must be managed for production of timber and other products. But how do we manage the natural forest for sustained production of timber or other products? We are far from achieving this sustainable management of tropical forests, although research efforts are being made to develop suitable management systems for achieving this goal. Although management of natural forests for multiple benefits has been advocated, the complexity of the operations required impose practical difficulties. Our present understanding of the functioning of tropical forest ecosystems is far from adequate for devising biologically and economically sustainable management systems. In the meantime, certification procedures are being developed by various institutions to ensure sustainable management. Managing pests in natural forests also requires an understanding of tropical forest dynamics.

Plantations

Historically, plantations in the tropics were established after cutting down the natural forests. Although at present plantations account for only 5% of the total forest area worldwide, the rate of their expansion in the tropics (1.7 million ha/yr), most often at the cost of destruction of natural forests, is cause for concern. Plantations produce the desired wood more efficiently than natural forests, but there are some negative impacts of plantations compared with natural forests.

Establishment of plantations in place of natural forests leads to loss of biodiversity, soil erosion, alteration of hydrological regime and loss of many of the non-timber products. Since harvesting of timber takes away some of the nutrients which are normally recycled in the ecosystem, plantations can lead to loss of soil fertility and deterioration of the site in the long run. Loss of biodiversity may also aggravate the development of pest problems and hinder our ability to manage them. In addition, there are some larger social and political issues.

The recent expansion of short rotation, fast-growing forest plantations across the tropics has been driven by the desire of some industries to ensure a steady supply of uniform pulpwood material for manufacture of paper, rayon and MDF (medium density fibreboard). A few multinational companies dominate this business and they have increasingly looked to the tropics for a cheap supply of the raw material. As noted earlier, aided by huge loans and other incentives, massive industrial-scale monocultures of fast growing species suitable for pulpwood have been established in several countries like Indonesia and Brazil, after cutting down the species-rich natural rain forests. These large-scale planting programmes, usually with exotic species, have altered the landscape in many tropical countries and deprived the indigenous human populations of their livelihood sources of food, firewood, fruits, medicines etc. They have also been alienated from their common property resources, as land untitled until then was usually taken over by national governments to lease out to the plantation companies. The culprits have not always been multinational companies. Sometimes, local governments entered into contracts with influential national companies to supply raw material for pulpwood industries in return for the benefit of industrialization and creation of jobs (and sometimes other unknown kickbacks!). Under these contracts, governments often supplied the industries with natural and plantation-grown wood at heavily subsidised prices. Thus large-scale plantations have usually worked against the interests of the poorer people of the tropical countries.

Exotics

Most of the large-scale plantations in the tropics are of exotic tree species, as already noted. For example, both Brazil and India have about 3 million ha of eucalypt plantations each (Brown and Ball, 2000; Goncalves *et al.*, 1999) and Indonesia has 500 000 ha of *Acacia mangium* plantations, 48 400 ha of *Falcataria moluccana* and 47 800 ha of *Gmelina arborea* (Cossalter and Nair, 2000). The area planted with exotics in the tropics is increasing and the genetic base of the planted species is decreasing. Although many exotics are now free of pests and diseases, there are instances of devastating pest outbreaks in exotics and there is lingering fear that massive pest outbreaks are just waiting to happen. Chapter 8 will analyse this issue critically.

Exotics have evoked a variety of negative responses in the tropics, sometimes with reason and sometimes without. It is argued that biodiversity is impoverished by exotic plantations; eucalypt plantations consume excessive water, causing drought; acacia pollen causes allergy etc. While some of these generalizations like loss of biodiversity are true, it is difficult to prove or disprove some others because the effects are species-specific and are common to some of the indigenous species. The effects often depend on the extent of plantations. Some of the alleged drawbacks are effects of monoculture plantations per se irrespective of whether the trees are exotic or indigenous. In some places, social activists have resorted to uprooting and burning of exotic plantations in protest, but it is often difficult to separate out the reasons for resentment because complex issues are involved: the negative impact of any large-scale plantation on local livelihood activities such as collection of firewood, fruits, fodder, honey and a variety of other non-wood products from a natural forest that was replaced by a plantation; the often unfair subsidies offered by governments to the industries that benefited from the plantation; restriction of entry into the area; the very different look of the landscape planted up with an unfamiliar exotic etc. These have often evoked feelings similar to that of patriotism in favour of indigenous species. It is beyond the scope of this book to discuss these issues in detail; interested readers may refer to the book entitled *Plantation Politics* edited by Sargent and Bass (1992).

2

An overview of tropical forest insects

2.1 History of tropical forest entomology and important literature

Apart from traditional knowledge about common forest insects such as bees, termites, lac insect and silkworm, which dates back to pre-historic times, scientific literature on tropical forest insects has started accumulating since the last quarter of the eighteenth century. In 1779, Koenig, a student of Linnaeus working in India, published the first scientific study of termites (Koenig, 1779) and in 1782, Kerr, also working in India, published a study on the lac insect (Kerr, 1781). A report on insect borers of girdled teak trees was published in 1836 and one on the beehole borer of live teak trees during the 1840s, both from observations made in Myanmar (then Burma) (Beeson, 1941). Forest entomology in India, in particular, produced prolific literature between the mid nineteenth and mid twentieth centuries. Accounts of immature stages of forest insects and of timber borers were published in India in the 1850s and 1860s. Several accounts on Indian forest insects appeared in *Indian Museum Notes* and *Indian Forester*, both published since 1875; and in the *Journal of the Bombay Natural History Society* published since 1883. The first volume of *Indian Forester* in 1875 contained an account of the toon shoot borer, even before the insect was scientifically named *Hypsipyla robusta* in 1886 (Beeson, 1941). In 1893–6, Hampson authored four volumes on *Moths* under the well-known *Fauna of British India* series, which contained taxonomic and biological information on many forest moths (Hampson, 1893–6). In 1899, Stebbing assembled all the available information on Indian forest insects in a publication entitled *Injurious Insects of Indian Forests* which included about 100 named species (Stebbing, 1899). At that period distinction into forest insects and others was not very relevant and the world-renowned book by Maxwell Lefroy entitled *Indian Insect Life*, published in 1909, contained information on many forest insects (Lefroy, 1909). A post of Forest Entomologist was established in British India in 1900 which, although short-lived, was revived in 1906 as Forest Zoologist in the Imperial Forest Research Institute established at Dehra Dun. In 1914, a masterly volume on forest beetles running 648 pages, entitled *Indian Forest Insects of Economic Importance – Coleoptera* was authored by Stebbing, the first Forest Zoologist in India (Stebbing, 1914). A. D. Imms, whose *General Textbook of Entomology* is well known, also served for a short period as Forest Zoologist in India and was succeeded in 1913 by C. F. C. Beeson who made substantial contribution to forest entomological studies in India.

European scientists, many of whom worked as officers of the Indian Forest Service, and others who worked elsewhere on collections made in India, pioneered studies on forest insects during this period and contributed substantially to our knowledge of tropical insects in general. Beeson's monumental work entitled *The Ecology and Control of the Forest Insects of India and the Neighbouring Countries*, published in 1941, is the most comprehensive and authoritative work on tropical forest insects, containing references to 4300 species of forest insects and continuing to be a very valuable reference book even today (Beeson, 1941). Study of insects associated with forest plants in India and adjacent countries was continued and a comprehensive nine-part list of 16 000 species of insects found on 2140 species of forest plants was published by 1961 (Bhasin and Roonwal, 1954; Bhasin *et al.*, 1958; Mathur and Singh, 1960–61). Recent research on Indian forest insects will be covered elsewhere.

In Myanmar between 1928 and 1940 entomological research concentrated on teak pests, notably on the beehole borer and the defoliators, with some attention to defoliators of *Gmelina arborea*, and borers of *Xylia dolabriformis* and bamboos (Beeson, 1941). Early work in Malaysia (then Malaya) initiated in 1933 by F. G. Browne concentrated on timber borers, particularly Platypodinae and Scolytinae. In 1968, Browne published a comprehensive reference book entitled *Pests and Diseases of Forest Plantation Trees: An Annotated List of the Principal Species Occurring in the British Commonwealth* running into 1330 pages (Browne, 1968). Some early work was also carried out on termites and wood-borers in Indochina (now Vietnam, Cambodia, and Laos) and on termites and pests of shade trees in tea gardens in Sri Lanka.

Forest entomological research in other tropical countries of Asia-Pacific is more recent. A valuable book in two volumes by L. G. E. Kalshoven (1950–51), entitled *Pests of Crops in Indonesia* and originally published in Dutch, was revised and translated into English by Van der Laan in 1981 and contains information on some forest pests of Indonesia. A SEAMEO-BIOTROP (Southeast Asian Regional Center for Tropical Biology) publication entitled *Forest Pests and Diseases in Southeast Asia*, edited by Guzman and Nuhamara (1987), gives an overview of forest insect pest problems in Indonesia, Malaysia, the Philippines and Thailand. A checklist of forest insects in Thailand was published more recently (Hutacharern and Tubtim, 1995). Other recent comprehensive publications in English on forest entomology in Asia-Pacific include *Asian Tree Pests – An Overview* by Day *et al.* (1994); *Forest Pest Insects in Sabah* by Chey (1996); *Insect Pests and Diseases in Indonesian Forests* edited by Nair (2000); and *Forest Entomology: Ecology and Management* (pertaining to India) by Thakur (2000).

Comprehensive publications on forest insects of Africa and Latin America are rare. A recent publication by Wagner *et al.* (1991) entitled *Forest Entomology in West Tropical Africa: Forest Insects of Ghana* covers the forest insects of Ghana and provides a brief history of forest entomology in West Africa. A West African Timber Borer Research Institute was established in Kumasi, Ghana in 1953, during the British colonial period, to focus on control of ambrosia beetles on logs for export. Also, research on termites was carried out in Ghana by a unit of the Commonwealth Institute of Entomology. It was only after the establishment of a Ghanaian national Forest Products Research Institute in 1964 that attention was paid to other areas of forest entomology.

Pest problems of pine and eucalypt in several Latin American countries were covered in a 1985 publication entitled *Noxious Insects to Pine and Eucalypt Plantation in the Tropics* assembled by Pedrosa-Macedo (1985). In 1992, the Tropical Agricultural Center for Research and Education (CATIE) published a field guide and a companion handbook, entitled *Forest Pests in Central America*, which dealt with pests of 18 common forest trees in the region and their control (CATIE, 1992a, 1992b).

A more recent book by Speight and Wylie (2001) covers the general aspects of tropical forest entomology for all the tropical regions of the world.

2.2 The diversity of tropical forest insects

2.2.1 *Structural diversity*

The insect orders

A forest insect is, to use Beeson's (1941) words, quite simply an insect which lives in a forest. Since forest comprises a variety of habitats, most insect groups except the highly specialized ones, though not all the species, are present in forests. Therefore it is instructive to look at the overall classification of insects to gain an insight into the structural diversity of forest insects. Insect groups more abundant in forests will be further considered below.

Insects are now classified into 30 orders, as listed in Table 2.1. The scheme of insect classification has undergone various changes over the past few years as

Table 2.1. *Classification of insects (class: Insecta)*[a]

Group	Group	Group	Order	Common names	Number of species
Primitively wingless insects			Archeognatha	bristletails	500
			Zygentoma	silverfish	400
Pterygota	Paleoptera		Ephemeroptera	mayflies	3 100
			Odonata	dragonflies, damselflies	5 500
	Neoptera	Polyneoptera	Dermaptera	earwigs	2 000
			Grylloblattodea	ice crawlers	26
			Mantophasmatodea	rock crawlers	15
			Plecoptera	stoneflies	2 000
			Embiodea	web spinners	500
			Zoraptera		32
			Phasmatodea	stick insects, leaf insects	3 000
			Orthoptera	crickets, grasshoppers	20 000
			Mantodea	mantises	1 800
			Blattaria	cockroaches	4 000
			Isoptera	termites	2 900

	Paraneoptera	Psocoptera	bark lice	4 400
		Phthiraptera	true lice	4 900
		Thysanoptera	thrips	5 000
		Hemiptera	bugs, aphids, scale insects	90 000
	Holometabola (Endopterygota)	Coleoptera	beetles	350 000
		Raphidioptera	snakeflies	220
		Megaloptera	alderflies, dobsonflies	270
		Neuroptera	lacewings	6 000
		Hymenoptera	wasps, bees, sawflies, ants	125 000
		Mecoptera	scorpionflies	600
		Siphonaptera	fleas	2 500
		Strepsiptera	twisted wings	550
		Diptera	true flies	120 000
		Trichoptera	caddisflies	11 000
		Lepidoptera	butterflies, moths	150 000

[a]*Based on Grimaldi and Engel (2005)*

new knowledge on phylogenetic relationships has accumulated from morphological, molecular and palaeontological data. The classification given here is based on the recent synthesis of information by Grimaldi and Engel (2005) in their book on evolution of the insects. The major recent changes are (i) removal of three orders, that is, Collembola, Protura and Diplura from the class Insecta and their placement under a separate class Entognatha (characterized by mothparts appendages recessed within a gnathal pouch on the head capsule) and (ii) the discovery in 2002 of a new insect order named Mantophasmatodea. The newly created class Entognatha, and the class Insecta (= Ectognatha) are grouped together under an epiclass Hexapoda (six-legged arthropods). Among the three orders of Entognatha, proturans and diplurans are minute organisms that occur in soil, rotting wood and leaf litter and are rarely encountered, while the collembolans, also minute, are very abundant soil organisms found on decaying organic matter in tropical forests and play a role in its recycling (see Chapter 3).

Among the 30 orders under Insecta, two major groups are recognized – the primitive, wingless insects comprising two orders, that is, Archeognatha (bristletails) and Zygentoma (silverfish), and the winged or secondarily wingless insects (Pterygota) which make up the rest (Table 2.1). The Pterygota has two major subdivisions – Paleoptera, consisting of the orders Ephemeroptera (mayflies) and Odonata (dragonflies and damselflies) in which the wings cannot be folded back over the body, a primitive condition, and Neoptera, consisting of the rest of the orders in which the wings can be folded. Neoptera, which comprises the bulk of the insect orders (26 out of 30), is further divided into three groups – Polyneoptera, Paraneoptera and Holometabola (Endopterygota), based on several considerations. Holometabola are those insects in which there is complete metamorphosis, with a larval, pupal and adult stage. In this group the wings develop internally and the immature stage, called larva, is different from the adult in structure and habits, as in the case of the butterfly. Other pterygote insects normally have a simple incomplete metamorphosis (hemimetabolous) and usually have no pupal instar. The wings develop externally and the immature insects, called nymphs, resemble the adults in structure and habits, as in the case of the grasshopper. In contrast, the primitively wingless insects display virtually no change from immature stages to adult (ametabolous). In all three groups of Neoptera some of the orders are grouped together to form superorders based on their closer relationships. Insects representing all the orders are illustrated in Fig. 2.1. Together, the 30 orders of insects display great structural diversity, unmatched by other living organisms which, coupled with their physiological and behavioural diversity, make insects successful in a wide variety of environments.

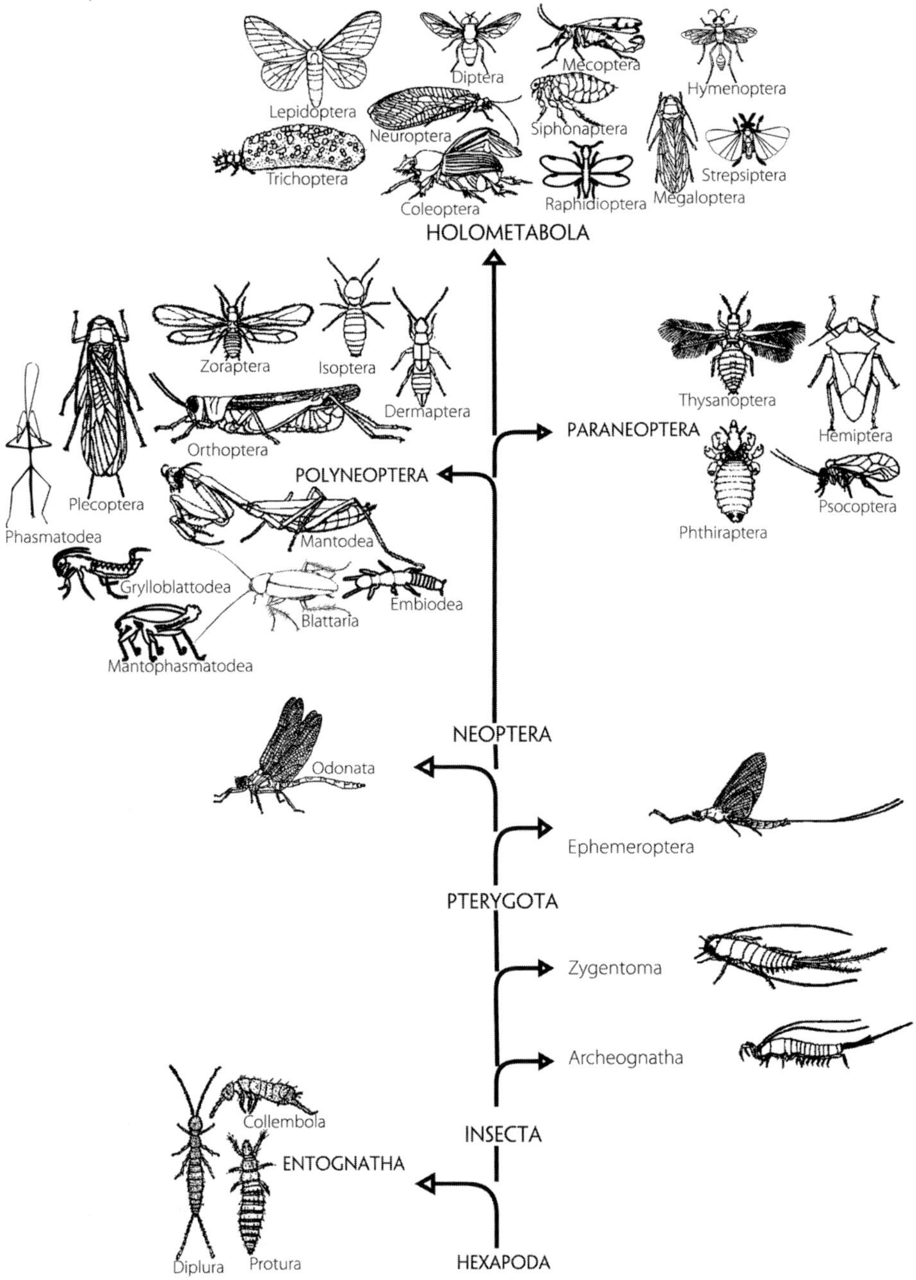

Fig. 2.1 The diversity of insect orders. Representatives of the various orders and their evolutionary relationships are shown. Modified from *Illinois Natural History Survey Circular 39* (Ross, 1962). Orders and groupings were updated as per Grimaldi and Engel (2005). Three orders of primitively wingless Hexapoda, now excluded from the class Insecta and placed under a separate class, Entognatha, are also shown.

Table 2.1 also shows the approximate number of species described throughout the world under each order. The order Coleoptera has the largest number of species, followed by Lepidoptera, Hymenoptera and Diptera. Together they account for about 80% of all insects, and it is interesting to note that the 'big four' orders are all holometabolous (Grimaldi and Engel, 2005).

Dominant orders of tropical forest insects

All insect orders are present in the tropical forest ecosystem, except Grylloblattodea (ice crawlers) which are confined to the cold temperate forests of the Northern Hemisphere and Mantophasmatodea (rock crawlers) confined to xeric, rocky habitats in southern Africa. However some orders are dominant, that is, more abundant, more conspicuous or more important, because of their negative impact on forest trees, particularly plantations. These dominant orders are discussed briefly below. As indicated in Chapter 1, tropical forests have a greater diversity of insects than temperate and boreal forests. However, this is not necessarily so for all groups of insects. For example, the orders Diptera and Hymenoptera have greater species diversity in temperate regions (Price, 1997). Similarly, among aphids (Hemiptera) 80% of species have been recorded in the temperate regions. Also some groups are more diverse in cooler regions within the tropics. For example, in India, more species of aphids and *Drosophila* have been recorded at higher elevations than at lower (Chakrabarti, 2001; Vasudev *et al.*, 2001). However, in general, insects are more numerous in the tropics. For example, out of about 2500 species of mosquitoes, 76% are found in the tropics and subtropics (Gillett, 1971) and out of 760 species of the carpenter bee *Xylocopa* 90% are in the tropics (Gerling *et al.*, 1989).

Order Coleoptera (beetles)

This is the largest order of insects worldwide, as well as in the tropical forests, in terms of the number of species. It is also of greatest importance in terms of damage caused to trees. Beetles are present everywhere, in all the major forest habitats, feeding on a variety of organic matter. A bewildering variety of beetles feeds on wood. They include the large beetles of the family Cerambycidae (longhorn beetles) that feed on freshly felled wood with intact bark, and small beetles of the families Anobidae, Bostrichidae, Brentidae and Curculionidae (Scolytinae, Platypodinae) that feed on drier wood. Passalidae, Anthribidae, Lucanidae and Oedemeridae feed on wet, rotten wood. The dominant leaf-feeding beetles belong to Chrysomelidae and Curculionidae, although some scarabaeids and buprestids also feed on leaves. There is even a buprestid leaf miner, *Trachys bicolor* on *Butea frondosa*. A large variety of beetles in the families Scarabaeidae, Tenebrionidae, Cucujidae and Elateridae feed on vegetable matter on the ground, humus and soil. There are carnivorous beetles in the families Carabidae, Cicindelidae, Cleridae, Coccinellidae and Staphylinidae. Species of the family Dermestidae feed on keratinous material of animal origin such as hide, hair and hoof. Species of Anthribidae and Bruchidae feed on seeds. Most of the above families are rich in species. For example, in the Indian region alone there are over 1200 species of Cerambycidae and 87 species of Bostrichidae (powder-post beetles) (Beeson, 1941).

Most of the earlier studies on insect fauna were based on ground surveys and light trap collections. Recent collections employing the technique of canopy fogging with insecticide have shown the very rich fauna of beetles in the canopy of tropical rain forests. For example, Erwin and Scott (1980) reported 1200 species of beetles representing at least 57 families, from the canopy of a single tree species *Luehea seemannii* in Panama. Of these the majority were herbivores and the rest predators, fungivores and scavengers (Table 2.2). The *L. seemannii* canopy beetle fauna may have included some species simply resting on the foliage or on bark. Erwin (1983b) also reported 1085 species of beetles belonging to at least 55 families from the canopies of four types of rain forests within 70 km radius of Manaus, Brazil. Curculionids and chrysomelids are the most dominant canopy beetles.

Order Lepidoptera (moths and butterflies)

This is the second largest order of insects in terms of number of species, both in the forests and outside. It is also second in importance economically, after Coleoptera, in terms of damage caused to trees. While the short-lived adults – the moths and butterflies – feed on nectar and other fluids, caterpillars of most species feed on foliage. Some species of Pyralidae, Gelechiidae, Blastobasidae and Oecophoridae bore into young shoots and some species of Cossidae, Hepialidae and Tineidae bore into branch wood. Indarbelidae and some species of Tineidae feed on bark. Caterpillars of some species of Pyralidae and Eucosmidae feed on seeds and fruits. Some species of Blastobasidae, Noctuidae, Tineidae, Lycaenidae etc., are carnivorous. Some 85 species of lepidopterans have been recorded on the teak tree alone (Table 2.4).

Some of the well-known forest pests such as the teak defoliator *Hyblaea puera* (Hyblaeidae), teak bee hole borer *Xyleutes ceramicus* (Cossidae) and mahogany shoot borers *Hypsipyla robusta* and *H. grandella* (Pyralidae) belong to this order. This order also includes some economically useful species such as the mulberry silkworm and the tassar silkworm (Saturnidae).

Order Hymenoptera (ants, bees, and wasps)

The order Hymenoptera which includes ants, bees and wasps is the third largest in the number of species worldwide, and its members play an important

Table 2.2. *Diversity of beetles (order Coleoptera) associated with the tree* Leuhea seemannii *in Panama*

Trophic group and Family	No. of species	Trophic group and Family	No. of species
Herbivores	*(682)*	*Predators*	*(297)*[a]
Anobiidae	14	Carabidae	41
Bruchidae	6	Cleridae	12
Buprestidae	14	Coccinellidae	36
Byturidae	1	Colydiidae	5
Cantharidae	19	Cucujidae	18
Cerambycidae	62	Dytiscidae	1
Chrysomelidae	205	Eucnemidae	11
Curculionidae	210[b]	Histeridae	3
Elateridae	12	Lampyridae	12
Helodidae	12	Lycidae	9
Languriidae	14	Melyridae	2
Limnichidae	1	Mycteridae	11
Monommidae	1	Orthoperidae	10
Mordellidae	43	Rhizophigidae	1
Phalacridae	28	Scydmaenidae	3
Ptilodactylidae	35	Staphylinidae	110
Rhipiphoridae	1	Trogositidae	7
Scarabaeidae	3		
Throsidae	1		
Fungivores	*(69)*	*Scavengers*	*(96)*
Anthribidae	11	Anthicidae	15
Biphyllidae	1	Cryptophagidae	9
Ciidae	8	Dermestidae	6
Curculionidae (Platypodinae)	2	Euglenidae	11
Endomychidae	5	Hydrophilidae	2
Erotylidae	9	Nitidulidae	22
Heteroceridae	1	Tenebrionidae	31
Lathidiidae	3	*Unidentifiable families*	
Melandryidae	14	Family 1	1
Pselaphidae	7	Family 2	1
Scaphidiidae	8		

[a]The subtotal is as given in Erwin and Scott (1980); the number of species given against the families adds up only to 292 but it was estimated that undetermined species under Staphylinidae (subfamily Aleocarinae) would add another 50 species to the total.

[b]Not given in Erwin and Scott (1980) but deduced from the total of 682 species of herbivores given in Erwin (1983a).

Data from Erwin and Scott (1980)

role in the ecology of tropical forests as pollinators, and parasitoids of injurious insects. The many species of tropical honeybees alone have provided subsistence and economic benefits to the tribal, rural and urban societies of the tropics since ancient times. The role played by parasitoids in the families Ichneumonidae, Braconidae, Chalcidae, Elasmidae, Eulophidae, Bethylidae, Trichogrammatidae etc. in keeping the populations of the several tree pests within bounds, by parasitising their eggs, larvae and pupae, is immeasurable. Although the leaf-cutting ants of tropical America are pests in general, large populations of ants are important predators and scavengers in the tropical forests. In the canopy of tropical forests in Panama, Erwin (1983b) found that among the 18 orders of canopy insects present 50.8% of the individuals were hymenopterans, of which 84% were ants. In addition to the leaf-cutting ants, a small number of hymenopterans such as sawflies, gall wasps and wood wasps are also pests of trees, although these are more important in temperate than in tropical forests. In addition, over 100 species of the genus *Tetramesa* belonging to the predominantly parasitic family Eurytomidae (subfamily Chalcidoidea) are phytophagous; *T. gigantochloae* infests the stem of some bamboos in Malaysia (Narendran and Kovac, 1995). Among the tropical sawflies (suborder Symphyta), *Shizocera* sp. (Argidae) is a defoliator of *Manglietia conifera* in Vietnam (Tin, 1990) and several species of *Sericoceros* (Argidae) feed on the leaves of some trees in tropical America (Ciesla, 2002).

Order Hemiptera (bugs)

This order includes bugs that can be distinguished into three main groups (suborders): Heteroptera or the 'true bugs' which includes water skaters, belostomatids, bed bugs, tingids, lygaeids, pentatomids etc.; Sternorrhyncha which includes whiteflies, scale insects, aphids and jumping plant lice (psyllids); and Auchenorrhyncha which includes the leaf hoppers, tree hoppers and cicadas. In Heteroptera, the forewings are thick and stiff at the basal half and thin and membraneous at the distal half, and the abdomen has scent or stink glands. In all bugs, the mouthparts are of a piercing and sucking type. Generally, the bugs suck the sap of plants, but members of some families such as Reduviidae and Pentatomidae are predators and suck the fluid of other animals including insects. The major families of importance to tropical forestry are Cicadidae, Coccidae, Psyllidae and Tingidae. Cicadas are well-known insects of the tropical forests and feed on the sap of tender shoots and twigs of trees. They feed gregariously for long hours and the copious fluid excreta ejected by them from the tree tops drops on the ground like an incessant spray. The shrill but loud noise produced by the male cicadas in chorus is characteristic of tropical forests. The Coccidae include the economically beneficial lac insect on which a whole industry has long flourished. Tingidae include the pest *Tingis beesoni* which causes dieback of *Gmelina arborea* saplings in plantations. Psyllidae include the well-known pests *Heteropsylla cubana*, that attacks leucaena, and *Phytolyma* spp. that attack *Milicia* spp. Some bugs are implicated in transmission of tree diseases such as sandal spike.

Recent studies have shown that the bug fauna of tropical tree canopies can be substantial. For example, Wolda (1979) recorded 332 species of bugs from the canopy of the tree *Luehea seemannii* in Panama, by canopy fogging collection over three seasons (Table 2.3). In a study in primary lowland rain forest in Sulawesi, Indonesia, Rees (1983) found 168 taxa of bugs in traps set at 30 m height.

Table 2.3. *Diversity of bugs (order Hemiptera) associated with the tree* Luehea seemannii *in Panama*

Family	No. of species
Membracidae	71
Derbidae	32
Deltocephalinae	23
Cicadellinae	22
Typhocybinae	21
Gyponinae	20
Issidae	19
Cixiidae	16
Coelidiinae	15
Flatidae	15
Achilidae	14
Neocoelidiinae	10
Idiocerinae	8
Psyllidae	8
Cercopidae	7
Xestocephalinae	6
Agalliinae	5
Delphacidae	3
Tropiduchidae	3
Cicadidae	2
Kinnaridae	2
Nirvaniinae	2
Fulgoridae	1
Nogodinidae	1
Total	332

Data from Wolda (1979)

Order Isoptera (termites)

Termites are characteristically tropical insects that feed on dead wood. Some 2900 species have been recorded. They are social insects, with caste differentiation among individuals. Some species live exclusively within wood (family Kalotermitidae), but the majority are ground dwellers. The nests of ground dwellers are either subterranean or project above ground in the form of small or large, conspicuous mounds. Some species make carton nests, attached to tree trunks. Generally, termites forage underground or under cover of mud tunnels, but a few species like *Acanthotermes* spp. (family Hodotermitidae) forage above ground in the open. They cut pieces of grass and carry them in procession to subterranean galleries much like the leaf-cutting ants of tropical America. Some species feed on the root of eucalypt saplings or other tree species. A few species attack the trunk of mature trees and hollow them out. Examples of trunk-feeding termites are *Neotermes* spp. which attack teak in Indonesia and mahogany in Fiji (Nair, 2001a) and *Coptotermes* spp. which attack eucalypts in Australia (Elliott *et al.*, 1998) and rain forest trees in central Amazonia (Apolinário and Martius, 2004).

The importance of termites in tropical forests is twofold; they are beneficial when they recycle wood and turn over the soil, but injurious when they destroy crops.

Order Orthoptera (grasshoppers and crickets)

Grasshoppers and crickets are common phytophagous insects of tropical forests. The major groups are the short-horned grasshoppers (family Acrididae) comprising about 9000 world species, the long-horned grasshoppers (family Tettigonidae) comprising about 5000 world species, the crickets (family Gryllidae) and the mole crickets (family Gryllotalpidae) (Hill, 1997). Locusts, although primarily agricultural pests, damage forest trees during outbreaks. Several species of locusts are known in the African and Asian regions and, although extensive outbreaks have occurred periodically in the past, the frequency and severity of outbreaks have been reduced substantially in recent times through international monitoring and control programmes. In general grasshoppers, although ubiquitous in tropical forests, do not increase in large enough numbers to cause serious damage. Exceptions are the indigenous *Plagiotriptus* spp. (Eumastacidae) which have become persistently severe defoliators of exotic pine plantations in parts of east Africa (Schabel *et al.*, 1999). The acridid *Zonocerus variegatus*, known as the variegated grasshopper, is also a serious pest of agroforestry crops in some parts of Ghana during the dry season (Wagner *et al.*, 1991). Crickets and grasshoppers sometimes cause extensive damage to forest tree seedlings in nursery beds by feeding on the succulent stems.

How many species?

No one knows how many species of insects are there in the tropical forests. The total number of insect species formally described throughout the world so far is close to a million, but the estimated total ranges from 3 to 30 million. At least 10 000 new species of insects are discovered and named each year worldwide. Many genera contain hundreds of species. For example, there are at least 600 species of *Culex* mosquitoes, 350 species of *Anopheles* mosquitoes, 1500 species of the fruitfly *Drosophila* (Wheeler, 1986) and 730 species of the carpenter bee *Xylocopa* (Gerling *et al.*, 1989). As in the case with plants, the number of species of insects is far greater in tropical forests than in temperate and boreal forests. Insects reflect and magnify the diversity of trees as each tree species offers a variety of niches. Also, in insects, which have a much shorter generation time than trees, speciation can be expected to be much faster, particularly in the warm tropics. The wide range in our estimates for the world total of insect species is mainly due to the uncertainty about their number in tropical forests. With about 80% of the world's insect taxonomists located outside the tropics (May, 1994), vast numbers of tropical insects remain unnamed.

One approach that has been taken to estimating the species totals is to thoroughly sample a taxonomic group, usually an order, in a relatively unstudied representative region in the tropics and determine what fraction of the species from this region have previously been recorded. Then, using the ratio between those previously described and the total determined by the intensive survey of the region, global totals are estimated. In such a study, Hodkinson and Casson (cited by May, 1994) found a total of 1690 species of bugs in a representative tropical rain forest in Sulawesi, Indonesia, of which only 37% had been previously recorded. This led them to estimate, by extrapolation, that the total number of insect species in the world is 2–3 million. A different approach was used by Gaston (1991) to estimate species totals: he surveyed insect systematists instead of forests, as Grimaldi and Engel (2005) put it, and arrived at an estimate of 5 million.

Identification of tropical insects is often difficult because of inadequate taxonomical knowledge of them. However, unless a species is formally described and named, it cannot be included in the species count. For example, in a study in the Silent Valley National Park in Kerala, India, about 400 taxa of moths were collected but only 246 could be identified with certainty to species level (Mathew and Rahamathulla, 1995), even after referring to experts in the International Institute of Entomology of the Commonwealth Agricultural Bureaux. Obviously many of the 154 unidentified taxa could turn out to be new species. Lepidoptera is a comparatively well-studied group; the situation in many other groups is worse.

For example, out of 295 species of Psocoptera collected from the forests of Panama, 264 (nearly 90%) were undescribed (Broadhead, 1983). Due to the paucity of specialist taxonomists our museums, especially in the developing, tropical countries, are glutted with collections representing new species (Cherian, 2004). According to Narendran (2001), about 60 000 insect species from India have been described but 6–10 times more Indian species are yet to be discovered.

Some tropical forest habitats are difficult to sample. Most available information comes from collections made at the ground level using nets or light traps. The comparatively recent technique of collecting insects by canopy fogging with insecticide has highlighted the richness of the insect fauna of tropical tree canopies. As noted earlier, in a lowland seasonal forest in Panama, insecticidal fogging of the canopy of a single tree species *Luehea seemannii* over three seasons, yielded about 1200 species of beetles (Erwin, 1983a) and 332 species of bugs (Wolda, 1979). Since then, several such studies have been made. One study of the vertical distribution of insects, using light traps set at 1 – 30 m above ground level in rain forests in Sulawesi, Brunei, Papua New Guinea and Panama (Sutton, 1983) showed marked concentration of insects of the orders Hemiptera, Lepidoptera, Diptera, Hymenoptera and Coleoptera in the upper canopy and of Ephemeroptera at mid-levels. Based on his study of the canopy beetle fauna of *L. seemannii* in Panama, and using a chain of extrapolations, Erwin (1982) estimated a global total of 30 million species of insects. His arguments rested on several assumptions, including a generalization that there are 163 beetle species specific to each tropical tree species, which obviously is a gross overestimate. Some of his other assumptions have also been shown to be unrealistic (May, 1994; Speight *et al.*, 1999).

May (1994) has reviewed the various methods used to arrive at estimates of insect numbers and has shown that all are based on some assumptions. He has concluded that a global total of fewer than 10 million insect species, and probably around 5 million, is a reasonable estimate. Hawksworth and Kalin-Arroyo (1995) put the 'reasonable' figure at 8 million. Nevertheless, the canopy insect surveys by Erwin and others have demonstrated the great diversity and abundance of insects present in tropical forest canopies, a habitat neglected for sampling in the past because of inaccessibility. Even 5 million species of insects represents an enormous diversity when compared with about 4500 species of mammals, 9000 of birds or even 21 000 of fish.

2.2.2 *Functional diversity: the feeding guilds*

The diversity of forest insects is also reflected in their feeding habits. Almost all organic matter in the forest is eaten by one or other insect species.

Although the feeding habits of the dominant taxonomic groups were indicated in Section 2.2.1, insects can be grouped into feeding guilds across the taxonomic groups. A group of species that all exploit the same class of resource in a similar way is called a guild and guild membership cuts across taxonomic groupings. This kind of grouping helps to focus attention on the ecological functions of insects as discussed in the next chapter and also on the impact of insects on the forest. Under each feeding type some examples are given, but the major pest insects will be discussed elsewhere.

Leaf feeders

Leaf feeders constitute a large proportion of forest insects. Members of the orders Lepidoptera, Coleoptera and Orthoptera are the common leaf feeding insects. Leaf is consumed in a wide variety of ways by different insects. The simplest is the wholesale consumption of leaf by groups such as caterpillars and beetles, of which there are thousands of species, together feeding on almost all species of trees. Defoliation by caterpillars often results in widespread damage to forest plantations. Some caterpillars, such as the teak leaf skeletonizer *Eutectona machaeralis* and the early instars of most caterpillars, feed only on the green leaf tissue between the network of veins which results in skeletonization of leaves. Some caterpillars tie the leaf together or roll the leaves and feed from within. Bagworms or case moths feed on leaves, hiding themselves within bags made of leaf or other plant material. Some lepidopteran caterpillars and some fly maggots (Agromyzidae) mine into the leaf between the upper and lower epidermal layers and feed on the green matter, creating mines, blisters or blotches of various shapes.

Sap feeders

Sap feeders constitute a comparatively small proportion of species, but some are of economic significance because of population outbreaks or because they act as vectors of disease. Most sap feeders belong to the order Hemiptera although some Diptera (fly maggots) and Thysanoptera (thrips) also feed by sucking. They feed on succulent plant parts such as tender leaf, shoot, fruit, flower or seed by sucking the sap or liquefied tissues. Cicada, leaf hoppers, psyllids, mealy bugs, scale insects and aphids are examples. Extensive outbreaks of the leucaena psyllid *Heteropsylla cubana* and the conifer aphids *Cinara* spp. and *Pineus* spp. have occurred across continents. Some bugs, like the tingid *Tingis beesoni* on *Gmelina arborea* saplings and the mirid *Helopeltis* spp. on eucalypt seedlings and saplings, inject toxic saliva during feeding causing necrosis of plant tissue and shoot dieback. Others like the sandal bug *Rederator bimaculatus* transmit pathogens to host trees (Balasundaran *et al.*, 1988). Yet others, like

Asphondyla tectonae (Diptera: Cecidomyiidae) on teak, feed from within stem galls and *Phytolyma* spp. (Hemiptera: Psyllidae) on *Milicia* spp. feed from within leaf galls.

Stem feeders

Stem feeders include shoot borers, bark borers, sapwood borers and sapwood cum heartwood borers. They constitute a fairly large group of tropical forest insects.

Shoot borers are mostly lepidopteran larvae of the families Pyralidae, Oecophoridae and Cossidae. They bore into the young, tender shoots of trees and saplings. Examples are the pine shoot borer *Dioryctria* spp., the mahogany shoot borer *Hypsipyla* spp. and the *Bombax* shoot borer *Tonica niviferana*. Some small beetles such as the curculionids bore into the shoot of seedlings.

Bark borers include the bark surface feeding caterpillar *Indarbela quadrinotata* as well as the more economically important 'bark beetles' of the family Curculionidae (Scolytinae). Although most scolytine bark beetles in the tropics do not cause damage as serious as their counterparts in the temperate forests, many species are present there and some, like the pine bark beetles in the Latin American countries, have caused occasional outbreaks. There is a wide variety of small beetles feeding on the bark and sapwood of many tree species. They multiply in large numbers when the trees are weakened by other causes. Many of them feed on dead wood under normal circumstances.

Sapwood cum heartwood borers of the coleopteran family Cerambycidae bore deep into the tree trunk and cause more serious damage. Examples are *Hoplocerambyx spinicornis* attacking *Shorea robusta* in India, *Aristobia horridula* attacking *Dalbergia cochinchinensis* in Thailand and *Xystrocera festiva* attacking *Falcataria moluccana* in Indonesia. Some lepidopteran caterpillars like *Xyleutes ceramicus* attack living teak trees in Myanmar and Thailand. Some species of termites also attack and hollow out the trunk of live trees.

Flower, nectar, pollen, and seed feeders

Many species of thrips (Thysanoptera) feed on the flowers of trees. Several insects in their adult stage feed on nectar or pollen and incidentally effect cross-fertilization of plants. Most members of this group belong to Hymenoptera, exemplified by the honeybee. Members of Lepidoptera, Coleoptera and Thysanoptera also feed on nectar and pollen. A large number of species belonging to Coleoptera and Lepidoptera feed on the young or mature seeds of trees while they are still on the tree or when fallen on the ground. The most common seed-feeding insects are listed in Chapter 3

Dead-wood feeders

The insect fauna that thrives on dead wood in the tropical forest is very rich and includes members of the coleopteran family Cerambycidae, which feed on freshly dead wood, as well as smaller beetles of several families and termites which feed on drier wood. In addition, there are insects that feed on decaying wood on the forest floor. The dead-wood feeders will be discussed in detail in Chapter 3.

Insects that feed on litter, fungi, algae, root, animal dung, and soil

This heterogenous group of insects constitute a large proportion of the total insect fauna, with members drawn from the orders Coleoptera, Collembola, Hemiptera, Orthoptera and Isoptera. They are involved in the breaking down of dead plant biomass. Litter-feeding insects are discussed in detail in Chapter 3. A study of British insect fauna showed that more than half of the insect species were carnivorous or saprophagous (Strong *et al.*, 1984). This must be true of tropical insects as well. A variety of coleopteran larvae feed on roots, animal dung and soil, as do many species of termites. Many insects feed on fungi associated with decaying matter. Even among the canopy insects, many species are scavengers and fungivores (Table 2.2). Trees usually have a large guild of Psocoptera feeding on fungal spores, algal cells and lichen present as micro-epiphytes on the bark and leaf surface (Broadhead and Wolda, 1985).

Predators and parasitoids

Predators and parasitoids constitute a large group of insects. They feed mostly on other insects. Predators belong to several orders – Hemiptera, Dictyoptera, Odonata, Dermaptera, Neuroptera, Coleoptera, Diptera and Hymenoptera. Most parasitoids belong to Hymenoptera as discussed in Section 2.2.1 and some to Diptera (family Tachinidae).

2.3 The concept of pests

The above discussion has shown that there is a great diversity of forest insects adapted morphologically, physiologically and behaviourally to feed on almost all forest vegetation and organic matter derived from it. By feeding on a variety of substances, they perform some ecological functions which are discussed in Chapter 3. As will be shown, the activities of some groups of insects such as decomposers and pollinators are beneficial to trees, but insects feeding on living trees have a negative impact on the growth and survival of individual trees. This impact becomes all the more serious in plantations.

Insects, as a group, are capable of feeding on almost all parts of a tree – the leaves, flowers, fruit, shoot, bark, sapwood, heartwood and the roots. Usually each tree species has a characteristic spectrum of associated insects comprising plant feeders, fungus eaters, detritivores, predators, parasitoids and even simply casual visitors. Phytophagous insects do not adversely affect the tree when the insect numbers are small, which is usually the case. Apparently a tree can dispense with some portion of its biomass without adverse effects on its growth.

To show the diversity of insects associated with a living tree, insects found on teak in India and the adjacent countries are listed in Table 2.4. Altogether there are 174 species. The list would be longer if species found on teak in all countries were included. The vast majority are leaf feeders, accounting for 137 species, followed by 16 sap feeders, 14 shoot/stem feeders, 5 root feeders and 3 seed feeders. Only a smaller number of species may be present in a given locality at a given time. Most species cause only slight or occasional damage and their impact on the tree is negligible. However, a few species are serious pests on teak. These include the leaf-feeding caterpillars *Hyblaea puera* and *Eutectona machaeralis* (or *Paliga machaeralis* in some countries) and the wood-boring caterpillar *Xyleutes ceramicus* (or *Alcterogystia cadambae* in India). An additional few like the hepialid sapling stem borers and the scarabaeid seedling root feeders are pests of lesser importance.

As with teak, a large but variable number of insect species is usually associated with each tree species. Sometimes, some of them increase in numbers enormously, creating a pest situation. Although some insects can cause economic damage even when present in small numbers (e.g. a worm in an apple or a borer in wood), generally it is a large increase in the number of individuals of a species that creates a pest situation. This may happen due to one or more of several causes which are discussed in Chapter 7. Usually, only very few of the insect species associated with a tree species will develop pest status as in the case of teak.

A pest is defined as an organism which causes economic damage or other negative impact on human well-being. It therefore reflects a human viewpoint. For example, termites are pests when they feed on the root of eucalypt saplings in plantations and kill them, or when they destroy valuable papers or woodwork in a building, but they are beneficial when they feed on wooden refuse in our backyard or on fallen logs in the natural forest. In the strict sense, only those insects which cause economic loss should be called pests, but in practice all insects that feed on a plant are called pests as the economic impact of many insects has not been determined. Again, an insect species may be a pest at one time but not at another. Thus it is improper to call an insect a pest; an insect is a pest only in some circumstances. Therefore, we can only talk of a pest situation;

Table 2.4. *Insects associated with the living teak tree in India and adjacent countries*

Plant part eaten	Insect order and family	Species
Leaf	**Coleoptera**	
	Chrysomelidae	*Aspidomorpha sanctae-crucis* Fabricius
		Aulacophora foveicollis Lucas
		Chrysochus nilgiriensis Jacoby
		Clitena limbata Baly
		Colasposoma asperatum Lefebvre
		C. downesi Baly
		C. rufipes Jacoby
		C. semicostatum Jacoby
		C. villosulum Lefebvre
		Corynodus peregrinus Herbst
		Hispa armigera Olivier
		Mimstra gracilicornis Jacoby
		Nodostoma bhamoense Jacoby
		N. dimidiatipes Jacoby
		Sebaethe brevicollis Jacoby
	Curculionidae	*Alcides scenicus* Faust
		Astycus aurovittatus Heller
		A. latralis Fabricius
		Attelabus feae Faust
		Crinorrhinus approximans Marshall
		Cyphicerinus tectonae Marshall
		Cyphicerus humeralis Marshall
		C. interruptus Faust
		Cyrtepistomus pannosus Marshall
		Episomus lacerta Fabricius
		Hypomeces squamosus Fabricius
		Myllocerus discolor variegatus Boheman
		M. dorsatus Fabricius
		M. echinarius Marshall
		M. lineaticollis Boheman
		M. sabulosus Marshall
		Peltotrachelus albus Pascoe
		P. pubes Faust
		Phytoscaphus fractivirgatus Marshall
	Scarabaeidae	*Adoretus epipleuralis* Arrow
		Apogonia clypeata Moser
		A. granum Burmeister
		A. nigricans Hope
		Autoserica insanabilis Brenske
		Holotrichia tuberculata Moser
		Lachnosterna serrata Fabricius (as adult)

Table 2.4. (cont.)

Plant part eaten	Insect order and family	Species
	Lepidoptera	
	Aganaidae	*Asota caricae* Fabricius var. *alciphron*
	Arctiidae	*Amsacta lactinea* Cramer
		Asura subricosa Moore
		Diacrisia flavens Moore
		D. obliqua confusa Butler
		Pericallia matherana rubelliana Swinhoe
	Cosmopterygidae	*Labdia callistrepta* Meyrick
	Epiplemidae	*Dirades adjutaria* Walker (syn. *D. theclata* Butler)
	Eupterotidae	*Eupterote germinata* Walker
		E. undata Blanchard
	Gelechiidae	*Deltoplastis ocreata* Meyerick
	Geometridae	*Ascotis infixaria* Walker
		A. selenaria Hubner
		A. selenaria imparata Walker
		A. trispinaria Walker
		Boarmia fuliginea Hampson
		Buzura suppressaria Guenee
		Cleora alienaria Walker
		C. cornaria Guenee
		Dysphania percota Swinhoe
		Ectropis bhurmitra Walker
		Hyposidra sp.
		H. successaria Walker
		H. talaca Walker
		Orsonoba clelia Cramer
		Problepsis vulgaris Butler
	Glyphiperrygidae	*Brenthia albimaculana* Snellen
	Gracilariidae	*Phyllocnistis tectonivora* Meyrick
	Hyblaeidae	*Hyblaea constellata* Guenee
		H. puera Cramer
	Lasiocampidae	*Cosmotriche* sp.
		Estigena pardalis Walker
	Limacodidae	*Macroplectra signata* Moore
	Lymantriidae	*Dasychira grotei* Moore
		D. mendosa Hubner
		D. pennatula Fabricius
		Euproctis bimaculata Walker
		E. fraterna Moore
		E. howra subsp. *rhoda* Swinhoe
		Euproctis sp.

Table 2.4. (cont.)

Plant part eaten	Insect order and family	Species
		Laelia sp.
		Lymantria ampla Walker
		Orgia postica Walker
	Noctuidae	*Beara dichromella* Walker
		B. nubiferella Walker
		Chilkasa falcata Swinhoe
		Falana sordida Moore
		Fodina pallula Guenee
		Heliothes armigera Hubner
		Maurilia iconica Walker *ab. instabilis* Butler
		Mocis undata Fabricius
		Paectes subapicalis Walker
		Phytometra albostriata Bremer & Gray
		P. chalcites Esper
		Prodenia litura Fabricius
		Tiracola plagiata Walker
	Nymphalidae	*Eriboea arja* Felder
		Telchinia violae Fabricius
	Pyralidae	*Acharana mutualis* Zeller
		Eutectona machaeralis Walker
		Hapalia mandronalis Walker
		Macalla plicatalis Hampson
		Margaronia glauculalis Gueneee
		M. vertumnalis Gueneee
		Sylepta sp.
		S. straminea Butler
	Sphingidae	*Acherontia lachesis* Fabricius
		Cephonodes hylas Linnaeus
		Herse concolvuli Linnaeus
		Macroglossum gyrans Walker
		Psilogramma menephron Cramer
	Thyrididae	*Theretra alecto* Linnaeus
		Striglina glareola Felder
	Tortricidae	*Cacoecia micaceana* Walker
		Homona coffearia Nietner
	Xyloryctidae	*Acria emarginella* Donovan
		Aeolanthes sagulata Meyerick
	Yponomeutidae	*Ethmia hilarella* Walker
	Orthoptera	
	Acrididae	*Aulacobothrus luteipes* Walker
		Aularches miliaris Linnaeus

Table 2.4. (cont.)

Plant part eaten	Insect order and family	Species
		Catantops innotabile Walker
		Ceracris deflorata Brunner
		Chlorizeina unicolor Brunner
		Choroedocus robusta Serville
		Dittoplernis venusta Walker
		Eucoptacra saturata Walker
		Pachyacris vinosa Walker
		Phlaeoba sp.
		Pyrithous ramachandrai Bolivar
		Schistocerca gregaria Forska
		Spathosternum prasiniferum Walker
		Teratodes monticollis Gray
		Trilophidia sp.
	Tettigonidae	*Conocephalus maculatus* Guillou
		Ducetia thymifolia Fabricius
		Mecopoda elongata Linnaeus
Sap	**Hemiptera**	
	Aphididae	*Aphis gossypii* Glover
	Cercopidae	*Phymatostetha deschampsi* Lethierry
		Ptyleus nebulosus Fabricius
		P. praefractus Distant
	Coccidae	*Drosichiella phyllanthi* Green
		D. tectonae Green
		Pseudococcus deceptor Green
		P. tectonae Green
		Icerya aegyptiaca Douglas
		I. formicarum Newstead
		Laccifer lacca Kerr
	Fulgoridae	*Eurybrachys tomemtosa* Fabricius
		Flara ferrugata Fabricius
	Jassidae	*Tettigonia ferruginea* Fabricius
	Membracidae	*Leptocentrus taurus* Fabricius
		Otinotus oneratus Walker
Shoot	**Coleoptera**	
	Curculionidae: Scolytinae	*Hypothenemus tectonae* Stebbing
Stem	**Coleoptera**	
	Cerambycidae	*Aristobia approximator* Thomson
		A. birmanica Gahan
		Dihammus cervinus Hope
		Nupserha variabilis Gahan

Table 2.4. (cont.)

Plant part eaten	Insect order and family	Species
	Chrysomelidae	*Sagra jansoni* Baly
		S. longicollis Lacordaire
	Curculionidae	*Alcides ludificator* Faust
	Isoptera	
	Kalotermitidae	*Neotermes tectonae* Dammerman
	Lepidoptera	
	Cossidae	*Zeuzera coffeae* Nietner
		Alcterogystia (syn. *Cossus*) *cadambae* Moore
		Xyleutes ceramicus Walker
	Hepialidae	*Endoclita chalybeata* Moore
		Sahyadrassus malabaricus Moore
Seed	**Coleoptera**	
	Anobidae	*Lasioderma sericorne* Fabricius
	Lepidoptera	
	Pyralidae	*Pagyda salvalis* Walker
		Dichocrocis punctiferalis Guenee
Root	**Coleoptera**	
	Scarabaeidae	*Lachnosterna serrata* Fabricius
		Clinteria clugi Hope
		Oryctes rhinoceros Linnaeus
	Cerambycidae	*Celosterna scabrator* Fabricius
	Isoptera	
	Termitidae	*Hospitalitermes birmanicus* Snyder

Data from Mathur and Singh (1961)

we cannot categorise a particular insect species as a pest. Many phytophagous insects are not pests when they occur in low densities. However, pest is a convenient term to refer to a phytophagous insect which appears to cause economic loss.

The major groups of forest pests are defoliators, shoot borers and wood borers. They can retard the growth of trees, cause deformation of the stem or even kill seedlings, saplings and trees. They also degrade and destroy harvested wood. Some pest outbreaks are spectacular, extending over thousands of hectares and causing enormous economic loss. The biology, ecology and economic importance of the major forest pests are discussed in Chapters 4, 5, 6 and 10.

3

Ecology of insects in the forest environment

3.1 The concept and functioning of ecosystem

To understand the status and role of insects in the forest environment, it is first necessary to briefly discuss the concept and functioning of an ecosystem.

Nature is a highly complex, interconnected system. Links exist not only between the living components but also between living and non-living components. The significance of this complex interrelationship has been well captured by the concept of **ecosystem**. An ecosystem can be defined as a functional unit or entity consisting of a community of living organisms and the physical environment in which they live, interacting with each other so that there is a flow of energy from plants to the consumer organisms, and the cycling of some materials between living and non-living components, with all the living components existing in a dynamically steady state. It denotes a level of organization above the living community, integrating it with its abiotic environment. It provides a framework to organize our thoughts, as well as facts observed from nature. In practical terms, a forest ecosystem consists of the community of living trees and other vegetation, animals and micro-organisms and their physicochemical (i.e. abiotic) environment which function together as an integrated unit or system. It is difficult to delimit the physical boundaries of an ecosystem because of the continuity of interconnections, but for practical purposes it can be delimited according to our convenience. Thus we can talk of a one-hectare patch of tropical forest as an ecosystem, or the entire forests of Sri Lanka, of Southeast Asia or of the tropical forests as a whole as an ecosystem. Each of these is itself part of a larger ecosystem. The largest ecosystem on earth is the **biosphere** or **ecosphere**, which includes all the living organisms on the earth interacting with the physical environment of the earth.

In the same way as a population of individuals of a species has properties not possessed by a single individual (such as reproductive potential, capacity to adapt to new environment, to evolve etc.), an ecosystem has properties not possessed by its components individually. Let us look at the properties of an ecosystem.

A generalized model of an ecosystem is shown in Fig. 3.1. The green plants capture a small part of the incident solar energy by converting it into chemical energy through the process of photosynthesis. This is accomplished by synthesizing glucose out of CO_2 obtained from the air, and water absorbed from the soil. Green plants that are capable of photosynthesis are therefore called **producers**. Trees in the forests, grasses, herbs, shrubs and algae in ponds and oceans are all producers. Using glucose, they also manufacture other organic compounds for their own metabolism and growth. All other life forms (except some micro-organisms capable of chemosynthesis) must obtain their energy directly or indirectly from the producers. Those that feed directly on the producers are called **herbivores** or **primary consumers**. Leaf-feeding insects form a significant group of primary consumers. Other primary consumers are herbivorous animals like hare, deer, elephant etc. Animals that feed on the

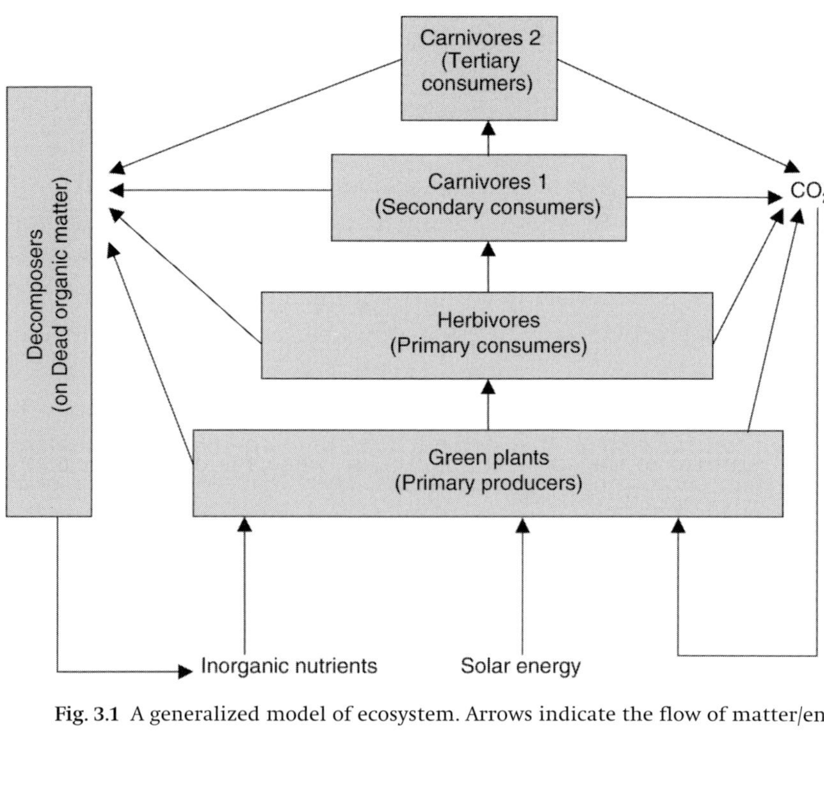

Fig. 3.1 A generalized model of ecosystem. Arrows indicate the flow of matter/energy.

primary consumers are called **secondary consumers** or **first-order carnivores** and include predatory insects, insectivorous birds and bats, frogs, reptiles and some mammals. At the next higher feeding (trophic) level are **tertiary consumers** or **second-order carnivores** like the eagle which feeds on other birds, or snakes which feed on frogs that eat insects. At the highest trophic level is the carnivore like the tiger, which has no predator.

Thus living organisms are linked to one another through feeding relationships or the **food chain** but these interconnections are often not as simple as the trophic levels described above would suggest. The same animal can be a herbivore as well as a carnivore (like some monkeys, or some leaf-eating but cannibalistic caterpillars like *Helicoverpa armigera*), or a first-order as well as second-order carnivore (like a snake which eats a herbivorous rat as well as a carnivorous frog). This makes it difficult to depict the feeding relationships in an ecosystem by simple food chains as these will have many cross-links, forming a complicated **food web**. Nevertheless, the simplified concept of trophic levels facilitates understanding of how an ecosystem functions. Energy is lost to the environment at each trophic level because life processes and activities require energy which is utilized and dissipated as heat. Also the energy efficiency of conversion of living matter from the lower to the higher trophic level is low. Therefore, if the energy contained at each higher level is plotted as a column on top of the energy at the lower level, it results in a characteristic pyramidal shape which is usually known as the **pyramid of energy**. Biomass and the number of individuals at higher trophic levels also follow a similar pattern.

While energy is lost as it passes through the ecosystem, nutrient elements are recycled. Life processes require a number of elements such as carbon, hydrogen, oxygen, nitrogen, phosphorus, sulphur, calcium and magnesium, which become constituents of living matter. If these were tied up in the living and dead organic matter forever, life would soon come to a halt because there is only a limited supply of these elements within the boundaries of the biosphere. These materials must therefore be cycled. Herbivores absorb these elements from the soil and the atmosphere, incorporate them into living matter and pass them on to the higher trophic levels through the food chain. However, they are returned to the soil and atmosphere when the dead living matter is broken down by another group of organisms called **decomposers**. Decomposers are largely bacteria, fungi, macroarthropods and microarthropods, including insects that inhabit the soil and litter. Some larger animals such as crows and vultures also aid in the process of decomposition. This process of recycling of nutrients through the ecosystem is called **nutrient cycling** or **biogeochemical cycling**. Water and carbon dioxide, the starting materials for photosynthesis, are cycled through the atmosphere and the forest. A typical nutrient cycle,

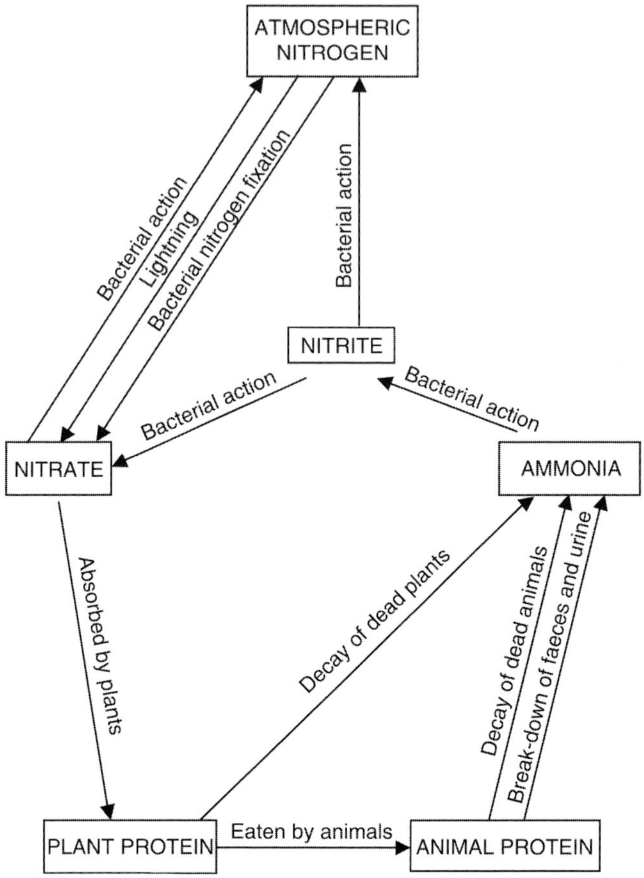

Fig. 3.2 The nitrogen cycle. Details from Andrews (1972).

that of nitrogen, an important constituent of protein and nucleic acids, is shown in Fig. 3.2 to illustrate the complex interrelationships between living and non-living matter. Note how the atmospheric nitrogen enters the living matter and returns to the atmosphere. Plants use nitrogen mainly in the form of the soluble nitrate which is absorbed through roots. It enters into plant proteins and other molecules from where it is transferred to animal tissues, and is liberated into the atmosphere in the form of gaseous nitrogen when they decompose, and then rebuilt into nitrate through various pathways to start the cycle again, as shown in Fig. 3.2. Similar cycles operate for many other inorganic nutrients such as phosphorus, potassium, calcium, magnesium and sulphur, although not all the chemical elements on earth are involved in the construction of biological materials. We have little knowledge of the cycling of many elements used in very minute quantities by different organisms.

3.2 Role of insects in ecosystem processes of tropical forests

Insects play key roles in ecosystem processes at two trophic levels – as primary consumers and as decomposers. They also play minor roles as secondary and tertiary consumers. In addition, they interact with many other life forms in innumerable ways. These direct and indirect effects of insects on trees, other organisms and the physical environment can influence primary production, succession and evolution of plant communities.

3.2.1 *Insects as primary consumers*

The phytophagous insect fauna of tropical forests is rich in species (i.e. diversity) as we saw in Chapter 2, although under normal conditions the number of individuals per species (i.e. abundance) remains low. In general, each plantation tree species has 10 – 200 species of associated insects (Chapter 5, Section 5.4). For trees in natural forests very little information is available. Published records for 20 species in the moist – deciduous forest of Kerala, India (Nair *et al.*, 1986a) show an average of 38 species (range, 2–188) of insects per tree species, but this is not based on a comprehensive search, most records being incidental. The richness of the canopy insect fauna of tropical forests was clearly brought out in several recent studies (see Chapter 2). In lowland seasonal forest in Panama, 1200 species of beetles and 332 species of bugs were recorded from the canopy of a single tree species, *Luehea seemannii*. The greater part of canopy insects are herbivorous, feeding on the leaves or sap. Some studies indicate that chewing insects consume 7–10% of the leaf area in tropical forest canopies (Wint, 1983) although higher levels of leaf consumption may occur in some seasons. For example, in the lowland rain forests of Panama and Papua New Guinea, Wint (1983) recorded 13–14% defoliation during the summer months. However, in a study in moist-deciduous and evergreen forests in Kerala, India, Nair *et al.* (1986a) found only about 2% annual foliage loss caused by insects. This estimate was based on monthly visual scoring of leaf loss, on five trees each of 38 representative species in the natural forests over a two-year period. The effect of sap-sucking, gall-forming and stem-boring insects was not assessed. Based on several studies made in temperate countries, Schowalter *et al.* (1986) estimated that insects normally consume less than 10% of annual foliage standing crop. The few studies in the tropics mentioned above suggest that this may be applicable to tropical forests as well, although quick post-defoliation regrowth of leaves in tropical trees, following insect feeding, complicates these estimates.

Leaf consumption may reach extremely high levels when insect outbreaks occur. Several examples of such outbreaks are described in Chapters 4 and 10. During these outbreaks, huge quantities of foliage of particular tree species are consumed by millions of larvae in a matter of weeks and defoliation may spread over hundreds or thousands of hectares. Such outbreaks may occur annually, as in the case of the teak defoliator *Hyblaea puera* in Asia-Pacific, or at irregular intervals in other cases. A regular periodicity of outbreaks, for example, the 9-year cycle of larch budmoth outbreaks in the European Alps (Baltensweiler and Fischlin, 1988) has been noted in some temperate forest insects. Although not a tree pest, the example of the African army worm, *Spodoptera exempta* illustrates the impact of such outbreaks. In a well-sampled infestation of *S. exempta* in Kenya, in May 1965, at a mean density of 28 sixth instar larvae per m^2 that spread over 65 km^2 southeast of Nairobi, Odiyo (1979) estimated that herbage consumption amounted to 50 tons dry weight per day for a week. Similarly, in a study of about 10 000 ha of teak plantation at Nilambur in Kerala, India, Nair *et al.* (1998a) showed that in the year 1993, between February and September, the foliage of about 7260 ha of plantations was almost totally consumed by the outbreaking caterpillar populations of the moth *Hyblaea puera*. It is obvious that such outbreaks have serious impact on primary production and also affect other ecosystem functions in various ways, by releasing the nutrients locked up in the trees into the soil, allowing penetration of light into lower canopy levels etc. These effects are discussed further in Section 3.2.7 below.

3.2.2 *Insects as secondary and tertiary consumers*

Insect predators and parasitoids function as secondary consumers. Predators include mantids, hemipterans, neuropterans (chrysopids), dermapterans, odonates, some beetles (Carabidae, Cicindelidae, Melyridae and Staphylinidae) and some hymenopterans (ants, wasps). Parasitoids include a wide variety of hymenopterans and some dipterans. The secondary consumers constitute a large group, feeding mostly on other insects but also on some other animals. For example, carabids feed on worms and snails and mosquitoes on the blood of mammals. Ants which act as predators and scavengers constitute an important group in tropical forests, often accounting for 20–40% of the arthropod biomass of the canopy. They also act as indirect herbivores when they feed on extrafloral nectar, specialized food bodies of some plants or on the liquid exudates of sap-feeding insects. Hyperparasites which feed on other parasitoids are tertiary consumers, but they are a small group.

3.2.3 *Insects as decomposers*

Insects play a vital role in nutrient cycling in tropical forests. A staggering diversity of insects is involved in the decomposition process. This is understandable because a large part of the biomass of the forest passes through the decomposer chain. In a temperate oak–pine forest, it was shown that while only 30 dry g/m^2 per year of the net primary production passed through the herbivore food chain, 360 dry g/m^2 per year passed through the decomposer food chain (Woodwell, 1970, cited by Price, 1997). The following account will illustrate how different types of organic matter on the forest floor are acted upon by various specialized groups of insects in the decomposition process.

Insects on litter

Litter fall is one of the main mechanisms by which cycling of nutrients between soil and vegetation takes place in the forest. Litter consists of dead plant material including leaves, twigs, bark, flowers, fruits and seeds that fall to the ground. Litter fall for a variety of forest types and localities in the tropics ranges from 5.7–13.3 tons of dry matter ha^{-1} yr^{-1} with a mean of 8.7 tons ha^{-1} yr^{-1} (Anderson and Swift, 1983) which is a substantial quantity. Litter decomposition involves a sequence of physical and biological processes by which litter falling on the ground is finally transformed into humus. Physical processes involve leaching of chemicals and mechanical disintegration. In the biological process, a complex community of fungi, bacteria, actinomycetes and invertebrates including insects take part, the action of one group of organisms making the litter suitable for action by the next group, as in an assembly line, along the vertical layers of litter. Fungi, bacteria and actinomycetes are the pioneers. They start growing on moist litter and initiate the process of biodegradation. As many as 32 genera of fungi have been recorded on teak litter (Mary and Sankaran, 1991). The chief role of insects is comminution of litter (breaking up into smaller particles) by feeding. This facilitates further microbial growth. Mixing of litter with the faecal pellets of insects also promotes microbial activity. Soil insects consume micro-organisms and thus regulate the microbial activity. Other groups of organisms involved include mites, symphylids, pauropods and earthworms.

The abundance and activity of soil and litter insects are influenced by a number of factors such as temperature, moisture level, nutrient composition and the chemical milieu determined by secondary plant chemicals. Usually, the fauna consists of detritivores, fungivores and their predators. Collembolans dominate all other groups in terms of number of individuals, followed by acarines, particularly the oribatid mites. In a study in Indonesian rain forest, Stork (1987) recorded 3000 individual organisms per m^2 of floor surface (600 in litter and 2400 in soil), consisting mostly of collembolans and mites. The species composition varies depending on the tree species that contribute to the litter, the stage of decomposition, climate, vertical position in the litter layer etc. For example, the groups of organisms associated with breakdown of teak litter in India are collembolans, scarabaeids, earwigs, staphylinids and termites (Ananthakrishnan, 1996). The topmost litter layer is dominated by the predatory food chain and contains groups such as Orthoptera, Hemiptera and Coleoptera; the intermediate layer consisting of partially decomposed litter contains Thysanoptera, Dermaptera, Coleoptera and Collembola; and the transitional, bottom humus layer contains mainly collembolans in addition to acarines (Ananthakrishnan, 1996). Table 3.1 shows the rich insect fauna associated with tropical forest litter in India. The food relationships among the major litter inhabiting organisms are depicted in Fig. 3.3. Note that while fungi attack fresh, semi-decomposed and decomposed litter, insects may feed on semi-decomposed litter (termites and some dermapterans), decomposed litter (termites, blattids, gryllids and some beetles) or fungi (thrips and some beetles) or may be predators (bugs, some dermapterans and some beetles).

The beetle fauna associated with decaying litter and humus is very large (Table 3.1). The more important families are Scarabaeidae, Nitidulidae and Staphylinidae. Larvae of the scarabaeid subfamilies/tribes Cetoninae, Dynastinae, Euchirinae, Melolonthinae and Rutelinae feed on rotting plant matter or humus and/or rootlets. A typical example is the ruteline genus *Anomala*, with over 200 species in India. The adult beetles swarm after the early heavy showers of rain and lay eggs in soil. The larvae tunnel through the soil eating the rootlets of plants and rotted vegetable matter and the life cycle usually takes a year (Beeson, 1941). Several species of the family Nitidulidae feed on fermenting or decaying vegetable matter, souring fruit and withering flowers, decomposing bark or sapwood, fungi, and pollen. Another important group is Staphylinidae that comprises at least 2500 species in the Indian region. They are small beetles with varied habits, mostly scavengers or predators, found on fresh and decaying fungi, pollen, rotting fruit, vegetable debris, under the bark of decaying trees etc. Larvae of many species of flies of the dipteran families, Muscidae and Drosophilidae, also live on decaying vegetables and fruits and hasten the process of decomposition.

Insects that feed on the fallen fruits and seeds on the forest floor include several species of beetles of the families Bruchidae and Curculionidae, and a few microlepidopterans of the families Blastobasidae and Pyralidae. A representative list of seed pests and their seed/fruit hosts is given in Table 3.2.

Insects on dead and fallen wood

In addition to the fine litter discussed above, wood of larger dimensions in the form of dead standing trees, fallen trees and the stumps and roots of dead trees make up a significant part of the dead forest biomass. Such woody material is attacked by a large variety of insects which aid its decomposition.

Table 3.1. *Insects and collembolans[a] associated with tropical forest litter in India*

COLLEMBOLA	HEMIPTERA	DERMAPTERA
Brachystomella terrefolia	**Anthocoridae**	*Dicrana kallipyga*
Callyntura ceylonica	*Amphiareus constrictus*	*Eparchus insignis*
Callyntra sp.	*Anthocoris* sp.	*Euborellia annulipes*
Cryptophygus sp.	*Buchananiella carayoni*	*Forcipula lurida*
C. thermophylus	*Cardiastethus* sp.	*F. quadrispinosa*
Cyphoderus sp.	*C. nagadiraja*	*Forminalabis sisera*
Entomobrya sp.	*C. pygmaeus pauliani*	*Gonolabidura nathani*
Folsomia sp.	*Orius maxidentax*	*Labia minor*
Folsomides sp.	*Physopleurella anathakrishnani*	*Labidura riparia pallas*
Hypogastrura communis	*Scoloposcelis* sp.	*Metisolabis bifoviolate*
Hypogastrura sp.	*S. asiaticus*	*Psalis castetsi*
Isotomodes sp.	*S. parallelus*	*Pygidicrana eximia*
Lepidocyrtus sp.	*Xylocoris (Proxylocoris) distanti*	*P. picta*
Lobella sp.	*X. clarus carayon*	*P. valida*
Onychiurus sp.	**Lygaeidae**	ORTHOPTERA
Paratulbergia indica	*Geocoris* sp.	**Blattidae**
Salina indica	*G. ochropterus*	*Blatella* sp.
S. quatturofasciata	*Graptostethus servus*	*B. germanica*
S. tricolour	*Lygaeus* sp.	*Theria* sp.
Sinella sp.	*Metochus uniguttatus*	**Gryllidae**
Sminthurinus sp.	*Naphilus dilutus*	*Gryllus* sp.
Sphaeridia sp.	**Reduviidae**	COLEOPTERA
Tulbergia sp.	*Acanthaspis coprolagus*	(families)
Xenylla sp.	*A. quinquespinosa*	**Anthicidae**
THYSANOPTERA	*A. pedestris*	**Bostrichidae**
Apelaunothrips madrasensis	*Catamiarus brevipennis*	**Carabidae**
Azaleothrips amabilis	*Ectomocoris ochropterus*	**Coccinellidae**
Boumieria indica	*E. tibialis*	**Cucujidae**
Dinothrips sumatrensis	*Echtrechotes pilicornis*	**Hydrophilide**
Ecacanthothrips tibialis	*Haematorhophus nigro-violaceous*	**Hydroscaphidae**
Elaphrothrips denticollis	*Lisarda annulosa*	**Nitidulidae**
Gastrothrips falcatus	*Lophocephala guerinii*	**Scarabaeidae**
Hoplandrothrips flavipes	*Piratus affinis*	**Staphylinidae**
Hoplothrips fungosus	*P. mundulus*	**Tenebrionidae**
Nesothrips acuticornis	*Rhaphidosoma atkinsoni*	
Neurothrips indicus	*Rhinocoris marginatus*	
Priesneriana kabandha		
Stictothrips fimbriata		
Stigmothrips limpidus		

[a]*Class Entognatha*

Data from Ananthakrishnan (1996)

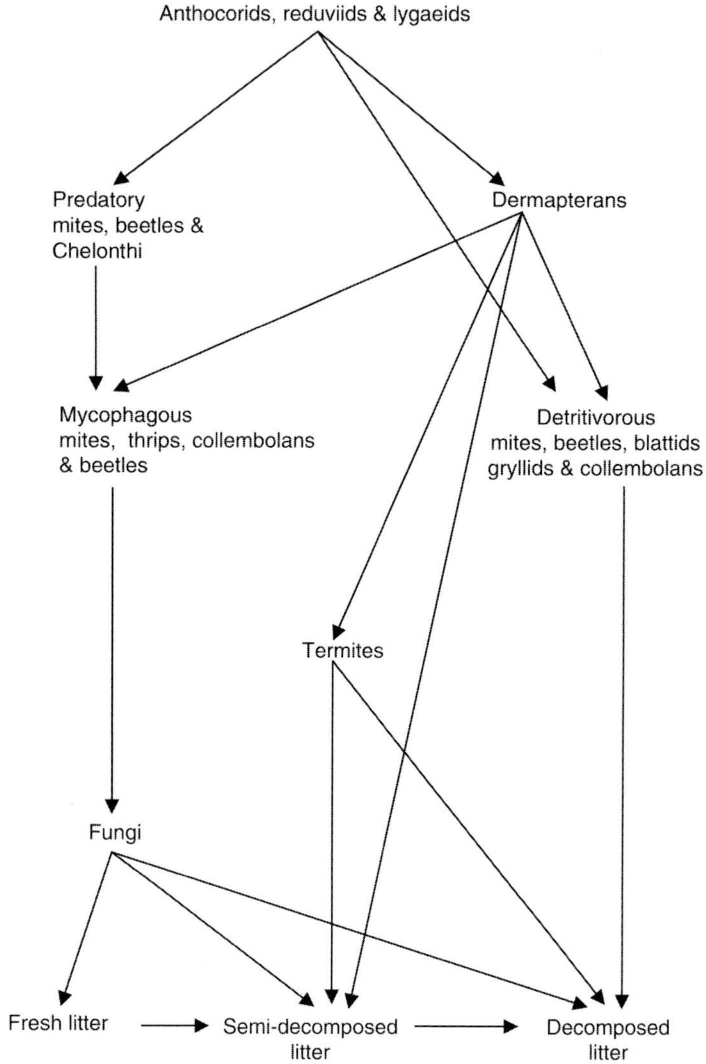

Fig. 3.3 Food relationships among the major litter-inhabiting organisms. Adapted from Ananthakrishnan (1996).

Beetles are the main group that infest freshly dead wood. The tropical wood-feeding beetle fauna is very rich and is dealt with in detail in Chapter 6. The large beetles of the family Cerambycidae attack freshly dead wood with intact bark while smaller beetles of several families, Anthribidae, Curculionidae and Lyctidae, attack wood devoid of bark. Together, they comprise thousands of species. In the Indo-Malayan region alone, for example, there are over 1200 species of cerambycids, 300 species of scolytines, 250 species of platypodines and 87 species of bostrichines (Beeson, 1941). The feeding activity

Table 3.2. *A short list of seed/fruit-feeding insects and their tree hosts*

Insect (order, species and family)	Host seed/fruit
Coleoptera	
Bruchus bilineatopygus (Bruchidae)	*Albizia procera*
B. pisorum	*A. lebbeck, Cassia fistula, Dalbergia sissoo*
Pachymerus gonagra (Bruchidae)	*A. lebbeck, C. fistula, Tamarindus indica*
Caccotrypes carpophagus (Curculionidae: Scolytinae)	*Diospyros quaesita, Polyalthia semiarum*
Stephanoderes cassiae (Curculionidae: Scolytinae)	*Cassia* spp.
Thamnugides cardamomi (Curculionidae: Scolytinae)	*Canarium strictum, Cullenia excelsa, Hardwickia pinnata, Vateria indica*
T. rubidus	*Dipterocarpus pilosus, Eugenia formosa, Mesua ferrea*
Nanophyes spp. (Curculionidae)	Dipterocarps
Lepidoptera	
Dichocrosis leptalis (Pyralidae)	*Pentacme suavis*
D. punctiferalis	*Tectona grandis*
Blastobasis crassifica, B. molinda and *B. ochromorpha* (Blastobasidae)	*Shorea robusta*
B. spermologa	*S. robusta, Dipterocarpus tubinatus, Ficus glomerata, Polyalthia longifolia*

Data from Beeson (1941), Elouard (1998)

of these beetles converts the wood to course fibres or fine dust. Some of these beetles (ambrosia beetles) have fungal associates which help in the degradation of cellulose and lignin. While various kinds of fungi act independently causing wood deterioration, fungal decay may also promote infestation by some species of beetles and termites.

Termites constitute another important group of insects that feed on dead wood, particularly in the drier tropics. The termite density in tropical forests may range from 390–4450 individuals per m^2 and the biomass from 0.7–9.4 g per m^2 (Sen-Sarma, 1996). Recent studies show that tropical wet forests of the Guinea-Congolese block in Africa are extremely rich in termites, predominantly soil feeders, and make up biomass densities of up to 100 g per m^2 (Eggleton *et al.*, 2002). They feed on a wide variety of dead plant material, including wood, bark and leaf litter. While the higher termites of the subfamily Nasutitermitinae produce cellulose-digesting enzymes themselves, others make use of symbiotic micro-organisms (flagellate protozoans or bacteria) that live in their gut to produce the enzymes necessary to digest cellulose. Species of the subfamily Macrotermitinae cultivate a fungus, *Termitomyces*, in fungus combs within their nests. The fungus breaks down the cellulose and lignin in dead plant material and the termites feed on the fungal spores. Thus termites make a significant contribution to the decomposition of woody biomass in tropical forests. Their activities also cause modification of the soil profile and fertility, by bringing large quantities of subsoil to the surface in the course of building mounds and runways and by mixing saliva, excreta, dead bodies of termites etc., in the soil. Many species do not build mounds but construct nests in the soil or buried wood. Some African species of the family Macrotermitinae build huge nests up to 9 m in height and 30 m in diameter (Lavelle *et al.*, 1994). Some species of the genera such as *Speculitermes, Anoplotermes, Pericapritermes* and *Procopritermes* also feed on soil and humus, and thus bring about modification of the soil (Sen-Sarma, 1996). The translocation and modification of soil by termites in tropical forests is enormous, perhaps similar to that of earthworms in tropical savannas. It has been estimated that 250–1250 tons dry weight of soil per ha passes through the guts of earthworms at Lamto in Côte d' Ivoire (Lavelle *et al.*, 1994). Some species of the family Rhinotermitidae even have the ability to fix atmospheric nitrogen using bacteria in their hindgut (Speight *et al.*, 1999).

Termites are replaced by beetles of the family Passalidae (Fig. 3.4) on the floor of wet tropical forests in Asia. These characteristic, fairly large (10–55 mm in length), flattened, black, shining beetles make tunnels inside the logs of fallen trees or stumps in which the larvae also live and feed, causing the disintegration of wood. Most species prefer wood that is moist or wet and rotting, and the infested wood can be easily broken up with the fingers. Passalids are generally gregarious (Beeson, 1941).

Insects on animal dung and carcasses

Animal dung and carcasses constitute another significant component of biological material on the forest floor. The dung-beetles of the scarabaeid subfamilies/tribes Aphodiinae, Coprinae, Geotrupinae and others are the main agents which act on animal excrement and cause its further decomposition (Beeson, 1941). Several species of the genus *Aphodius* feed on fresh animal dung as both adults and larvae, while some others feed on fresh dung as beetles but lay eggs on dry dung in which the larvae make tunnels and feed. Beetles of the genus *Geotrupes* dig a tunnel under a heap of fresh dung and store a quantity of dung in it. The tunnel is later interconnected with a main chamber and side tunnels in which more dung is stored for the larvae. The genus *Onthophagus* comprising about 200 species in India has similar habits. Species of *Copris*, of which there are about 40 in India, breed on the dung of ruminants. Large quantities of dung are taken into a chamber in the ground where it is triturated

Fig. 3.4 A passalid beetle, *Pleurarina brachyphyllus*, an inhabitant of wet decaying wood on tropical evergreen forest floor. Length 45 mm.

by the female and built up into a large mass, then cut up into several balls, each of which receives an egg. In addition, there are the dung rolling beetles such as the coprine genera *Gymnopleurus*, *Scarabaeus* and *Sisybus* which remove a quantity of dung from the main mass, make it into a ball and roll it some distance before it is buried. The dung ball is buried in soil and an egg is deposited in it. Most dung beetles complete their life cycle in three to five weeks. Large beetles of the genus *Heliocopris*, some more than six centimetres long, bury large dung balls which are covered by a layer of hard cemented earth. They have an annual life cycle and the new generation of beetles emerges when the wall of the brood chamber is softened by the monsoon rains. *Heliocopris dominus*, the Indian elephant dung beetle (Fig. 3.5) and *H. dilloni*, the African elephant dung beetle, breed on elephant dung. In *H. dominus*, the brood ball (a dung ball covered by a soil layer) weighs 530–1750 g and two to four such balls may be placed by a single female beetle in a cluster at the end of a slanting tunnel about 40 cm below ground (Joseph, 1998). Several females may work on one dung pat and each may move up to 2.5 kg of fresh dung. Thus the extent of dung removal and soil excavation by these beetles is substantial.

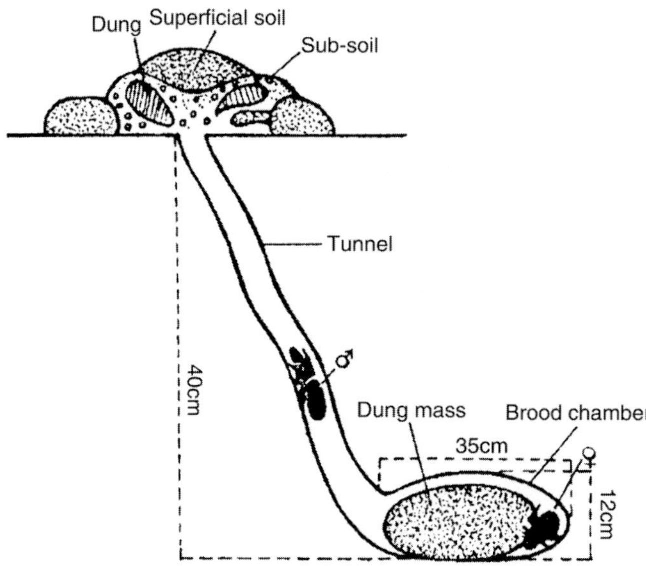

Fig. 3.5 Vertical section (diagrammatic) through dung pat, tunnel and brood chamber of the elephant dung beetle *Heliocopris dominus*. From *Entomon* (Joseph, 1998).

Some species of scarabaeid beetles, known as carrion beetles (e.g. some *Onthophagus* spp.), as well as some staphylinid beetles feed on carrion as do many species of the dipteran families Calliphoridae (flesh-flies), Muscidae and Phoridae. Several species of the beetle family Dermestidae (e.g. *Dermestes vulpinus, Anthrenus flavipes*) feed on dry meat, hide, skin, hoof, horn, hair, wool etc. and cause their decomposition.

3.2.4 *Insects as food*

While many insects form the food of other insects, insects also serve as food for a wide variety of other animals – amphibians, reptiles, birds and mammals. Because of their large number and variety, insects constitute a quantitatively important link in the food chain. In some countries, insects form part of the human diet also. In addition to honey from honeybees which is a prized food item worldwide, many kinds of insects like locusts, grasshoppers, termites and lepidopteran larvae and pupae form part of the human diet, particularly for tribal people. In Nigeria the larvae of *Anaphe venata* (Lepidoptera: Notodontidae), a defoliator of *Triplochiton scleroxylon* in the high forests, are roasted in dry sand and eaten by local tribes (Ashiru, 1988, cited by Wagner *et al.*, 1991) and in Uganda the grasshopper *Homorocoryphus nitidulus* which periodically swarms in large numbers is eaten either raw or cooked

(Hill, 1997). In some parts of Indonesia pupae of the teak defoliator *Hyblaea puera* are eaten. Roasted grasshoppers are often available at roadside food stalls in Thailand.

Insects form part of the food for some plants also. An example is the pitcher plant (*Nepenthes* spp.) which lives in nutrient poor soil and uses trapped insects as a dietary supplement. The plant has modified leaves holding a liquid in a cavity. Insects such as flies that fall into this cavity, attracted by the plant's odour, colour etc., cannot escape and are drowned in the liquid which contains digestive enzymes secreted by the plant. The digested nutrients are absorbed by the plant.

3.2.5 *Insects as pollinators*

Another important ecological role of insects is pollination. A wide range of tree species are insect pollinated. Examples are *Acacia*, eucalypts, *Ficus*, *Mesua* and many dipterocarps. Insect pollination of trees appears to be more common in the tropics, particularly the humid tropics, than in the temperate regions. It is believed that in humid climates wind pollination is ineffective (Price, 1997). Cross-pollination is the general rule in tropical trees although a few are self-fertile. Pollination is usually incidental, but during the course of evolution intricate adaptations have been developed by plants and insects for effecting pollination. As is well known, honey bees collect pollen in the pollen baskets on their hind legs and store it in their hives. Several species of fig trees (*Ficus* spp.) have specially adapted fig wasps as pollinators, a different species of fig wasp for each species of fig, and the complex interrelationship developed for pollination of figs through coevolution is almost unbelievable. Orchids are dependent on bees for pollination and produce chemicals that mimic the sex pheromones of bees to attract them for pollination.

Some species of most of the insect orders may be involved in pollination, but the most important groups of pollinators belong to the orders Hymenoptera (solitary and social bees), Diptera (flies), Lepidoptera (butterflies and moths), Coleoptera (beetles) and Thysanoptera (thrips). While the honey bee is well known as a pollinator, there are also a large number of solitary bees which effect pollination. There are a total of about 20 000 species of bees worldwide, of which more than 85% are solitary (Batra, 1984, cited by Hill, 1997). Many *Shorea* species in Malaysia are pollinated by thrips.

Since pollination is a major ecological function of insects in tropical forests, sustenance of the tropical forest ecosystems is dependent on a critical minimum level of insect biodiversity. Without them as agents of cross-pollination, many trees would not be able maintain their genetic heterogeneity.

3.2.6 *Other ecological interactions*

In addition to those discussed above, insects are involved in innumerable interactions with other organisms. Some examples are given below.

Insects often act as vectors of plant and animal diseases in the tropics and thus influence the dynamics of plant and animal populations. Transmission may be effected either through mechanical transfer of the disease-causing organism or through biological transfer in which the disease organism replicates in the insect vector. Most vectors of animal diseases are Diptera. Examples are mosquitoes transmitting malaria, yellow fever, dengue fever, elephantiasis and encephalitis, tsetse fly transmitting sleeping sickness, flea transmitting bubonic plague etc. Tree diseases may be transmitted by Hemiptera, Coleoptera and Hymenoptera. For example, the bug *Rederator maculatus* transmits spike disease to the sandal tree through a mycoplasma-like organism and *Helopeltis* sp. transmits inflorescence blight and dieback of cashew. While there are many serious tree diseases transmitted by insects in temperate forests, such as Dutch elm disease (caused by a fungus) transmitted by the scolytine bark beetle, *Scolytus* sp.; pine wilt (caused by a nematode) transmitted by a cerambycid beetle, *Monochamus* sp.; and the wood rot of pines transmitted by the wood wasp, *Sirex* sp., many tree diseases transmitted by insects in the tropics are probably unrecorded. Even when insects do not act as vectors of tree diseases, injury caused by them may provide a port of entry to pathogenic organisms.

Mutually beneficial ant–plant associations are well known (Huxley, 1986). Many plants possess specialized structures or chambers for housing ants, called *domatia* (little houses). These structures develop independently without the influence of ants and may be swollen thorns, hollow stems or tubers. Plants which have such structures are called *myrmecophytes*, and myrmecodomatia have been described for over 250 plant species from 19 families. In *Acacia* the domatia provide living or nesting space for the ants while extrafloral nectaries and small nodules at the tip of the leaflets (called Beltian bodies) provide food. In return, the plant benefits from the protection afforded by the ants from herbivorous insects as the ants predate or drive them away. In *Macaranga* trees (Euphorbiaceae), the domatia are inside internodes in the stem of the plant which becomes hollow due to degeneration of the pith. In the ant tree *Tachigali myrmecophila* (Fabaceae: Caesalpinioideae) in Amazonia, the hollow leaf axis and petiole are inhabited by the stinging ant *Psuedomyrmex concolor* (Psuedomyrmecinae). The ant preys on a colony of coccids kept inside the domatia. The coccids, *Catenococcus* sp., produce honeydew which is used by the ants as their main energy source. Ant exclusion experiments showed that removal of the ants increased the herbivore density 4.3 fold and the level of leaf damage tenfold indicating the beneficial role of ants to the tree (Fonseca, 1994). In some cases, waste material produced by the ant colony is absorbed by the plant through the inner lining of domatia and used as a source of mineral nutrients and nitrogen.

On many non-myrmecophytic trees, ants tend or farm mealy bugs, aphids or other sap-sucking bugs and feed on the sugar-rich honeydew secreted by the bugs. Ants protect these bugs from natural enemies and transport them to new shoots or plants.

In tropical America, an exclusive group of ants of the family Attinae cut the foliage of trees into small pieces and carry it to their nest to cultivate a fungus. In the nest, the ants cut the leaf into smaller fragments, 1–2 mm in diameter, chew them along the edges to make them wet and pulpy, mix them with their own faecal exudate, place them in the fungus garden and then add tufts of fungal mycelia picked up from the substratum (Wilson, 1971). The ants eat the inflated tips of the growing fungal hyphae, called 'gongylidia', which are also fed to the larvae. Thus the ant acts as an agent promoting direct decomposition of green leaves. The fungi cultivated by leaf-cutting ants have been identified as species of Agaricaceae (Basidiomycetes). Leaf-cutting, fungus-growing ants are present only in the New World, distributed mainly in the tropics and subtropics. About 200 species of fungus-growing ants have been recognized in 11 genera and some genera make use of the corpses of other arthropods, insect frass etc. for cultivating fungi. Several species of the genera *Atta* and *Acromyrmex* are the dominant leaf-cutting ants. For example *Atta cephalotus* is found in tropical forests from Mexico in the north to Bolivia in the south. Like other ants, leaf-cutting ants are also social insects and they make large nests on the ground. The nest of *A. cephalotes* may cover up to 250 m^2 in surface area and be several metres deep, with hundreds of chambers (Cherrett, 1983), while that of the subtropical Texas leaf-cutting ant *A. texana* may have a central nest mound 30 m in diameter, with numerous smaller mounds extending outwards to a radius of 80 m and may occupy a 30–600 m^2 area (Kulhavy *et al.*, 2001). A nest may have several million workers and they gather several kilograms of leaves per day. In *Eucalyptus* plantations in Brazil, it has been estimated that an adult ant colony uses about one ton of leaves per year (Lima and Filho, 1985). Leaf-cutting ants are generally polyphagous and Lugo *et al.* (1973) estimated that in tropical wet forest in Costa Rica, *A. colombica* took about 0.2% of the gross productivity of the forest.

Termites of the family Hodotermitidae, known as harvester termites, are also reported to forage for grass which they cut and carry to their underground nests. In the moist-deciduous forest of Kerala, India, workers of an unidentified termite species were seen in procession, in the open, each carrying a cut piece of grass, much like the leaf-cutting ants of tropical America (unpublished observations). It is not known whether these termites use the grass to cultivate fungi like the leaf-cutting ants.

The regeneration of some trees is facilitated by the activity of termites. It has been shown (Chacko, 1998) that feeding of termites on the mesocarp of fallen teak fruits (seeds) on the forest floor induces germination of the recalcitrant seeds. On the other hand, insect seed predators may adversely affect regeneration of some tree species. For example, Curran and Leighton (1991) reported that in one year a dipterocarp seed crop of about 100 000 seeds ha^{-1} in the lowland forest of West Kalimantan, Indonesia was entirely destroyed by seed-feeding insects. The phenomenon of mass fruiting of dipterocarps in some years is thought to be a strategy to escape complete seed destruction by satiating the seed pests (Janzen, 1974).

Only some of the known relationships between plants, insects and their environment have been discussed above and several other intricate relationships remain little known. Because of the manifold interactions, forest fragmentation and degradation which lead to loss of biodiversity will adversely affect the proper functioning of tropical forest ecosystem processes.

3.2.7 *Influence on forest primary production, succession and tree evolution*

Mattson and Addy (1975) have argued that phytophagous insects function as regulators (in the cybernetic sense) of primary production in forest ecosystems. According to them, the activity and abundance of phytophagous insects is dependent on the vigour and productivity of the forest ecosystem. When the vigour and productivity of the ecosystem is lowered due to tree age, stressful climatic conditions, low fertility of the site or bottlenecks in the flow of certain vital nutrients, the insects respond by increase in their numbers, leading to population outbreaks. This ultimately results in rejuvenation of the ecosystem as insect grazing stimulates the host's physiological system, increases the penetration of sunlight, increases soil fertility through increased litter fall (including insect excrement and cadavers) and kills weakened or old trees, leading to the growth of more vigorous younger plants of the same species or individuals of other species. Thus insect outbreaks help to maintain nutrient cycling and primary production at optimal rates for a particular site. Each tree species and forest ecosystem supports a variety of insects whose composition varies with the seasonal and ontogenic development of the plants and at least a few of these insects are thought to be capable of making dramatic population changes in response to subtle changes in individual plant or ecosystem processes (Mattson and Addy, 1975). In this manner, insects are thought to act as regulators of forest primary production. Although some authors hold the view that insect outbreaks are too infrequent and their effect too ephemeral to cause substantial and enduring top-down effects on plant communities, there is increasing evidence to show that insect outbreaks are common in many community types worldwide, particularly in large, dense and continuous host stands and that outbreaking insects function as keystone species by reducing the abundance of the dominant species and increasing diversity (Carson *et al.*, 2004).

Insect outbreaks can regulate forest succession by changing the composition of forest stands. When some species of trees are killed by insect outbreaks, the growth and regeneration of other trees are favoured. For example, very heavy and widespread outbreaks of the sal heartwood borer *Hoplocerambyx spinicornis* occur periodically in natural stands of the sal tree *Shorea robusta* in India (see Chapter 4, Section 4.2.2). In a recent epidemic during 1994–99, over three million sal trees spread over half a million ha of forest in Madhya Pradesh were affected and most of the trees were killed (Dey, 2001). While the cause of some of these outbreaks cannot be determined with certainty, most pest outbreaks in natural forest have occurred in tree species that occur gregariously, like in a monoculture, and indications are that at least in some species, outbreaks begin in epicentres where the trees are under stress due to ageing, drought or other causes. Like the scolytine bark beetle outbreaks in temperate pine stands, these outbreaks seem to aid in thinning high density stands of some species to facilitate regeneration of a more balanced mixture of tree species. Thus phytophagous insects may have an ecological role in regulating forest succession. Alteration of plant community dynamics by periodic outbreaks of a chrysomelid beetle, *Microrhopala vittata* was demonstrated experimentally in a herbaceous perennial, *Solidago altissima* (goldenrod) in New York, USA (Carson and Root, 2000). Even at less than outbreak levels, insect herbivores promoted plant species diversity and co-existence through their effects on litter accumulation and light penetration below the canopy of the dominant plant species. Increasing evidence is now accumulating to indicate that insect herbivores exert a major influence in regulating the plant community structure.

Two theories have been proposed to explain the regulatory effect of insects on plant communities. According to the 'Resource Supply Theory', supply of resources (nutrients) to the plants determines primary plant production as well as resource allocation to defences, which in turn determine herbivore population size. This theory suggests that when resource supply to the plants is not uniform, it affects plant defences against insects, leading to insect population outbreaks. In other words, resources at the base of the food web are of primary importance in precipitating herbivore outbreaks and this kind of regulation of a plant community is often referred to as bottom-up control (Chase *et al.*, 2000). Alternatively, the 'Host Concentration Theory' proposes that specialist insect herbivores will exert strong regulatory effects on plant communities whenever their hosts form large, persistent dense stands (Carson and Root, 2000; Long *et al.*, 2003; Carson *et al.*, 2004). Host concentration is believed to promote pest build-up and outbreak by providing a larger absolute supply of food, greater ease in host location due to the physical proximity of the host trees as well as the absence of interfering non-host chemicals and reduced dispersal of pests out of the dense host patch. This kind of regulation of plant community from above (i.e. a higher trophic level) is called top-down control. The host concentration theory is discussed further in Chapter 8, in connection with pest incidence in monocultures versus mixed stands.

Insects may sometimes influence tree evolution. Insect herbivory can affect many aspects of tree performance – growth, form, seed production, seed germination, competitive ability and survival. These effects often exert a negative or positive influence on the success of individual plants or groups of plants which exhibit genetically controlled deviations from the rest of the conspecific population, and can drive evolutionary change in the plants. To illustrate this, consider a simplified example of interaction between a plant and an insect. Assume that a teak tree develops, through mutation, a heritable capacity to produce on bark injury a chemical (inducible defence) that is lethal to newly hatched larvae of the beehole borer, *Xyleutes ceramicus* (see pest profile under teak, in Chapter 10). When the larva attempts to bore into the bark of the tree, the chemical is released and the larva is killed. This chemical will protect such a tree from the pest and increase its survivorship in comparison with other teak trees. Therefore, in course of time, the proportion of individuals which carry this novel borer defence mechanism will increase by the process of natural selection. This process can go on and lead to evolutionary change in the host. Sometimes, an evolutionary 'arms race' will result, with the insect developing new strains that can detoxify the harmful chemical. It is logical to assume that herbivorous insects may influence the population dynamics and evolution of plants in the manner described above, but it is difficult to come up with conclusive evidence because of the complexity of interactions involving a multitude of physical and biotic factors and the many other unknown functions a mutation may serve. Perhaps such effects may operate more effectively on plants which have a shorter life cycle, while on trees they may be limited to instances where particular insect species have the propensity to cause premature death of trees. In general, it may be the insects that adapt and evolve according to tree characteristics because of the very short generation time of insects compared to that of trees. Nevertheless, insects can drive plant evolution (Leather, 2000).

In spite of the many roles that insects fulfil in forest ecosystems, their role in plant evolution is not generally recognized, partly because the total biomass of insects in the forest appears to be small comparison with the tree biomass or the biomass of other animals, except on some occasions. Our understanding has also suffered due to lack of manipulative studies where the insect populations in an ecosystem are experimentally altered (Weisser and Siemann, 2004). Much more remains to be learnt of the role of insects in ecosystem processes.

4

Insect pests in natural forests

4.1 Introduction

It is generally believed that tropical forests, characterised by high species diversity, are free of pest outbreaks, although the trees may support small populations of phytophagous insects. In keeping with this view, mixed tropical forests are usually cited as examples that demonstrate the strong correlation between diversity and stability in relation to pest outbreak. The following statements highlight this conventional wisdom.

> No biologist who has penetrated and explored a truly virgin forest in the tropics has ever reported the occurrence of insect epidemics or has seen evidence of extensive defoliation and borer damage. In tropical evergreen forests with their numerous species of trees and still more numerous hordes of insect species, the absence of epidemics is not surprising. (Beeson, 1941, p. 633)

> Mixed stands are much safer from insect injury than are pure stands. In fact, we may safely say that the greater the diversification of tree species, the less frequent will be insect outbreaks. This is an illustration of the general principle that other things being equal, the degree of environmental stability is in direct proportion to the number of species living together in an environment. (Graham and Knight, 1965, with reference to temperate forests, p. 213)

> It can be generally stated that extensive outbreaks of defoliating insects are uncommon in the high forests of Ghana. This is true because the forests have a high degree of species diversity and most insects have a narrow host range. . . . When, however,

> single species plantations are established, the probability of an outbreak of a defoliating insect increases substantially. (Wagner *et al.*, 1991, p. 24)

> The potential for pest outbreak is nil in primary forest with high diversity of >200 species per acre and in secondary forests with medium to high degree of diversity; low in regenerating forest with mainly pioneer and non-pioneer light demanders; low to medium in enrichment plantings in degraded forests with one to few tree species; and high in forest plantations which are mainly monocultural plantations. (Cobbinah and Wagner, 2001, summarised from Table 1)

> A virgin forest, almost undisturbed by anthropogenic interferences, represent a climax state, where no insect epidemics are known to occur. (Thakur, 2000, p. 473)

> Tropical forests [show] a tendency for more pest problems as they become more disturbed or perturbed away from natural situations. It is somewhat of a dogma these days to state that more diverse ecosystems are also more stable; to put it another way, species-rich communities tend not to exhibit large fluctuations in the abundances of one or more of their constituent species . . . Proponents of this dogma thus stress the need to promote biodiversity in crop systems to avoid the development of pest outbreaks . . . There is nothing wrong with this philosophy in principle, though it may be too simplistic and unreliable in some cases . . . (Speight and Wylie, 2001, p. 40)

> The higher an ecosystem has been simplified, the higher the risk for the outbreak of an insect pest. (Foahom, 2002, p. 40)

While detailed studies are rare, such general statements linking absence of pest outbreaks with high tree species diversity and occurrence of outbreaks with simplification or disturbance of the natural ecosystem are thus common in tropical forestry literature. It is believed that in mixed forests, other tree species associated with a host tree may mask or interfere with its attractiveness for a pest as well as provide nectar and pollen sources or shelter for the natural enemies of the pest. On the other hand, proximity of host trees in a host-dense stand, as in a plantation, is believed to favour the build up of pests by reducing dispersal mortality and providing abundant food.

4.2 Empirical findings

Before we consider pest incidence in natural forests, we may recall the discussion in Chapter 2 on the concept of pests. We defined pests as organisms that cause economic damage or adversely affect human welfare. In the natural forest, it is often difficult to judge whether an insect causes economic damage or not. This is due to several reasons. First, pest incidence in natural tropical forests has not received enough research attention. Second, it is not easy to carry out economic analysis of a pest situation in natural forests because of a large number of variables and uncertainties, and therefore this has seldom been done. Third, all gradations of insect incidence may occur in natural forests, from mere presence of phytophagous insects in small numbers, to occasional local eruptions, to widespread outbreaks. Therefore there is uncertainty as to what conditions might qualify for calling an insect a pest in the natural forest. We will discuss this question further after examining the empirical findings.

Systematic investigations on pest incidence in natural forests are rare and most available information is of an anecdotal nature, i.e. based on unplanned, incidental observations. It is convenient to discuss these empirical findings under two headings, general pest incidence and pest outbreaks, although this is an arbitrary separation of the continuum ranging from minor insect feeding to large-scale outbreaks.

4.2.1 General pest incidence

In a specific pest incidence study in natural forests in Kerala, India, Nair *et al.* (1986a) observed 20 tree species in moist deciduous forests and 18 tree species in evergreen forests, at monthly intervals over a two-year period (Fig. 4.1). All the 38 tree species suffered some insect damage. The most common damage was leaf feeding, noticed on all tree species at some time. Sap-sucking, gall-forming and wood-boring insects were also recorded on some species. The annual defoliation percentage ranged from 0.1–6.7 for the different tree species. The mean monthly defoliation value did not exceed 21% for moist deciduous species and 17% for evergreen species, although individual trees of some species suffered more than 50% defoliation at times. For many species, the mean monthly defoliation never exceeded 5% (Fig. 4.1b). In general, evergreen tree species suffered less damage than moist deciduous species. One species in which greater than 50% defoliation was noted in some individual trees was *Tectona grandis* (teak) in the moist deciduous forest. This defoliation was caused by the caterpillar *Hyblaea puera*, which is a well-known outbreak species in plantations of teak (see Chapter 10). In the above study, not all the species that caused damage were collected and identified due to difficulty in gathering

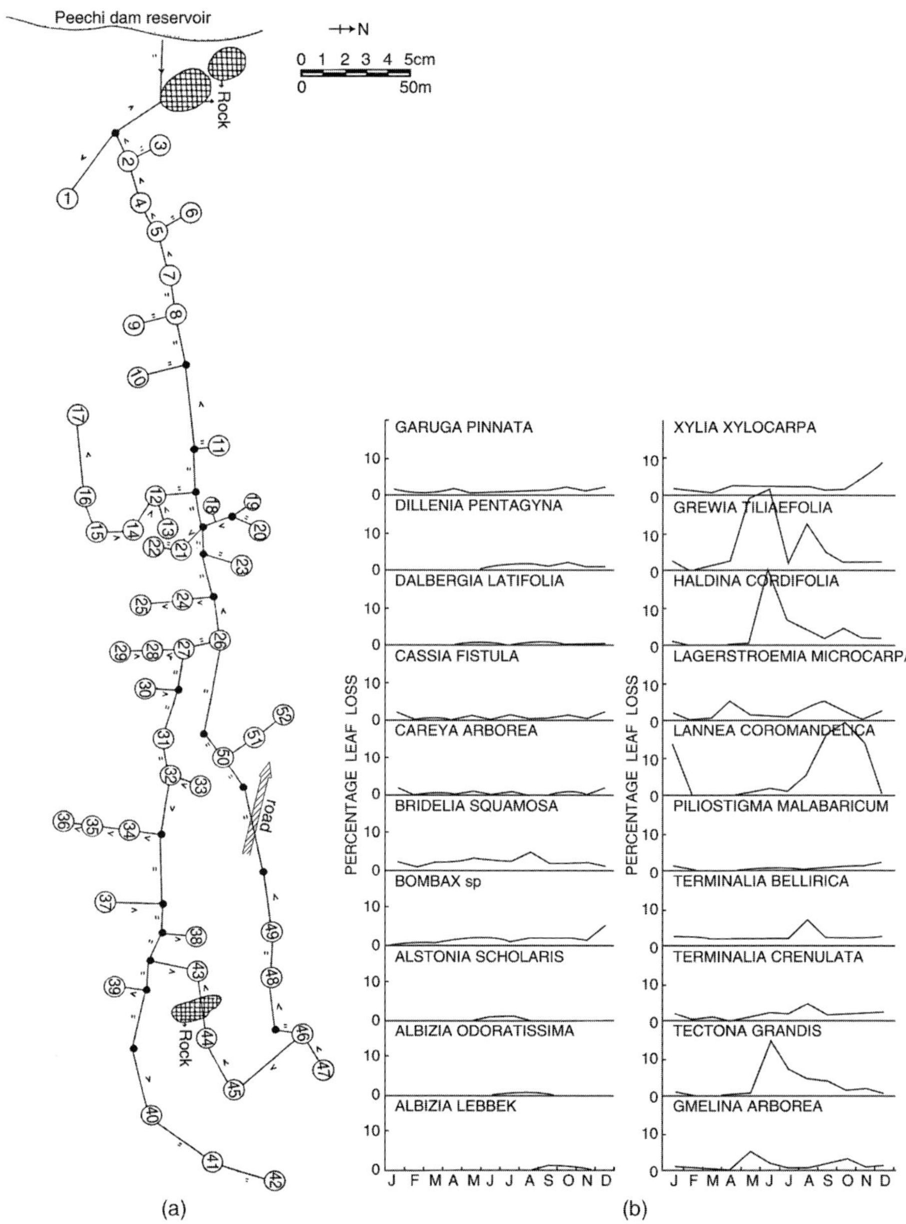

Fig. 4.1 Pest incidence in natural forest in Kerala, India. From *KFRI Research Report No. 44* (Nair *et al*., 1986a). (a) A typical study plot in moist-deciduous forest, which consisted of a walking path along which the selected tree species (circles) were situated. All the sampled trees are numbered serially; species identity is given in the Tree Identification Key overleaf. Other tree species are not shown. The symbols along the path indicate slope. A sample of five trees of each species was sampled over two plots. (cont.)

the insects from the tall canopy. Out of the 85 insect species collected from the lower canopy levels in the moist deciduous forest, 60% were new records for the respective hosts in India, indicating the meagre knowledge of insects associated with trees in the natural forest. Knowledge is particularly poor for insects associated with evergreen trees where collection is more difficult because of the lofty nature of the trees. Only eight species (six leaf feeders and three wood borers) could be collected from the evergreen forests although defoliation was noted in all the trees. This reflects the difficulty of collection, not the paucity of insect fauna.

In a similar study in the Guinea-Congolian domain of dense humid evergreen forests of Cameroon in Africa, Foahom (2002) observed saplings and young trees (<5 m tall) of seven species (*Lophira alata*, *Nauclea diderrichii*, *Celocarium preussii*, *Pycnanthus angolensis*, *Staudtia kameroonensis*, *Anthrocaryum klainianum* and *Uapaca guineensis*) in undisturbed and disturbed sites at monthly intervals over a year. He found very low level of leaf feeding, shoot boring, sap sucking and wood boring damage in the undisturbed forest, while in the disturbed (logged and liana cut) forest the damage was of higher intensity. It is noteworthy that all types of damage occurred in the undisturbed forest, although at very low intensity. He also studied a total of 22 species in disturbed forest and found that all species suffered damage, individual trees of nine species showing defoliation exceeding 50% and one among them, *Irvingia gabonensis* suffering 100% defoliation, leading to the death of trees. However, the monthly mean defoliation was below 15%. Other types of damage were less severe except sap sucking in *Milicia excelsa* and shoot boring in *Lophira alata*.

In a study in natural dipterocarp forests in East Kalimantan, Indonesia, Rahayu *et al.* (1998) reported damage to *Shorea* spp. caused by leaf-feeding caterpillars. Seed pests also make a significant impact in natural dipterocarp forests. Curculionid beetles and larvae of some small moths attack the fruits on the tree and on the ground, and prevent seed germination (Elouard, 1998).

Caption for Fig. 4.1 (cont.)

Tree Identification Key: *Albizzia lebbek* – 10, 15; *A. odoratissima* – 41, 46, 47; *Alstonia scholaris* – 33, 36, 49; *Bombax* sp. – 21, 51; *Bridelia squamosa* – 12, 19, 22; *Careya arborea* – 1, 3; *Cassia fistula* – 16, 23, 34; *Dalbergia latifolia* – 13, 18; *Dillenia pentagyna* – 20, 24, 26; *Garuga pinnata* – 5, 32, 50; *Gmelina arborea* – 7, 48; *Grewia tileaefolia* – 29, 45; *Haldina cordifolia* – 31, 35, 43; *Lagerstroemia microcarpa* – 14, 27, 28; *Lannea coromandelica* – 6, 52; *Piliostigma malabaricum* – 17, 25, 42; *Terminalia bellirica* – 30, 39, 40; *T. crenulata* – 4, 11, 37; *Tectona grandis* – 38, 44; *Xylia xylocarpa* – 2, 8, 9.

(b) Mean monthly defoliation (mean of five trees over two years) in 20 tree species in the moist-deciduous forest.

Dirzo (1982) reported that in a study in Mexican tropical rain forest up to 60% of the seedling populations of trees were damaged by herbivorous insects although only less than 25% of leaf tissue was lost in the affected seedlings.

In the pristine subtropical mixed conifer forest in Baja, Mexico, with a mean tree density of 160 trees ha^{-1} with no species dominating, Maloney and Rizzo (2002) reported widespread incidence of the bark beetle *Scolytus ventralis* (fir engraver) on white fir (*Abies concolor*). Other pests encountered included the bark beetles *Dendroctonus jeffreyi*, *D. valens* and *Ips* spp. and the sawfly *Neodiprion* spp. on Jeffrey pine (*Pinus jeffreyi*); *D. ponderosae* and *Ips* spp. on sugar pine (*P. lambertiana*) and *D. ponderosae* on lodgepole pine (*P. contorta*). Apart from fire which affects all trees, fungal diseases and insects were the primary cause of mortality of older trees in these forests. Most bark beetle infestation occurred on dead trees, with much less incidence in living trees. For example, the borer *Scolytus ventralis* was found on 87% of dead but only on 10% of live white fir, and *Dendroctonus jeffreyi* was found on 71% of dead and less than 2% of live Jeffrey pine.

It can be seen from the above that low-level pest incidence is common in mixed tropical forests.

4.2.2 *Pest outbreaks*

In spite of the general belief to the contrary, there are many examples of pest outbreaks in natural forests in the tropics. Based on observations in Barro Colorado Island, Panama, Wolda and Foster (1978, p. 454) even stated "it seems that outbreaks of insects in a good tropical forest are by no means rarer than they are in a temperate forest." Examples, with brief details where available (as noted earlier, many examples are of anecdotal nature), are given below, arranged by insect order.

Lepidoptera

Eulepidotis spp. (Noctuidae) in Panama and Brazil

Eulepidotis superior (Noctuidae) is an insect whose larvae feed on the young leaves of *Quararibea asterolepis* (Bombacaceae), a canopy tree species common in the tropical moist forest on Barro Colorado Island, Panama. An outbreak of *E. superior* on *Q. asterolepis* was observed in a 50-ha plot at the above site, in late May to early June 1985 (Wong *et al.*, 1990; Pogue and Aiello, 1999). During the outbreak, thousands of caterpillars descended from defoliated crowns on silken threads. Among the *Quararibea* trees in the 50 ha plot, about 20% suffered near total defoliation, 5% suffered no defoliation and the rest were in between. *E. superior* is a tender-leaf specialist; variation in leaf phenology at the time of the outbreak explained the variations in defoliation level among

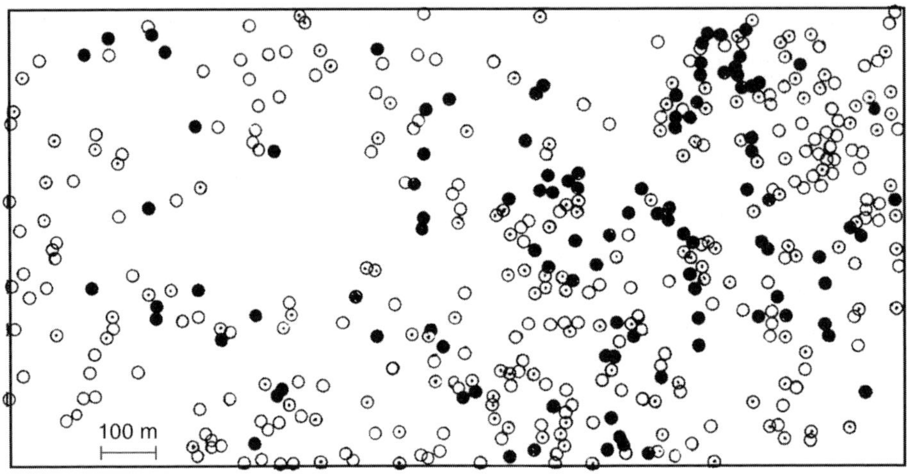

Fig. 4.2 Spatial pattern of defoliation of *Quararibea asterolepis* trees, caused by the caterpillar *Eulepidotis superior*, on a 50 ha plot in Barro Colorado Island, Panama. Trees are categorized into three levels of defoliation: light (0–20%, light circles), moderate (21–80%, circles with dot) and heavy (81–100%, dark circles). Adapted from Wong *et al.* (1990).

the trees. Also, the level of defoliation was higher where the tree density was higher, suggesting preferential oviposition by the moths in dense patches of the host tree. The spatial pattern of defoliation in the 50 ha plot is shown in Fig. 4.2. There was no subsequent outbreak in the same area that year, although pupae were found in great numbers on the underside of fallen leaves in the outbreak area. In the following two years, *E. superior* pupae were found in low density in the same area but no outbreak occurred. However, two major outbreaks have been observed since then (unpublished observations by Wright, Condit, Hubbell and Foster, Center for Tropical Forest Science (CTFS), Smithsonian Tropical Research Institute (STRI)). It is not known what causes the infrequent outbreaks. Although *Q. asterolepis* is the second most common tree species in the 50-ha observation plot, its density is not high. Out of 4276 stems per ha of all the tree species, *Q. asterolepis* accounted for 3–100 stems per ha, with an average of 44, constituting about 1% of the stems (Lao and Aiello, CTFS, STRI, personal communication, 2002). Thus the outbreak is not associated with high host density.

Outbreak of a related species, *E. phrygionia*, has been reported on the monodominant rain forest of *Peltogyne gracilipes* (Caesalpineaceae) in Maracá Island, Brazil (Nascimento and Proctor, 1994). This species also feeds on tender leaves and virtually all trees with tender leaves suffered heavy defoliation during outbreaks, which occurred during the early flushing season. The *Peltogyne* forest forms strips, each up to several hundred hectares in area, on Maracá Island where the outbreak was observed. The level of damage was lower in stands where the host trees were less dense. Two waves of outbreaks occurred in the first year of observation, but none in the following year.

Ophiusa spp. (Noctuidae) on *Palaquium* and mangrove in Indonesia

Kalshoven (1953) reported that outbreaks of the caterpillar *Ophiusa serva* occurred on *Palaquium* sp. which often constitutes 50% or more of the crop in some primary forests in South Sumatra, Indonesia. Another species, *O. melicerta* (syn. *Achaea janata*) is reported to have caused near total defoliation of a mangrove species *Excoecaria agallocha* over a stretch of 500–1000 ha of forest south of Belawan in North Sumatra, where the tree occurs essentially as single species stands (Whitten and Damanik, 1986).

Cleora injectaria (Geometridae) on the mangrove *Avicennia alba* in Thailand

Piyakarnchana (1981) reported that on one occasion a vast area of the mangrove species *Avicennia alba* in the Gulf of Thailand was defoliated by the larvae of *Cleora injectaria* (Lepidoptera: Geometridae). Although this insect is known to feed also on *Rhizhophora mucronata*, during the outbreak *R. mucronata*, as well as *R. apiculata*, which were mixed within the avicennia forest were not attacked.

Teak pests (Hyblaeidae and Pyralidae) in India and Myanmar

Hyblaea puera is a well-known defoliator of teak in plantations in many countries in Asia, but outbreaks are known to occur in natural forests as well. Nair and Sudheendrakumar (1986) reported heavy defoliation of isolated teak trees or small groups of trees in natural forests in the Kerala and Karnataka States in India. Fairly high-density infestations over larger patches of teak-bearing natural forests have also been observed in Myanmar, in Nagalaik Reserve Forest where teak trees occur at greater densities (Nair, 2001a). Outbreaks of *H. puera* have also been reported in natural stands of the mangrove *Avicennia marina*, an alternative host, on the Bombay coast of India (Chaturvedi, 1995). The biology and dynamics of defoliation of this insect are discussed in detail in Chapter 10.

Outbreaks of another caterpillar *Eutectona machaeralis* (Pyralidae) periodically occur on teak in India. Extensive outbreaks of this insect in natural teak areas in central India were reported as early as 1892–98 (Thompson, 1897; Fernandez, 1898). Thompson wrote that the whole forest where teak predominates had a sombre brown appearance due to skeletonization of leaves caused by the insect. He added "on 27th July, I traversed 32 miles, principally through teak forest and

I cannot recollect seeing a single tree that had entirely escaped." (Thompson, 1897, p. 325)

Voracia casuariniphaga (Lasiocampidae) on *Casuarina junghuhniana* in Indonesia

Kalshoven (1953) reported that occasional severe outbreaks of the caterpillar *Voracia casuariniphaga* occur in natural stands of *Casuarina junghuhniana* growing on mountain ridges and peaks in East Java. In an outbreak in February 1938 on Mt. Lawu, 800 ha were totally stripped.

Lymantria galinaria (Lymantriidae) on *Sonneratia acida* in Indonesia

Kalshoven (1953) reported that on one occasion a caterpillar provisionally identified as *Lymantria galinaria* caused defoliation of all trees of the mangrove *Sonneratia acida* in an estuary at Barito River in Southeast Kalimantan.

Zunacetha annulata (Dioptidae) on *Hybanthes prunifolius* in Panama

Although a pest of forest shrub rather than tree, *Zunacetha annulata* (Dioptidae) provides an example of pest outbreak in tropical forest. Larvae of *Z. annulata* feed on the leaves of *Hybanthes prunifolius* (Violaceae), a common shrub in the understorey of the lowland tropical monsoon forest at Barro Colorado Island in Panama. Outbreaks of the insect were noticed in two years (1971 and 1973) out of seven years of observation (1967–74) (Wolda and Foster, 1978). During outbreaks, most of the plants were almost completely defoliated, often repeatedly, by successive generations of the insect. With a total developmental period of about 33 days, up to seven generations occurred per year in the outbreak years. The insect is not traceable during the dry season from January to April. The beginning of an outbreak is abrupt and the insect is possibly a seasonal immigrant from some unknown area. The moths were caught in light traps at 27 m above ground at the canopy level and they are believed to come down through gaps in the canopy where the host shrubs are prevalent. The first few generations have the largest densities during outbreak years and the outbreaks were very effectively ended by fungus and/or a bacterial or viral disease (Wolda and Foster, 1978)

Bagworms (Psychidae) and *Miliona basalis* (Geometridae) on pine in Indonesia

Natural stands of *Pinus merkusii* cover an area of about 100 000 ha in North Sumatra in Indonesia. Severe outbreaks of a bagworm *Pteroma* sp. occurred over large areas in these stands in the years 1924, 1933 and 1934–38, the last one continuing over a four year period during which repeated defoliation occurred month after month (Kalshoven, 1953). The insect was probably *P. plagiophleps*, a polyphagous bagworm known to outbreak in plantations of

Falcataria moluccana in Indonesia (Nair, 2000) and India (Nair and Mathew, 1992) and known also to attack other trees such as *Tamarindus indica, Delonix regia, Emblica officinalis, Syzygium cumini, Populus deltoides, Tectona grandis* and *Trema orientalis* (Mathew and Nair, 1986). The affected pine stands were subjected to resin tapping and the insect attack was reported to be more serious on poorer sites. The biology and infestation characteristics of *P. plagiophleps* are described under *Falcataria moluccana* in Chapter 10. The adult female is wingless and the outbreaks are usually clumped.

Outbreaks of another species of bagworm *Eumeta* (= *Clania*) *variegata*, a common polyphagous pest, also occurred on these stands but were less frequent. Repeated outbreaks of a third pest, *Miliona basalis* (Geometridae) have also been recorded, smaller outbreaks developing simultaneously in different places all over the pine stands.

Anaphe venata (Notodontidae) on *Triplochiton scleroxylon* in Ghana

In the natural high forests of Ghana, *Anaphe venata* (Notodontidae) causes extensive defoliation of the valuable timber tree *Triplochiton scleroxylon* (Wagner *et al.*, 1991). The insect also occurs in Nigeria and Cameroon. Moths lay eggs on the leaves of tall trees and the larvae often strip the trees. The larvae, when mature, descend to the ground in long processions. On the ground, they form communal cocoons on the underside of the leaves of shrubs and low trees and remain in the prepupal stage for 2–3 months before pupating in separate cocoons within the communal cocoon. Repeated annual defoliations have been recorded during the months of August and September.

Coleoptera

Hoplocerambyx spinicornis (Cerambycidae) on *Shorea robusta* in India

Shorea robusta (sal) is a dipterocarp of commercial importance, distributed in over 10 million ha of forest in central and northern India, and extending into the subtropical zone. The tree occurs gregariously and attains a total height of about 30 m under favourable conditions. Severe epidemics of a cerambycid beetle *Hoplocerambyx spinicornis* have occurred on this tree repeatedly almost throughout its range. The beetle, which has an annual life cycle, lays eggs under the bark of the trunk. The larvae bore into the sapwood and heartwood, creating extensive galleries, and causing partial or complete girdling, eventually killing the tree when the infestation is severe. Large, extensive epidemics are common and may last for a few years before they subside. During an epidemic in 1923–28, about seven million sal trees were killed in Madla Forest Division in Madhya Pradesh in India (Roonwal, 1978). Another epidemic in the same state which started in 1994 after a gap of about 30 years, covered over half a million ha of sal forests and killed about three million trees by 1998, before it subsided naturally in 1999 (Dey, 2001). Several large and small epidemics have been recorded in the States of Assam, Bihar, Himachal Pradesh, Uttar Pradesh, West Bengal and, more frequently, in Madhya Pradesh. The infestation is endemic and chronic throughout the sal region but periodically flares up into an outbreak covering an extensive area and causing severe mortality of trees. The exact cause of the outbreaks is not known but high host density and unfavourable growth conditions for this gregarious tree species are thought to trigger outbreaks. Outbreak populations of the beetle can overcome the defences of healthy trees through mass attack. Although *H. spinicornis* is widespread in South and Southeast Asia and has several other host trees including *Duabanga sonneratioides, Hevea brasiliensis, Parashorea robusta, Pentacme suavis, Shorea assamica* and *S. obtusa*, outbreaks are known to occur only on sal. The biology of the insect and characteristics of outbreaks are discussed in detail under *Shorea robusta* in Chapter 10.

Dendroctonus frontalis (Curculionidae: Scolytinae) on pine in Honduras and other Central American countries

Dendroctonus frontalis, called the southern pine beetle, is a small (3–4 mm long) bark beetle that infests pine. The beetle bores through the bark of the tree trunk and feeds and oviposits in the phloem. Normally it attacks trees weakened by various causes, but when the beetle population is large, even healthy trees can be overcome. Tunnelling by the adult beetles and development of the broods in larval galleries result in girdling of the tree and tree death is hastened by invasion of fungi that are carried by the beetle. In Honduras, *D. frontalis* has a life cycle of less than a month.

Outbreaks of *D. frontalis* have occasionally occurred in the natural pine forests of Honduras, Nicaragua and other Latin American countries. In an outbreak in Honduras during 1962–65, more than 2 million ha of forest were affected (Billings and Espino, 1995). The affected area contained overmature trees, but as infestations expanded, trees of all ages above five years were infested and killed. The outbreak eventually subsided due to natural causes. Another outbreak occurred during 1982–1991 which was controlled at great effort by felling infested trees, in order to prevent continued expansion of active infestations on expanding outbreak fronts. It started in 1982 in about 1700 ha of forest consisting primarily of *Pinus oocarpa* and *P. caibaea*, located at 600–1000 m elevation, within about 112 530 ha of pine forests in north central Honduras. The stand was weakened by resin extraction, wounds, recent fire and a prolonged drought (Billings and Espino, 1995). With little or no effort at control, the infestation had affected over 8000 ha by 1983 but the affected area declined subsequently, in line with the control effort.

A more extensive *D. frontalis* outbreak erupted again in 2000–2002 and covered pine forests in Belize, Guatemala, El Salvador and Nicaragua, in addition to Honduras (Billings *et al.*, 2004; Billings and Espino, 2005). In Honduras young stands 18–25 yrs old were affected. The stands were dense and weakened by overcrowding, resin extraction wounds, frequent fires and prolonged drought. In spite of suppression efforts by means of cut-and-leave and cut-and-remove operations (which often suffered due to lack of adequate funds and delays), large areas were affected (1743 ha in 2000, 9078 ha in 2001 and 9500 ha in 2002).

D. frontalis is distributed in the south eastern United States, parts of Mexico and Central America, and is one among several bark beetles that are well known as very destructive pests of coniferous forests in the temperate and subtropical regions. Typically, bark beetles are pests of the temperate and subtropical forests, but their occurrence in Honduras is not surprising because Honduras, although classified as a tropical country (lying between latitude 13° N and 16° N) has a subtropical climate with a mean annual temperature of about 21.1 °C in the cooler highlands where most forests are located. The forests are dominated by oak and pine.

Agrilus kalshoveni (Buprestidae) on *Actinophora fragrans* in Indonesia

In lowland forests in Java, an outbreak of a small buprestid wood borer, *Agrilus kalshoveni* caused large-scale mortality of scattered trees of all sizes of *Actinophora fragrans* (Tiliaceae) in 1926–28 (Kalshoven, 1953).

Buprestid and curculionid borers on chir pine in India

In the subtropical mixed forest at Morni Hills (30° 32″–30° 45″ N latitude) in Haryana, India, where chir pine (*Pinus roxburghii*) occurs in scattered patches comprising young and middle-aged trees, heavy mortality of pine trees of all ages has been reported (Singh *et al.*, 2001). The dead, dying and live trees were found heavily infested by a complex of borers. Four species of beetles were encountered; *Sphenoptera aterrima* (Buprestidae), *Cryptorhynchus rufescens* (Curculionidae), *Platypus biformis* (Curculionidae: Platypodinae) and *Polygraphus longifolia* (Curculionidae: Scolytinae). Although the immediate cause of tree mortality was borer attack, occurrence of recurrent fires and past resin tapping had rendered the trees weak, making them susceptible to borer attack. Similar mortality of fire-scorched chir pine trees in the adjoining state of Himachal Pradesh had also been reported earlier.

Hemiptera

Phytolyma spp. (Psyllidae) on *Milicia* spp. in Africa

The psyllids *Phytolyma* spp. attack *Milicia* (syn. *Chlorophora*) spp. in Africa. *P. lata* on *M. regia* is the most damaging (Wagner *et al.*, 1991). The insect lays eggs on buds, shoots or leaves of the host plant. The newly hatched nymph bores into the plant tissues causing the formation of a gall which completely covers the nymph. Galls may develop on bud, shoot or leaf and may occur singly or in clumps. Several galls will often coalesce and become a bunched mass affecting the growth of the shoot, particularly of young plants. When the infestation is heavy, the shoots and leaves become a putrefying mass, the stem dies back and the seedling may eventually die. While damage by *Phytolyma* has been noticed in natural forests, the injury is more severe in nurseries and young plantations below one year old; 100% failures have sometimes been reported in nurseries and plantations in Ghana (Wagner *et al.*, 1991). A pest profile of *Phytolyma* spp. is given under *Milicia* species in Chapter 10.

Udonga montana (Pentatomidae) on bamboos in India

The pentatomid bug, *Udonga montana* feeds on the developing seeds of bamboos. Very heavy build up of this bug has occurred periodically in bamboo forests in India and Myanmar, coincident with gregarious flowering of bamboos. Huge swarms of the insect assemble on all sorts of trees and vegetation during these periods. Characteristics of these outbreaks are described more fully under bamboos in Chapter 10.

Hymenoptera

Shizocera sp. (Argidae) on *Manglietia conifera* in Vietnam

Larvae of the sawfly *Shizocera* sp. feed on the leaves of the Mo tree, *Manglietia conifera* (Magnolaceae) in mixed natural forests in Vietnam. Outbreaks have occurred often in pure stands of the tree in the northern temperate region of Vietnam with an average temperature of 21–24 °C (Tin, 1990). The insect passes through one or two generations per year depending on the temperature conditions. Additional details are given in Chapter 10.

4.3 Discussion and conclusion

The empirical evidence shows that contrary to conventional wisdom tropical forests are not free of pest outbreaks. All gradations of insect damage, from minor feeding with no significant impact, to large-scale outbreaks resulting in massive tree mortality have been observed.

Some authors have argued, however, that practically all contemporary forests have been subject to human impact at sometime and that outbreaks do not occur in truly virgin forests where the insects and trees have coevolved over a long period of time. Some of the forests in which outbreaks have been recorded have indeed been subject to human interference, and it is difficult to prove otherwise in other cases. But disturbance is also a natural event and part of the dynamics of the forest ecosystem (see Chapter 1). Therefore it is safe to conclude that pest outbreaks do occur in natural forests in the tropics although they may be less frequent and less severe than in plantations.

Two types of outbreaks can be distinguished among those described above – those triggered by host stress and those which are not. The periodic outbreaks of the bark beetle on pines in Honduras and other central American countries is believed to be triggered by host stress, in stands weakened by overcrowding, resin extraction, frequent fires or drought. Under normal conditions, a small population of bark beetles thrives on a small number of unhealthy hosts such as senescent trees, trees growing on poor sites etc., as shown in the Mexican study (Maloney and Rizzo, 2002). These beetles multiply in large numbers when healthy trees become weakened by adverse conditions, precipitating an outbreak. The case of beetle borer attack on chir pine in India is similar. Sal borer (*Hoplocerambyx*) outbreaks in India are also believed to be caused by multiplication of beetles in particularly favourable local epicentres. The large number of beetles thus produced overcomes the defences of healthier trees by mass attack. Host stress may also be the cause of outbreak of the bagworm defoliators in pine stands in North Sumatra, Indonesia, although in general host-stress induced outbreak is characteristic of boring and sucking insects (Koricheva *et al.*, 1998) and not of leaf feeders. Many insect outbreaks however, are not caused by host stress. The causes of insect outbreaks will be discussed in detail in Chapter 7.

Another notable feature of the insect outbreaks in natural forests is that many of them, but not all, have occurred in stands where the host density was high. This is the case in the outbreaks of *Eulepidiotis phrygiona* on *Peltogyne gracilipes* in Brazil, bagworms on pines in Indonesia, *Ophiusa* spp. on *Palaquium* and on *Excoecaria agallocha* in Indonesia, *Hoplocerambyx* on sal in India, bark beetle on pines in Honduras and sawfly on *Manglietia glauca* in Vietnam. Host concentration has been proposed as one of the causes of insect pest outbreaks, as discussed in Chapter 8. However, not all pest outbreaks occurr in stands of high host density.

In the natural forest, gradation in the severity of pest attack is very wide. For a given insect species, pest status varies in time and space and with respect to the host tree species. In other words, an insect may become a pest some times but not always, at some locations but not at others and on some of its host trees but not on all. Also, only some species of phytophagous insects, not all, may become pests. Outbreaks are apparently less frequent and less severe in natural forests. In addition, in the absence of routine damage survey in natural forests, it is very rare that an infestation receives our attention, even when it occurs, because in mixed forests with a large number of tree species pest damage is not conspicuous. For example, compare the visibility of shoot borer (*Hypsipyla grandella*) attack in mahogany trees which are distributed at a density of about one mature tree per hectare in the Brazilian natural forest with its visibility in a young monoculture plantation. On the other hand, infestations are more visible in the temperate forests dominated by single tree species, as well as in some types of subtropical forests of similar nature, like eucalypts in Australia and sal (*Shorea robusta*) in India.

It is obvious that all forest pests had their origin in natural forests and are still present there. But due to a variety of natural control factors, both biological and physical, the populations of most insects remain small in natural forests. Thus the natural forest, far from being free of pests, is a reservoir of pests. However pest outbreaks are rare and their impact is therefore minimal. Our economics-based definition of pests is not adequate for natural forest situations. 'Pest' is primarily an agriculture and plantation-related concept. In the continuum of insect damage scenarios ranging from minor feeding to large-scale outbreaks in the natural forest, it is difficult to determine what constitutes a pest situation, particularly when it is granted that minor insect damage may even have a stimulatory effect on plant productivity (Mattson and Addy, 1975). Population outbreak is the result of uncontrolled increase in the population. The factors which control the dynamics of insect populations and the circumstances under which outbreaks develop are discussed in Chapter 7, and the possible reasons for greater pest incidence in plantations are discussed in Chapter 8.

5

Insect pests in plantations: General aspects

5.1 Introduction

Plantation forestry is now a major activity in the forestry sector in the tropics, with a large number of species grown in plantations to serve a variety of purposes (see Chapter 1, Section 1.5). For example, about 170 species have been tried in plantations in India (Ghosh, 1977), 80 in Malaysia (Appanah and Weinland, 1993) and 24 in Indonesia (Cossalter and Nair, 2000). Increasing numbers of species are now being put on plantation trials as most commercially exploited species are potential candidates for plantations, and their numbers are large. For instance, in Cameroon alone there are 400 commercially exploited species (Foahom, 2002). Because of the large number of plantation species, it is impracticable to draw up a list of all species planted and deal with their pests. Such a treatment would be encyclopaedic and would not permit us to see the forest for the trees. A smaller number of species such as eucalypts, tropical pines and acacias have dominated the plantation scenario in the tropics, mainly for the production of pulpwood, but they are not representative as there are many other valuable tree species that are locally important and planted over smaller areas. In order to get a balanced view, we shall consider a representative group of plantation species. Trees commonly planted in the tropics are chosen, irrespective of the extent of area planted and whether they suffer from serious pest problems or not. The list includes selected species of *Acacia*, *Agathis*, *Ailanthus*, bamboo, *Casuarina*, *Dalbergia*, *Eucalyptus*, *Milicia*, *Pinus*, *Shorea* and *Swietenia*, and *Falcataria moluccana*, *Gmelina arborea*, *Leucaena leucocephala*, *Manglietia conifera*, *Neolamarckia cadamba* and *Tectona grandis*. These tree species are representative of the tropical plantations, although there is a dominance of species from the Asia-Pacific region which is understandable as this region accounts for the greater part of the tropical planted forests.

Pests associated with plantation tree species fall under three major categories – nursery pests, sapling pests and pests of older, established plantations. It is convenient to consider them separately. As indicated in the title of this chapter, only general aspects of the pest problems in plantations are discussed here. Specific details of the pests associated with each of the selected tree species are given in Chapter 10 where, following a brief tree profile, an overview of its pest problems and detailed pest profiles of the important pests are presented.

5.2 Nursery pests

Nursery pests are those insects which attack the tree seedlings in nursery beds. They do not generally attack older trees, although there are exceptions. Generally, forest tree seedlings are raised in nursery beds and planted out in the field when they are 6–12 months old. For many trees like eucalypts, small seedlings are pricked out from the nursery bed and raised in soil-filled polythene bag containers kept in secondary nurseries, before they are planted out. In nursery beds and container beds the plant density is high, as in agricultural fields. Nursery pests include root-feeding whitegrubs and termites; shoot-cutting caterpillars, crickets and grasshoppers; leaf-feeding caterpillars and beetles; sap-sucking bugs; and shoot-boring scolytine beetles. Some of these groups of insects are generalists which attack a wide range of tree species (e.g. whitegrubs and termites) while others are host specific (e.g. moth caterpillars which attack teak, mahogany or *Ailanthus*). The major groups of nursery pests, which usually attack seedlings of more than one tree species, are briefly discussed below. More host-specific nursery pests are dealt with in Chapter 10, under the respective tree species.

Whitegrubs

Whitegrubs (Fig. 5.1) are the immature stages of some beetles of the family Scarabaeidae (mostly of the subfamilies Melolonthinae and Rutelinae). They have a soft, white, curved body, brown or black head and three pairs of prominent thoracic legs. They live in soil and feed underground on the roots of seedlings, grass etc. The adult beetles feed on the foliage of trees. Generally, they have an annual life cycle, with the beetles emerging from April to June, following the monsoon showers. Eggs are deposited in moist soil, rich in organic matter, or near the roots of plants. The larvae usually pass through three instars. Whitegrubs are serious pests of teak nurseries in some localities. *Holotrichia* (syn. *Lachnosterna*) *consanguinea* and *H. serrata* (Melolonthinae) are the common

Fig. 5.1 Whitegrub, larva of a scarabaeid beetle.

species found in teak nurseries in India and Sri Lanka. The larvae bore into the fleshy taproot of teak seedlings and cause their death. Other tree species subject to whitegrub damage in the nursery include neem (*Azadirachta indica*), pines, sal (*Shorea robusta*), acacias etc. In whitegrub-affected nursery beds, loss of 20–30% of the seedlings is common. Control of whitegrub damage in nursery stock has generally been achieved by use of chemical insecticides. In areas prone to severe whitegrub infestation, an insecticide is mixed with the soil at the time of preparation of the nursery bed as a prophylactic measure (Oka and Vaishampayan, 1981).

Termites

Some species of termites attack the roots of tree seedlings, killing the plants. Eucalypt seedlings are particularly prone to termite attack, but other species such as pines, acacias, casuarina, dipterocarps and *Falcataria moluccana* are also attacked. In eucalypts, most damage is caused to newly transplanted seedlings; mortality of up to 80% of plants, within a few months of planting out of container-raised seedlings has been reported (Nair and Varma, 1985). Typically, the taproot of seedlings is eaten up a few centimetres below the soil surface, severing it from the rest of the root system. Because of the underground activity of the termites, the attack is recognizable only when the sapling is almost dead. The termite problem is discussed in more detail under the tree species, *Eucalyptus*, in Chapter 10. Effective control of termite attack can be obtained by prophylactic insecticidal treatment applied to seedlings in polythene bag containers prior to planting them out in the field.

Shoot-cutting caterpillars, crickets and grasshoppers

Caterpillars of some noctuid moths characteristically cut off the shoots of small seedlings at ground level. They are therefore called 'cutworms' (Fig. 5.2). They hide in the soil in shallow burrows under litter or stones during the day and become active at night. The larvae drag portions of the cut shoots into their burrows to feed, but cause considerable wastage as a single caterpillar may cut off several shoots per night. The most common species are the cosmopolitan *Agrotis ipsilon* (greasy cutworm) and *A. segetum* (syn. *Euxoa segetis*) (black cutworm). Ali and Chaturvedi (1996) reported the loss of 10–20% of seedlings of *Albizzia lebbek* and *Eucalyptus tereticornis*, due to an attack of *A. ipsilon* in Bihar, India.

Crickets and mole crickets (Orthoptera: Gryllidae and Gryllotalpidae) also cause damage in forest nurseries. These insects, which nest in tunnels made in the ground, come out at night and feed on seedlings, cutting them and dragging pieces to their tunnels. Seedlings of *Casuarina equisetifolia, Tectona grandis, Dalbergia sissoo*, eucalypts etc. are damaged. Common species causing damage are *Brachytrupes portentosus, Gymnogryllus humeralis, Nisitrus vittatus* and *Gryllotalpa africana* (mole cricket). Wylie and Brown (1992, cited by Speight and Wylie, 2001) reported that in a 10-ha plantation in China about 40% of 3-month-old *Eucalyptus urophylla* were damaged by the cricket, *Brachytrupes portentosus*.

Fig. 5.2 Cutworm, larva of a noctuid moth, in curled resting posture.

Several species of grasshoppers (Orthoptera: Acrididae) also feed on the foliage of seedlings and saplings, often cutting off the shoots. They often appear in swarms, when they cause the most damage. Small swarms of the grasshopper *Valanga nigricornis* are common in *Acacia mangium* nursery sites in Thailand (Hutacharern, 1993) and Indonesia (Nair, 2001a). *Stenocatentops splendens* is a common defoliator of *Acacia mangium* and *Falcataria moluccana* nurseries in Sabah, Malaysia (Chey, 1996). *Aularches miliaris* and *Chrotogonus* spp. are other grasshoppers injurious to forest tree seedlings. In Paraguay, an unidentified grasshopper of the genus *Baeacris* has been reported as causing severe damage to *Eucalyptus grandis* transplants during the early establishment phase, by feeding on the bark close to the ground level (Speight and Wylie, 2001).

Leaf-feeding caterpillars and beetles

Several species of leaf-feeding caterpillars damage seedlings in forest nurseries. *Diacrisia obliqua* (Arctiidae) and *Spodoptera litura* (Noctuidae) are polyphagous. *Eutectona machaeralis* (Pyralidae) and *Hyblaea puera* (Hyblaeidae) attack teak (Ambika-Varma *et al.*, 1996) and *Eligma narcissus* (Noctuidae) attacks *Ailanthus* spp. (Sivaramakrishnan and Remadevi, 1996). These insects cause extensive, often near-total, defoliation of seedlings in the nurseries. *Strepsicrates rhothia* (Tortricidae) is a cosmopolitan species which attacks seedlings of many species of eucalypts, sticking together the tender terminal leaves and feeding on them. It has been reported as damaging about 20% of seedlings in eucalypt nurseries in Sabah, Malaysia (Chey, 1996) and 50% of seedlings at some places in Ghana (Wagner *et al.*, 1991). The pyralid *Lamprosema lateritialis*, which is widespread in the lowland rain forest of Nigeria, Ghana and Ivory Coast, is a serious pest of nursery seedlings of the valuable indigenous timber species, afromorsia (*Pericopsis elata*). Its larvae feed gregariously in nests made by sticking the leaves together. It is reported to cause loss of 30–40% of seedlings in nurseries in Ghana, by repeated defoliation (Wagner *et al.*, 1991).

Many species of chrysomelid and curculionid beetles also cause damage to forest nursery seedlings. The chrysomelids *Chrysomela populi* and *Nodostema waterhousie* in poplar nurseries in Kashmir and Himachal Pradesh, respectively, are examples from India (Khan and Ahmad, 1991; Singh and Singh, 1995).

Sap-sucking bugs

Several species of psyllids (Hemiptera: Psyllide) cause serious damage to nursery seedlings. They lay eggs on the leaves and buds of the seedlings and the nymphs burrow into the plant tissues, often causing formation of galls. Total failures of nursery crops of *Milicia* spp. have often been caused by *Phytolyma* spp. in Ghana (Wagner *et al.*, 1991) (see details under the tree species *Milicia*

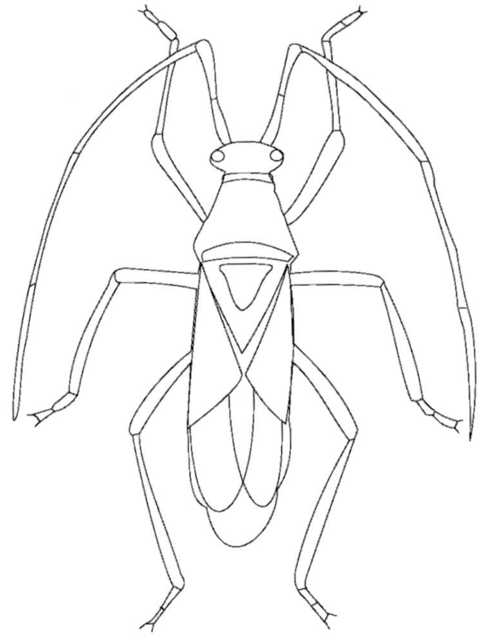

Fig. 5.3 The bug *Helopeltis antonii*. Length 6–8 mm. After Nair (1989).

in Chapter 10). In India, *Arytaina* sp. causes serious damage to seedlings of *Albizzia lebbek* in Karnataka (Sivaramakrishnan and Remadevi, 1996) and an unidentified species attacks seedlings of *A. odoratissima* and *Pterocarpus marsupium* in Kerala (Mathew, 1993). A few species of the mirid bug *Helopeltis* (Fig. 5.3) cause serious damage to *Acacia mangium* seedlings in Malaysia, the Philippines and Indonesia (Nair, 2001a). They cause dieback of shoots, probably as a result of injection into the plant of toxic saliva or pathogenic organisms. The lace bug *Dictyla monotropidia* (Homoptera: Tingidae) is a chronic pest of young *Cordia alliodora* in the neotropics (Schabel *et al.*, 1999).

Shoot-boring scolytine beetles

Some species of the small scolytine beetles (Curculionidae) attack seedlings of forest trees. They lay eggs in galleries made in the shoot of seedlings, and feeding of the larvae causes death of the seedlings. The species recorded, their hosts, countries of occurrence and severity of damage where known are shown in Table 5.1.

5.2.1 *Impact of nursery pests*

Generally, insect damage to tree seedlings in the decentralized forest nurseries in the tropics is not serious, although substantial loss of seedlings may

Table 5.1. *Scolytine beetles attacking tree nurseries*

Insect species	Host species	Country	Comments
Hypothenemus birmanus	*Casuarina equisetifolia*	Malaysia	
H. dimorphus	*Acacia auriculiformis*	Malaysia	
H. eruditus	*Swietenia macrophylla*	Malaysia	
H. pusillus	*Cedrela odorata* *Gmelina arborea* *Tectona grandis* *Terminalia ivorensis*	Ghana	Attack appears to be heavy on seedlings weakened by other causes
Xylosandrus compactus (syn. *X. morstatti*)	*Acacia auriculiformis*	Indonesia	
	Swietenia macrophylla	Indonesia, Malaysia, Sri Lanka and Thailand	
	Khaya grandifoliola and K. senegalensis	India	Introduced African tree species
Unidentified	*Acacia mangium*	Malaysia	Killed 40% of seedlings

Data from Browne (1968), Tho (1987), Natawiria (1990), Zakaria (1990), Wagner *et al.* (1991), Meshram *et al.* (1993), Day *et al.* (1994), Mayhew and Newton (1998), Nair and Sumardi (2000)

occur sporadically. As noted above, loss of up to 20–30% of seedlings of teak due to whitegrubs, 80% of eucalypts due to termites, 10–20% of *Albizia lebbek* or eucalypts due to cutworms, 40% of eucalypts due to crickets, 40% of *Acacia mangium* due to scolytines, 30–40% of *Pericopsis elata* due to a pyralid etc. have sometimes been recorded. However, these are exceptional cases. Serious damage can usually be prevented by application of prophylactic or remedial control measures as discussed above and in Chapter 9.

5.3 Sapling pests

Some pests attack trees only during the sapling stage. They include root, stem, or terminal shoot borers, leaf-feeding caterpillars and sap-sucking bugs. Root-feeding termites attack saplings of eucalypts, pines, casuarina etc. particularly during their establishment phase after transplanting into the field. Larvae of the cerambycid beetle *Celosterna scabrator* bore into the root-shoot portion of the saplings of *Acacia nilotica*, eucalypts etc. often causing their death. Larvae of the hepialid moths *Sahyadrassus* and related species, as well as the cossid moth *Zeuzera coffeae* bore into the stem of saplings of teak, eucalypts etc. and some bostrichid beetles bore into the stem of saplings of *Acacia mangium* and *Falcataria moluccana*. The larvae of the beetle *Acalolepta cervina* bore into the shoot of teak saplings and cause cankers. Larvae of the moths *Hypsipyla robusta* or *H. grandella* bore into the terminal shoot of saplings of mahogany and other meliaceous trees, causing severe growth retardation. Leaf-feeding caterpillars of the moth *Eligma narcissus* cause defoliation of saplings of *Ailanthus* species. The gregarious sap feeding bug *Tingis beesoni* attacks saplings of *Gmelina arborea* and causes dieback of shoots. Details of these sapling pests are described in Chapter 10, under their main host tree species.

The reasons why these pests confine their attack to saplings is not understood. Stray instances of the mahogany sapling shoot borer infesting the branches of older trees are on record. Stem borers of saplings, such as hepialids and cossids apparently have a preference for succulent and small diameter stems of saplings.

5.3.1 *Impact of sapling pests*

Some sapling pests such as the hepialid and cossid stem borers cause a small percentage of the attacked saplings to break in the wind, but generally the impact is negligible. Similarly, the defoliating caterpillars do not usually cause serious damage. On the other hand the mahogany shoot borer causes severe growth retardation due to dieback of the leading shoot and multiple shoot growth, which has often led to abandonment of plantation programmes. The *Gmelina* bug also causes similar damage. Thus, depending on the tree species, sapling pests can cause serious economic loss, particularly where no effective control methods have been developed as in the case of the mahogany shoot borer.

5.4 Pests of older plantations

A large number of insect species are usually associated with each tree species in a plantation, as indicated in Chapter 2. Detailed information on insects associated with representative tropical tree species is given in Chapter 10. A summary of this information is given in Table 5.2. In this table, the tree species are grouped by genus in some cases (as in eucalypts) or by a still higher category (as in bamboos), as the pests are mostly common to the group. The approximate number of insect species associated with a tree species, genus or group is given, rounded off to the nearest 10. This number gives only a rough indication of the associated insect fauna as the research effort spent in collecting and identifying

Table 5.2. *Overview of the number of insect species associated with common plantation tree species in the tropics*

Tree species	Minimum no. of associated insect species[a]	No. of major pest species
Acacia auriculiformis	10	Nil
A. catechu	10	Nil
A. mangium	80	Nil
A. mearnsii	200[b]	3
A. nilotica	70	2
A. senegal	20	Nil
Agathis spp.	10	Nil
Ailanthus spp.	40	2
Bamboos	240[b]	Nil
Casuarina equisetifolia	70	Nil
C. junghuniana	10	Nil
Dalbergia cochinchinensis	20	1
D. latifolia	40	Nil
D. sissoo	130	1
Eucalyptus spp.	920[c]	4
Falcataria moluccana	40	2
Gmelina arborea	100	2
Leucaena leucocephala	10	1
Milicia spp.	10	1
Neolamarckia cadamba	10	1
Tropical pines	30	4
Shorea spp.	150	1
Swietenia spp.	20	1
Tectona grandis	170	3

[a]Rounded off to the nearest 10
[b]Including some from the temperate region
[c]World total, including those from the temperate region

the insects associated with the different tree species varies greatly. The following generalizations can be made from the table.

The number of phytophagous insect species associated with a tree species ranges from 10–200 in general, with a mean of 65. An exceptional 920 species are associated with *Eucalyptus*, including those in the temperate region; no separate estimate is available for the tropics. *Eucalyptus* is indeed an exception because (1) there are more than 600 species represented in the genus, and (2) the area and range of natural and introduced distribution of the genus is wide, encompassing both tropical and temperate climates. In fact, the majority of eucalypt insect fauna have been recorded in the temperate region. The comparatively higher number of insects in the case of bamboos and *Acacia mearnsii* is also due to inclusion of some insects from the temperate regions. About half of the tree species examined had 40 or fewer number of associated insect species and about a quarter had 10 or fewer. A smaller percentage of trees had higher numbers of associated insects. Obviously the number of species of insects associated with a plantation tree species will be influenced by several factors – the chemical profile of the species, the extent and climatic diversity of the geographical area covered, the period over which the species has been cultivated on a large scale etc.

Although fairly large numbers of insects are associated with all tree species, most of them are casual or minor pests. Only a few species have acquired major pest status on any given tree species. While some of these are chronic pests causing serious damage every year, others cause serious damage occasionally. The number of serious pests listed in Table 5.2 is therefore based on subjective judgement. Serious pests include defoliators, sap suckers and stem borers. Leaf-feeding insects occur on all tree species, with serious pests occurring on *Ailanthus, Dalbergia sissoo, eucalypts, Falcataria moluccana, Gmelina arborea, Neolamarckia cadamba* and *Tectona grandis*. Sap-sucking insects are not major pests except in *Leucaena leucocephala, Milicia* and pines. Among the trees not included in the detailed case studies in Chapter 10, a sap-sucking bug, *Rederator bimaculatus* (and possibly other bugs) is responsible for transmitting a serious disease known as spike disease, caused by a mycoplasma-like organism in the sandal tree, *Santalum album*. Stem borers are major pests of *Dalbergia cochinchinensis, Falcataria moluccana*, pines and *Shorea robusta*. Detailed accounts of these pest problems are given in Chapter 10.

In summary, most tree species raised in plantations are attacked by one or more serious pests; freedom from pests is exceptional. This is in contrast to the situation in natural tropical forests where serious pest attack is exceptional.

5.4.1 *Impact of pests in older plantations*

For some tree species, pests have a devastating impact in plantations, much more serious than in the mixed species natural stands. The details are covered in Chapter 10. In Asia, annual defoliation caused by the caterpillar *Hyblaea puera* in teak plantations has ben shown to result in loss of 44% of the wood volume increment. This pest is becoming increasingly important in exotic plantations of teak in Latin America, but has not so far become serious in Africa. Chronic defoliation caused by other insects on other trees must also be causing serious economic loss, but the losses have not been quantified in most cases. The sap-sucking leucaena psyllid has had a devastating impact on the cultivation of *Leucaena leucocephala*, an important exotic fodder and fuel wood tree introduced from Mexico and grown extensively in Southeast Asia, as described in Chapter 10. The enormous loss caused by periodic outbreaks of the stem borer *Hoplocerambyx spinicornis*, which has killed millions of the valuable timber tree *Shorea robusta* in India is also described in detail in Chapter 10. In general, effective control methods are not available for most of the pest problems afflicting older plantations and losses continue to occur in the form of chronic loss of growth increment and occasional large-scale mortality of trees. At the same time, many plantation tree species such as *Acacia, Agathis*, bamboos, eucalypts, *Casuarina* and *Dalbergia* in most places, and *Gmelina*, pines, *Shorea* etc. in some places, are comparatively free of serious insect pests. Various aspects of the plantation pest problems and the influences of monoculture and exotics are discussed in Chapters 7 and 8.

6

Insect pests of stored timber

6.1 Introduction

A large variety of insects attack timber during various stages of its utilization from trees felled in the forest to manufactured articles in use. The resulting waste of this valuable raw material is enormous although no reliable estimate of the loss is available. An indication of the potential for damage can be obtained from the fact that about 130 species of insect borers have been recorded from sal (*Shorea robusta*) timber alone in India (Beeson, 1941). The species of insects found vary depending on the geographical region, species of timber, season and stage of processing of the timber. However, unlike many pests of living trees which attack only a single or a narrow range of hosts, the timber borers in general attack a large number of timber species; that is, they are less host-specific.

Insects attack wood mainly for food although the wood serves as a place of shelter also. The insect's basic food requirements are surprisingly similar to that of humans, with minor exceptions – they require proteins, carbohydrates, fat and vitamins, although there are variations in the specific requirements of different insect species. Sapwood generally contains more nutrients like carbohydrates and proteins and for this reason most insects feed on the sapwood, and bore into the heartwood only for shelter. Like us, most insects are unable to digest the cellulose and lignin of wood, but some insects like termites accomplish this with the help of their intestinal symbionts, that is, some kinds of protozoa and bacteria. Some insects known as ambrosia beetles cultivate a type of fungus, called ambrosia, inside their tunnels in the wood and feed on it, and another insect feeds on microscopic algae (vide infra) inside its tunnel.

The damage caused by insects ranges from wholesale consumption of timber as in the case of termites to consumption of or tunnelling in localized areas. Localized damage is sometimes very extensive because of the large population of insects so that the attacked wood is reduced to a crumbling mass of tissue. The type of damage depends on the kind of insect and its feeding habit. In general, large beetles cause large tunnels in the wood, and small beetles cause pinholes or shotholes and black staining of the wood due to growth of fungi. Feeding by some groups of beetles reduces the wood to fine powder. Unfortunately a number of colloquial, often undefined terms such as powder-post beetles, pinhole borers, shothole borers, ambrosia beetles, engraver beetles etc. have entered the literature on wood-destroying insects. These designations are vague and often confusing, because the same insect can be classified under different categories. For example, a shothole borer also produces wood dust during tunnelling and therefore could be called a powder-post beetle, and an ambrosia beetle may belong to either the subfamily Scolytinae or Platypodinae of the family Curculionidae.

6.2 Categories of wood-destroying insects and their damage

The wood-destroying insects fall into two major groups – termites (order Isoptera) and beetles (order Coleoptera). In addition, in some parts of Africa, the nymphs of the mayfly *Povilla adusta* (Ephemeroptera: Polymitarcidae) bore into wood under fresh water. This unique insect causes considerable damage by tunnelling into wood that is used as support for fishing docks or stilt houses (Wagner *et al.*, 1991). It does not, however, feed on wood, but makes silk-lined galleries in it and feeds on microscopic algae. Some timbers like *Chlorophora excelsa, Nauclea diderrichii* and *Chrysobolanus ellipticus* are resistant to its attack.

6.2.1 *Termites*

Termites constitute a large group of insects, with more than 2900 species, as discussed in Chapter 2, and they play a dominant role in the recycling of wood, as discussed in Chapter 3. In India alone, about 64 species attack wood and 16 are regarded as major wood-destroying species. Termites live in colonies with a complex social organization, and the damage is caused by the worker caste which forages for food. Termites attack a large variety of timber and other woody materials and are divisible into two groups based on the nature of the damage and their habits – subterranean termites and wood-dwelling termites.

Subterranean or soil-dwelling termites

Most damage to timber is caused by this group which constitutes the majority of termite species. They build nests in soil and forage for food through underground tunnels or mud tubes built above ground. They may either have large nests which project above ground (mounds) or live in small, diffuse, below-ground nests. They attack only wood in contact with the ground. Unprocessed logs or sawn timber stored on the ground as well as timber used in the construction of buildings, bridges, furniture etc. may be attacked. Termites enter the buildings by working their way through the earth and crevices in the foundation and walls, particularly through damp spots. Since many species consume the wood from inside, their presence becomes detectable only after major damage is done. The damage caused by termites to wood work in buildings in the tropics is enormous and the literature on termites attacking wood is vast. Some important wood destroying termites include species of *Amitermes, Ancistrotermes, Coptotermes, Heterotermes, Macrotermes, Microtermes, Microcerotermes* and *Odontotermes*. Many of the species have a wide distribution in the tropics.

While most species of timber are susceptible to termite attack, the degree of susceptibility varies. Usually the durability of timbers against termites is assessed by what is known as the 'graveyard test' in which stakes of standard dimension, of different species of timber, are arranged vertically in soil in a termite infested site, with the top of the stakes protruding above ground like the headstones in a graveyard, and assessing the damage level at intervals. Based on this, timbers are rated as susceptible, moderately resistant, and highly resistant. Termite resistance ratings are available for most timbers in many countries. Although the durability varies among timbers, there are some exceptional timbers such as teak (*Tectona grandis*) and ironwood (*Eusideroxylon zwageri*) that are highly resistant. Even in resistant timbers, generally only the heartwood portion is resistant, although in exceptional cases like *Anopyxis klaieana, Diospyros sanza-minika* and *Klainedoxa gabonensis*, the sapwood is more resistant than the heartwood (Ocloo and Usher, 1980, cited by Wagner *et al.*, 1991)

Wood-dwelling termites

A small group of termites comprising a few genera (family Kalotermitidae) attacks comparatively dry wood and builds small nests entirely within the attacked timber. The colony is established by winged reproductives landing on the wood. Concealed, internal feeding often hollows out the timber from within, leaving only a papery thin outer layer. Some species of timber like *Artocarpus hirsuta* are very susceptible to dry-wood termites even when the wood is used in construction.

6.2.2 *Beetles*

Beetles consume wood from inside after boring into it. Generally they can be detected by the presence of frass outside the timber. Wood-feeding beetles fall into two major groups – large borers belonging to the families Cerambycidae and Buprestidae and small borers belonging to the families Bostrichidae, Curculionidae and Anthribidae. The large borers show some specificity to timber species while the small borers in general attack a wide variety of timber. Some species of low-density timbers are particularly prone to heavy damage by the small borers. In a survey carried out mainly in government-owned timber depots in Kerala State, India, and covering 46 timber species, Mathew (1982) found that all the timbers were attacked by one or more of 53 species of beetles.

Usually different groups of beetles attack the timber in succession, at various stages from freshly felled to dry, processed material. The first group to attack is usually Buprestidae, Cerambycidae and Curculionidae (Scolytinae and Platypodinae) and the second group, Bostrichidae. Some of these families may sometimes occur together. The decisive factor appears to be the moisture content of the wood. By far the most economically damaging groups of wood-destroying insects are the bostrichid beetles for wood used indoors and subterranean termites for wood used outdoors in contact with the ground. Characteristics of the various groups of wood-destroying beetles are described below.

Large borers

Family Cerambycidae

Most large wood destroying insects belong to the family Cerambycidae, commonly called longicorn beetles or longhorn beetles as they possess antennae that are about as long as or longer than the body. Some representative cerambycid timber borers are shown in Fig. 6.1. They generally attack freshly felled timber. Adult beetles discover newly felled timber and lay eggs in crevices in the bark. Newly hatched larvae feed initially under the bark and then tunnel into the sapwood. Typically, the larva is cylindrical and elongate, with a large head and thorax like that of *Hoplocerambyx spinicornis*, shown in Fig. 10.27(b) under *Shorea robusta* in Chapter 10. The mature larva bores into the heartwood and makes a shelter, thus damaging the heartwood, where it transforms into a pupa and then an adult. The cerambycids are usually large insects and cause extensive tunnelling damage. The damage is typically characterised by extrusion of coarse wood fibres although in some species the wood fibres are loosely packed within the larval tunnel (e.g. *Remphan hopei* attacking logs of *Dipterocarpus turbinatus* in Southeast Asia) and in others fine floury dust is tightly packed in

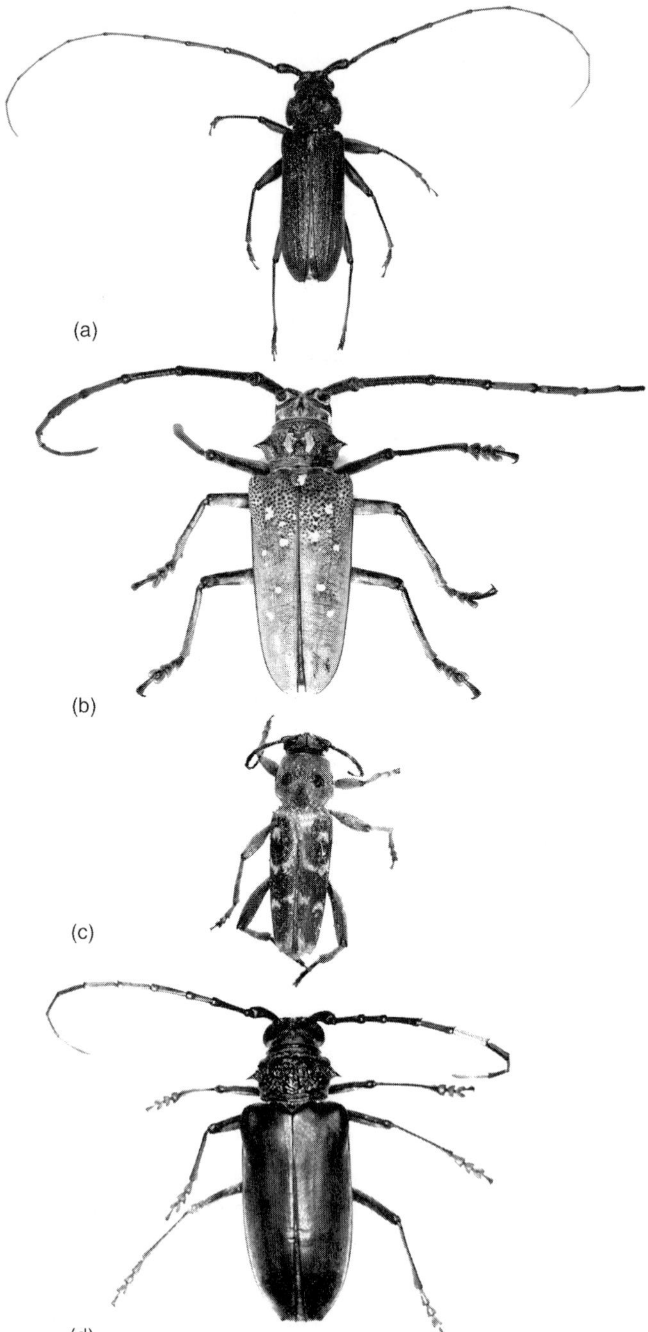

Fig. 6.1 Some representative cerambycid wood borers. (a) *Stromatium barbatum* (length 24 mm); (b) *Batocera rufomaculata* (length 45 mm); (c) *Xylotrechus* sp. (length 14 mm); (d) *Plocaederus ferrugineus* (length 37 mm).

the tunnel (e.g. *Stromatium barbatum* attacking a wide variety of timbers). In timber species without heartwood, the damage is generally more extensive. Most insect species of this group have an annual life cycle but some species complete a generation in less than a year and some take more than a year. For example *Xylotrechus smei* which attacks more than 40 timbers in India, including *Dalbergia* spp., *Gmelina arborea, Mangifera indica, Shorea robusta* and *Tectona grandis*, can complete a generation in about 2.5 months in summer but later generations might hibernate and emerge in the second year. As the cerambycid beetles lay eggs only when bark is present, debarked, sawn and seasoned timber are not attacked. However, some exceptional species like *Stromatium barbatum* attack debarked timber also.

The number of wood-boring cerambycid beetles in the tropics is very large. India alone has more than 1200 species of cerambycids, although some of them attack only slender stems, bark, cones or roots. Also, some cerambycids attack the wood of living trees. Examples are *Aristobia horridula* on *Dalbergia cochinchinensis*, *Celosterna scabrator* on *Acacia nilotica*, *Hoplocerambx spinicornis* on *Shorea robusta*, and *Xystrocera festiva* on *Falcataria moluccana*, pest profiles of which are given in Chapter 10. These insects usually continue their damage in felled timber. Wagner *et al.* (1991) have listed 60 species of cerambycids which attack various timbers in Ghana. In the survey carried out in timber depots in Kerala, India, referred to previously, Mathew (1982) found that 18 out of 46 timber species were attacked by cerambycid borers (Table 6.1). More than 15 species of insects, including some unidentified species were involved. Some low density timbers like *Anacardium occidentale, Artocarpus heterophyllus, Bombax ceiba, Hevea brasiliensis, Mangifera indica* and *Polyalthia fragrans* suffered major damage.

Family Buprestidae

This family of beetles (Fig. 6.2) range in size from 6 mm to 70 mm in length and are usually brightly coloured, with a metallic lustre. The buprestid larva has a characteristic appearance, with a large, flat prothorax into which the small head is withdrawn. The larvae are usually known as flatheaded borers. In general, buprestid borers attack sickly standing trees or recently felled trees and continue their damage in stored logs, although some species attack dry wood. The female beetle lays eggs on the bark and the young larvae bore irregular galleries between the bark and sapwood, and later penetrate into the sapwood. The galleries are usually packed with fine wood dust. Some species of dry wood borers such as *Buprestis geometrica* which attack pine wood and *Chrysochroa gratiosa* which attack *Sterculia alata* in India, penetrate deeper into the wood and tunnel extensively, reducing it to a mass of dust with flakes of wood left here and there (Beeson, 1941). *Belionota prasina* (Fig. 6.2) is a widely

Table 6.1. *Cerambycid borers infesting stored timber in Kerala, India*

Insect species	Host timbers
Acalolepta cervina	*Gmelina arborea*
A. rusticatris	*G. arborea*
Batocera rubus	*Careya arborea*
	Mangifera indica
B. rufomaculata	*Anacardium occidentale, Artocarpus heterophyllus, Bombax ceiba, Careya arborea, Ceiba pentandra, Mangifera indica, Syzygium cumini*
Celosterna scabrator	*Eucalyptus* spp.
Eucommatocera vittata	*Eucalyptus* spp.
Glenia homonospila	*Bombax ceiba, Zanthoxylum rhetsa*
G. indiana	*Z. rhetsa*
Olenecamptus bilobus	*Lagerstroemia microcarpa*
Plocaederus ferrugineus	*Anacardium occidentale*
P. obesus	*A. occidentale*
Serixia sp.	*Garcinia indica*
Xylotrechus buqueti	*Tectona grandis*
X. quadripes	*T. grandis*
Xystrocera globosa	*Albizia odoratissima, Falcataria moluccana, Haldina cordifolia*
Unidentified species	*Hevea brasiliensis, Polyalthia fragrans*

Data from Mathew (1982)

distributed buprestid borer of several commercially important timbers such as *Hopea parviflora, Mangifera indica* and *Terminalia* spp., distributed throughout the Oriental region and Africa. Different species of *Chrysobothris* attack various timbers like *Anogeissus pendula, Mimusops elengi, Shorea robusta, Terminalia tomentosa*, eucalypts etc., in India, *Triplochiton scleroxylon* in Ghana and *Mimosa scabrella* in Costa Rica. Buprestid borer infestation usually occurs in the forest when the log is still moist and is carried to the storage depots. Generally only the sapwood is affected.

Small borers

Family Curculionidae: Scolytinae

Popularly known as bark beetles, scolytines are among the first to attack newly felled trees. The beetles bore through the bark and make galleries either between the bark and sapwood or within the sapwood, depending on the insect species. The gallery system is picturesque (Fig. 6.3). Typically it consists of

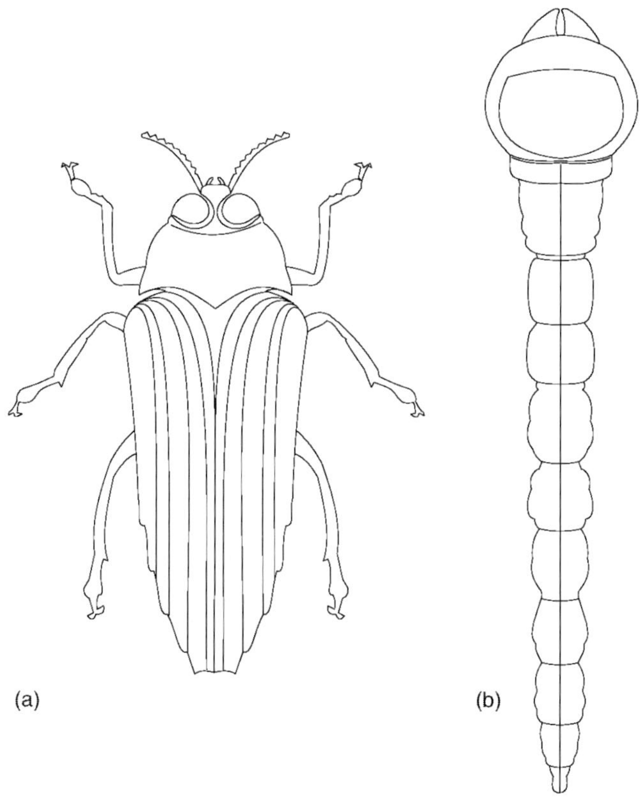

Fig. 6.2 A buprestid wood borer, *Belinota prasina*. (a) adult (length 26 mm); (b) larva (length 53 mm). After Beeson (1941).

a mother gallery between sapwood and inner bark, bored by the adult female, from which larval galleries radiate outward at regular intervals. The female beetle lays eggs on pits made on either side of the mother gallery and the larvae bore their galleries at about right angles to the mother gallery. The mature larva pupates at the end of the larval gallery, in a pupal cell, from which the adult emerges through an exit hole in the bark. The scolytines are also called engraver beetles because of the pattern they make on the wood. The gallery pattern is quite variable in the different species, with different combinations and spatial arrangements of the essential gallery elements. For example, in polygamous species, several mother galleries may be interconnected and the larval galleries may become crowded and irregular.

Scolytinae is a large group, with more than 2000 world species. At least 300 species occur in India and 68 in Ghana. They are small beetles, 1 mm to 9 mm in length. Many of them attack small branches and twigs. Some species cultivate a fungus known as ambrosia in their tunnels and feed on it, and are therefore

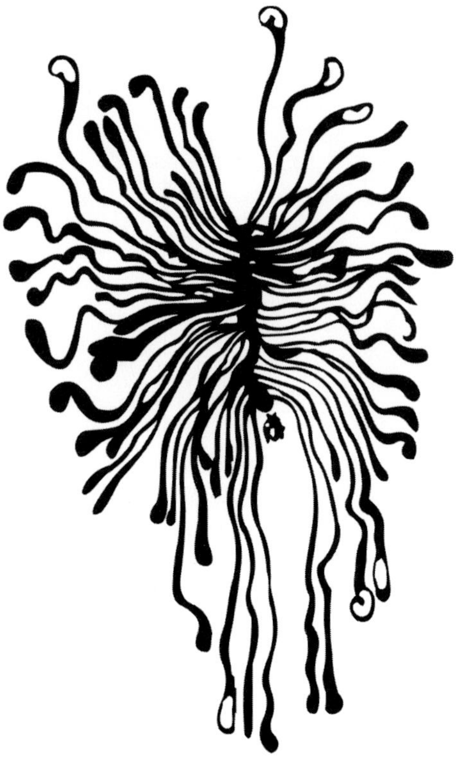

Fig. 6.3 Gallery system of the scolytine bark beetle *Scolytus major*, on the inner surface of bark of a log of *Cedrus deodora*. Note the central mother gallery (length 25 mm) and the radiating larval galleries. After Beeson (1941).

called ambrosia beetles. The ambrosia beetles construct their gallery system within the woody tissue. As a result, they cause small holes, and staining of the sapwood. The structural strength of the timber is not seriously affected but the infested timber shows defects such as pinholes, shotholes, black spots or lines on the sawn surfaces which spoil the appearance of plywood and ornamental veneers. An important genus causing serious damage to timber is *Xyleborus*, comprising over 100 species distributed throughout the world. Most species of *Xyleborus* are polyphagous, attacking 30–40 species of timber, with some like *X. ferrugineus* breeding in 74 timbers (Browne, 1962, cited by Wagner *et al.*, 1991) and *X. testaceous* breeding in over 100 timbers (Beeson, 1941). In this genus, the males are flightless and do not generally leave the parent nest. In the study in timber depots in Kerala, India, referred to earlier, Mathew (1982) recorded 12 species of scolytine borers infesting 15 timbers (Table 6.2).

Some bark beetles of the genera *Dendroctonus, Ips* and *Scolytus* are very serious pests of living coniferous trees in the temperate regions where their outbreaks

Table 6.2. *Small beetle borers infesting stored timber in Kerala, India*

Insect species	Host timbers
Family Curculionidae: Scolytinae	
Cryphalus carpophagus	*Hevea brasiliensis*
Cryphalus sp.	*Mesua ferrea*
Euwallacea fornicatus (syn: *Xyleborus fornicatus*)	*Gmelina arborea*
Hypocryphalus mangiferae	*Mangifera indica*
Hypothenemus birmanus	*Falcataria moluccana*
Phloeosinus tuberculatus	*Knema attenuata*
Phloeosinus sp.	*Knema attenuata*
Scolytomimus assamensis	*Palaquium ellipticum*
Sphaerotrypes sp.	*Syzigium cumini, Lagerstroemia speciosa, Vateria indica*
Xyleborus interjectus	*Artocarpus heterophyllus, Bombax ceiba, Ceiba pentandra, Dysoxylum malabaricum*
X. similis	*A. heterophyllus, Dysoxylum malabaricum, Erythrina indica, Hevea brasiliensis*
Xyleborus sp.	*Erythrina indica*
Family Curculionidae: Platypodinae	
Crossotarsus indicus	*Erythrina indica*
C. nilgiricus	*Canarium strictum*
C. saundersi	*Vateria indica*
Diacavus assamensis	*V. indica*
Platypus andrewesi	*Lophopetalum wightianum*
P. cavus	*Bombax ceiba*
P. latifinis	*B. ceiba, Ceiba pentandra, Hevea brasiliensis, Knema attenuata, Mangifera indica*
P. solidus	*Aglaia elaeagnoidea, Ailanthus triphysa, Ceiba pentandra, Elaeocarpus tuberculatus, Hevea brasiliensis, Machilus macrantha, Mangifera indica, Syzygium cumini*
P. uncinatus	*Lagerstoemia reginae, Mangifera indica*
Other Curculionids	
Aclees birmanus	*Artocarpus heterophyllus*
Cossonus divisus	*A. heterophyllus*
Mecistocerus mollis	*Erythrina indica*
Mecopus sp.	*Artocarpus heterophyllus, Grewia tiliaefolia*
Myocalandra exarata	Bamboo (*Ochlandra* spp.)
Pagiophloeus longiclavis	*Toona ciliata*
Phaenomerus sundevalli	*Aglaia elaeagnoidea, Hopea parviflora, Machilus macrantha*
Sipalinus gigas	Bamboo (*Bambusa* sp.)
Sipalus hypocrita	Bamboo (*Bambusa* sp.)

Table 6.2. (cont.)

Insect species	Host timbers
Family Bostrichidae	
Dinoderus bifoveolatus	*Albizzia procera, Bombax* sp., *Ficus hispida*
D. minutus	*Falcataria moluccana, Bombax ceiba, Toona ciliata,* bamboos
D. ocellaris	bamboos
Heterobostrychus aequalis	*Bombax ceiba, Calophyllum elatum, Grewia tiliaefolia, Hevea brasiliensis, Vateria indica,* bamboos
Lyctus brunneus	*H. brasiliensis*
Minthea rugicollis	*H. brasiliensis, Tetramelus nudiflora*
Rhizopertha dominica	*Albizia odoratissima,* bamboos
Sinoxylon anale	*A. odoratissima, Anacardium occidentale, Dalbergia latifolia, Hevea brasiliensis, Lagerstoemia reginae*
S. atratum	*Falcataria moluccana, Bombax ceiba*
S. conigerum	*Erythrina indica, H. brasiliensis, Lagerstroemia microcarpa*
S. crassum	*Albizia odoratissima, Terminalia bellerica*
S. pygmaem	*Grewia tiliaefolia*
Xylothrips flavipes	*A. odoratissima, Artocarpus hirsutus, Alstonia scholaris, Bombax ceiba, Hopea parviflora, Vateria indica*
Family Anthribidae	
Eucorynus crassicornis	Bamboo (*Bambusa* sp.)
Phloebius alternans	Bamboo (*Bambusa* sp.)
P. lutosus	Bamboo (*Bambusa* sp.)
Sintor sp.	*Artocarpus heterophyllus*

Data from Mathew (1982)

occur periodically, particularly in single species forest stands, killing trees over extensive areas. In the tropics, however, scolytines are not important pests of living trees, for reasons not well understood, except in pines in the cooler regions of Honduras and the Philippines, as discussed under *Pinus* in Chapter 10.

Family Curculionidae: Platypodinae

Platypodines are another group of small beetles that attack freshly felled timber. They are small, elongate and cylindrical beetles, 2–12 mm in length, and are known as pinhole or shothole borers based on the damage they cause. They are also called ambrosia beetles because they grow the ambrosia fungus in their tunnels. They generally attack only unseasoned wood with high moisture content. The life cycle may last from about five weeks to six months or more, depending on the species and the season. The larvae are soft and legless.

An entrance tunnel, usually drilled by the female beetle, extends radially into the sapwood from which the main galleries continue left and right, parallel to the circumference of the log (Fig. 6.4). From the main galleries, secondary galleries branch off and run to variable distances. In logs without distinct heartwood, the tunnels run deeper, in a sinuous or spiral course. In some species, the galleries may be interconnected. The galleries are bored by the adult beetles. Eggs are usually laid in the entrance tunnel and later transferred to the branch tunnels. The tunnels are kept clean by pushing out the wood dust and waste products through the tunnel entrance. The pupal chamber, excavated by the mature larva, is vertical to the larval gallery and the new adults escape through the original parental entrance tunnel.

Platypodines, like scolytine ambrosia beetles, spoil the appearance of timber, with pinholes, shotholes and black staining, although the structural strength

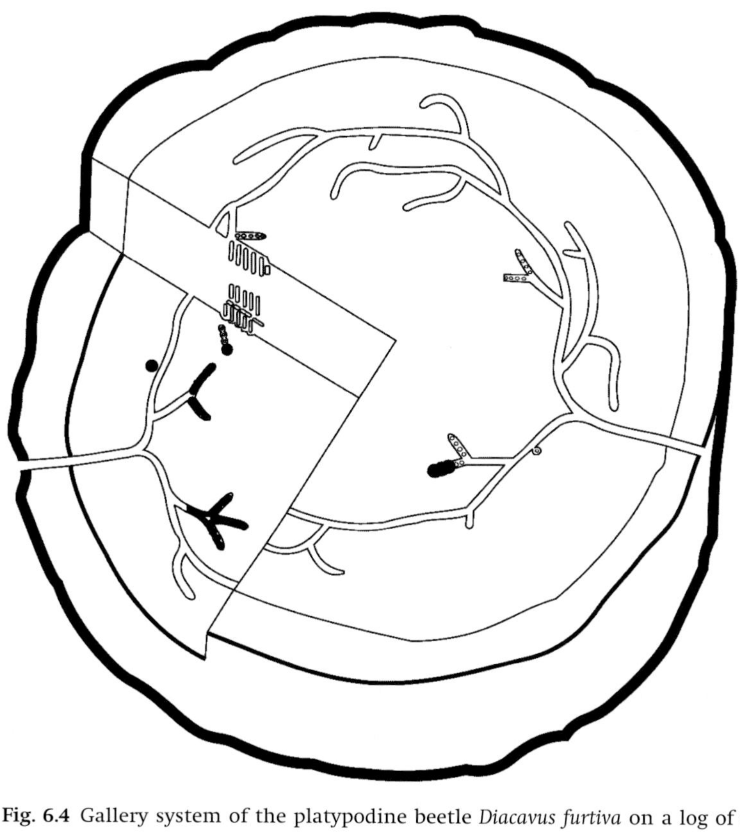

Fig. 6.4 Gallery system of the platypodine beetle *Diacavus furtiva* on a log of *Shorea robusta*. Diameter of the log is 13.5 cm. After Beeson (1941).

is little affected. Platypodines are generally polyphagous. They are often very abundant as they breed in felling refuse. Dry, seasoned timber is not attacked and the beetles usually leave the host wood when the moisture content falls below about 50%. *Crossotarsus* and *Platypus* are two common genera, with a large number of species. The biology of several platypodines of the Indian region is described by Beeson (1941) and of the West African region, by Browne (1968). In the Kerala study mentioned earlier, Mathew (1982) recorded nine species of platypodines, most belonging to the above two genera, attacking 15 timber species (Table 6.2).

Other Curculionidae

There are a few other larger curculionid wood borers, generally 10–20 mm in length. Some species have galleries confined to the bark, some bore superficially in the sapwood and others bore deeper into low-density wood. The larvae are legless, soft and curved. Most have annual generations and are not of much economic significance. In the Kerala study mentioned earlier, Mathew (1982) recorded nine species attacking nine timbers, including bamboo (Table 6.2).

Family Bostrichidae

Popularly called 'powder-post beetles' bostrichids attack dry timber. Low-density timbers as well as the sapwood portion of hardwoods are susceptible. They attack all kinds of wood, including stored logs, tent poles, sawn timber, plywood, manufactured articles such as furniture, packing cases, matchwood, tool handles etc. Their attack is characterised by copious extrusion of wood dust from the infested wood, and hence the popular name. Bostrichid beetles are of great economic importance because of the heavy damage they cause, although the family is small, with a little over 100 species.

The beetles are generally about 5–10 mm in length, although some like the African *Apate* spp. which bore into live trees can be over 20 mm long. Usually the beetles are cylindrical, with the head directed downward and covered by a hood-like thorax (Fig. 6.5 a,b,c). The larva has a curved body, enlarged at the thorax, with three pairs of well developed thoracic legs and is an active borer like the adult. Most species have three to four generations per year and successive generations attack the same piece of wood, reducing it to dust. Generally the adult female bores radially into the wood for a short distance, then parallel to the wood surface, to make a branched tunnel in which the eggs are laid. The larvae then bore through the wood. The tunnels are circular in cross-section and packed densely with fine wood dust. Usually the larval tunnels are close and crowded, leaving an outer skin of wood intact, but reducing the inner wood to powder.

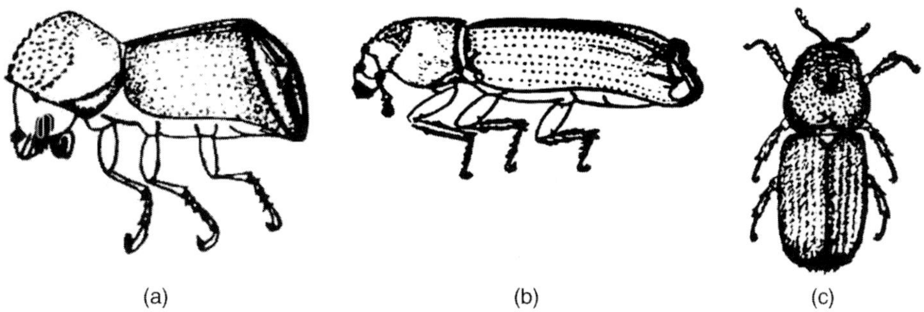

Fig. 6.5 Some bostrichid wood borers. (a) *Sinoxylon anale* (length 6 mm); (b) *Heterobostrychus aequalis* (length 10 mm); (c) *Dinoderus minutus* (length 3.5 mm). From *KFRI Research Report No. 10* (Mathew, 1982).

Bostrichids are generally very polyphagous. Presence of starch is believed to make the wood attractive to them. Debarking increases the damage by exposing the sapwood. Some timbers like rubber wood (*Hevea brasiliensis*) are highly susceptible to attack. Common bostrichid genera are *Dinoderus*, attacking a variety of bamboos worldwide (Fig 6.6); *Heterobostrychus*, attacking a variety of timbers in Asia and Africa and *Sinoxylon*, several species of which commonly occur in timber depots and saw mills in Asia (Fig. 6.5). *Lyctus brunneus* and *Minthea rugicollis* occur in Asia and Africa. In the Kerala study mentioned earlier, Mathew (1982) recorded 13 species of bostrichids, attacking over 20 species of timber (Table 6.2).

Family Anthribidae

Some members of this small family of beetles bore into wood. They lay eggs in the bark and the larvae tunnel mainly in the bark, or into the sapwood. The larval tunnels run parallel to the axis of the stem and are filled with fine wood dust. Life cycle is generally annual. They are not of much economic significance. In the Kerala study mentioned earlier (Mathew, 1982), four species were recorded, mostly from bamboo (Table 6.2).

General observations on small borers

As seen above, a large number of small beetles attack wood on storage. Generally they confine their attack to the sapwood, but in low-density timbers with no distinct heartwood the damage extends deeper. Since these borers are numerous and attack a wide variety of timber, with little host specificity, no timber's sapwood escapes their attack. Consequently, as a precaution, the sapwood layer of all timbers is discarded when the timber is processed for use, except when it is used for pulping.

Fig. 6.6 Stored, dry bamboo culms showing damage caused by *Dinoderus* beetles. Courtesy: R.V. Varma, Kerala Forest Research Institute.

Among the borers, the bostrichids are the most damaging as they attack even dried and converted timber. Some timbers are extremely susceptible to bostrichids, and these include the rubber wood *Hevea brasiliensis* in Asia and *Triplochiton scleroxylon* in West Africa. Some borers of the family Curculionidae, notably *Xyleborus* spp. (Scolytinae) and *Platypus* spp. (Platypodinae), are also highly damaging.

7

Population dynamics: What makes an insect a pest?

7.1 Introduction

As seen in Chapters 2 and 5, a large number of insect species is usually associated with each tree species, but only a few of them become serious pests. For example, out of over 174 species of insects that can feed on the living teak tree, only three have become pests. Usually insects do not cause serious damage to trees unless the number of individuals, i.e. the population size, becomes large. Under what circumstances do insect populations increase to damaging levels? And why do some insects build up in large numbers while others do not? Our ability to control pests depends on the answers to these questions.

A group of individuals of a species living together in a defined area is called a population, and the study of the changes in the size or density of populations over time is known as population dynamics. It tries to predict these changes and explain the causes. A population has certain group characteristics, in addition to those possessed by the individuals constituting the group. It has a genetic composition, sex ratio, age structure, density and dispersion (clumped, random, etc.), each of which influences its behaviour. It is obvious that a pest problem is essentially a population dynamics problem. So we shall examine in some detail the circumstances under which insect population densities change.

7.2 Characteristics of population growth

Under ideal conditions insects, like other organisms, have the capacity to increase exponentially. For example, the female moth of the teak defoliator *Hyblaea puera* lays an average of 500 eggs and the life cycle is completed in about 20 days. Therefore one female moth can produce 500 moths in less than a month.

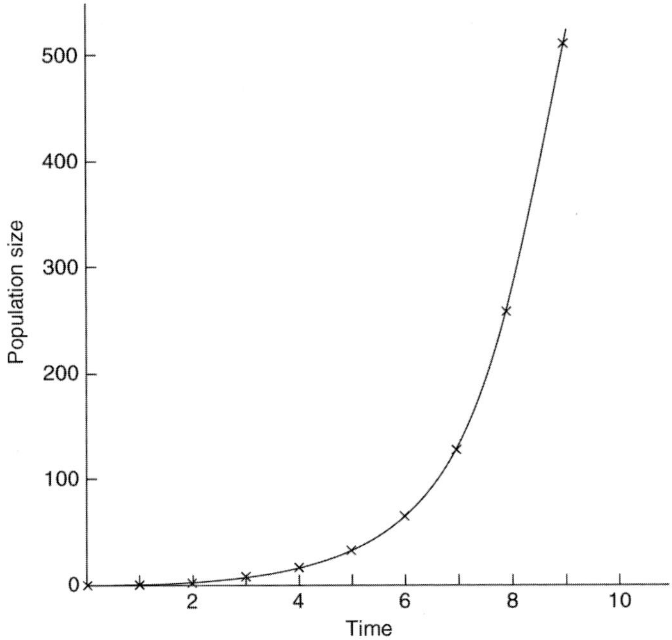

Fig. 7.1 Exponential growth of population under ideal conditions. Here the population increases by a constant factor per unit of time.

Half of them would be females and they could each produce another 500 moths over the next month. Thus if all of them survived, there would be 125 000 moths in two months, 31 250 000 moths in three months, 7 812 500 000 moths in four months, 1 953 125 000 000 moths in five months and so on, ad infinitum, all originating from a pair of moths. This kind of population growth is known as geometrical or exponential growth and is characteristic of all populations growing under ideal conditions (Fig. 7.1). In this type of growth, the population increases by a constant factor per unit of time. However, in nature conditions are seldom ideal. There are a large number of mortality factors affecting the survival of insects so that in a real situation, only a small fraction of the progeny survives.

The size of a population is usually measured in terms of the number of insects present per unit area of habitat, or *density*. Thus we speak of caterpillars per leaf or shoot, termites per m^3 of soil or insects per hectare of forest. We measure the population density by sampling the insects in small units such as leaf, shoot or tree and then extrapolate it to larger areas. The rate at which a population changes is determined primarily by three variables – births, deaths and movement of insects. The movement is very important because most insects are highly mobile and individuals may move in and out of an area rapidly and in large numbers. In the simplest sense, population change can be defined as follows (Berryman, 1986).

Population change per unit of time equals birth rate minus death rate plus immigration rate minus emigration rate.

This can be expressed symbolically as

$$\Delta N = (b - d + i - e)\, N,$$

where N is the initial density of population, b is the average birth rate per individual and d, i and e are rates of individual death, immigration and emigration. In a given area, the population density increases when the rates of births and immigrations exceed the rates of deaths and emigrations, i.e. when $b+i > d+e$. The density declines when $b+i < d+e$, and remains unchanged, i.e. in equilibrium when $b+i = d+e$. One useful measure of population growth potential is the *net per capita finite growth rate* (g) which is defined as $(b - d + i - e)$. The population growth equation then becomes

$$\Delta N / \Delta t = gN,$$

where t is time. Here g is the growth rate over a finite interval of time, when free from environmental constraints. Although these simple equations describe how populations change over long periods of time, they are applicable only when the important parameters, the rates of birth, death, immigration and emigration remain constant over that period. This condition is seldom met in natural situations because there are a large number of biological and environmental factors that alter these rates from time to time and place to place (Berryman, 1986). Consequently it is difficult to predict the changes in population density of an insect although significant advances have been made in theory and methods. The literature on insect population dynamics is extensive and good insight has been gained through the use of mathematical models. Berryman (1999) gives a concise account of the principles of population dynamics and their applications, and chapters in Cappuccino and Price (1995) provide an overview of the vast literature on various aspects of the subject. Since this is a specialized area of study which requires the use of mathematical methods, we will not go into the details. What is given here is an overview of the topic to gain an appreciation of the importance of population dynamics in understanding the origin of pest problems and how we can control them.

7.3 Factors affecting population change

Any factor which affects the rates of birth, death and movement of insects is potentially capable of influencing population growth and such factors are legion. They can be broadly categorised into physical (or abiotic) factors and biotic factors.

7.3.1 *Physical factors*

The physical factors may exert their influence directly or indirectly through their effects on other organisms, including the host plants and natural enemies. Temperature, humidity, rainfall, wind, soil properties etc. all exert their influence on insects in various ways. The higher temperatures of the tropics, for example, are conducive for the growth of the poikilothermic insects and consequently most tropical insects can have continuous generations throughout the year, unlike the temperate insects. Therefore the populations of tropical insects can grow faster. Extremes of temperature, however, may induce aestivation or hibernation. Temperature can have different effects on different stages of the life cycle or on different life processes such as survival, mobility, rate of development and reproduction, and consequently the effect of temperature on population growth is often difficult to predict. Furthermore different species of insects have different temperature optima and tolerance limits. Population outbreaks of the leucaena psyllid, for example, are suspected to be prevalent in places with lower optimal temperatures in their introduced habitats (see Chapter 10).

As with temperature, moisture also exerts an influence on the growth and survival of insects and there is an optimal range of moisture which may vary between insect species and stages of development of the same species. Some species become dormant in the absence of adequate moisture as in the case of the eulophid parasitoid *Sympiesis hyblaea* of the teak defoliator *Hyblaea puera*. This can impact on the populations of *H. puera*. High atmospheric humidity often favours the survival and spread of fungal pathogens of insects with effect on the abundance of insect populations. High rainfall has been found to favour the development of populations of the cerambycid borer *Hoplocerambyx spinicornis* on *Shorea robusta* and of the cossid borer *Xyleutes ceramicus* on *Tectona grandis* (see Chapter 10).

Similarly, various parameters of light such as photoperiod, intensity and wave length also affect the life of insects in various ways. For example, UV radiation can kill viral pathogens of insects. Weather also influences the movement of some insects very significantly, with important consequences on population development. The monsoon-linked migratory movement of the teak defoliator *H. puera* (Chapter 10) is one example where weather has a decisive influence on the population dynamics of an insect. Dispersed populations of some flying insects are also known to be concentrated by wind convergence (Pedgley, 1990). Many other components of the physical environment like gaseous composition of the air, pollutants, electromagnetic radiation, chemical composition of the soil etc. affect the life of insects either directly or indirectly in various ways.

As mentioned above, physical factors can influence insect populations either directly or through their effects on host plants or natural enemies. When the effect is indirect, for example when it adversely affects host foliage quality, it can lead to a delayed response from the insect population.

7.3.2 *Biotic factors and logistic growth*

The biotic factors influencing a given species include other individuals of the same species as well as other species of animals and plants.

Interactions among members of the same species, i.e. intraspecific interactions, may have beneficial as well as inhibitory effects on population growth, depending on the population density. Moderately high density favours mate finding and offsets the impact of natural enemies. The beneficial effect of cooperation reaches its peak in social insects where there is division of labour among castes. High population density can also be beneficial when it breaks down the host tree defences, as seen in the cases of the sal borer *Hoplocerambyx spinicornis* attacking *Shorea robusta* and bark beetles attacking pines, or helps to overwhelm the parsitoids and predators as during teak defoliator outbreaks (see Chapter 10). On the other hand, high population density can also lead to competition among individuals for limited resources of food and shelter. This competition inhibits population increase. It retards the birth rate and/or enhances the death and emigration rates through various mechanisms. The net per capita growth rate, g, decreases progressively as population size increases, until it reaches a constant value, resulting in a logistic population growth curve (Fig. 7.2). The population growth curve levels off when the *carrying capacity* (K) of the environment is reached. K represents the maximum population size that can be supported by a given environment. For example, a patch of grassland has a maximum number of grazing deer that can be supported, depending on the regenerating capacity of the grass, although there are many other factors that prevent populations from reaching the carrying capacity of the environment. Population growth under these conditions is described by the equation

$$\Delta N/\Delta t = g[1 - N/K]N,$$

where N is the initial density of the population, t is time, g is the net per capita finite growth rate, and K is the carrying capacity of the environment.

Among interspecific interactions, the most important are the insect–host tree interaction and the insect–natural enemy interaction.

Host quality in terms of nutrients, secondary plant chemicals, physical and chemical deterrents etc. have important implications for pest population

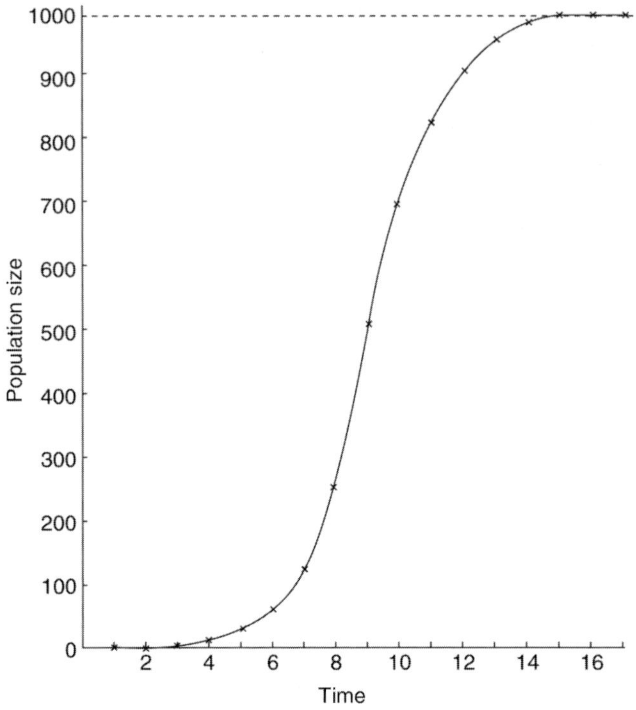

Fig. 7.2 Logistic population growth. When the population size increases beyond a certain limit, the growth rate decreases until it is stabilized to maintain the carrying capacity of the environment.

build up. Drought stress, for example, favours the build up of some insects like the bark beetles of pines (see Chapter 10). Seasonal and spatial variations in the quality of plants (e.g. flushing) alter the rates of birth, death and movement of pest insects. There are innumerable ways in which the host plants influence the growth, reproduction and survival of pest insects.

Natural enemies are thought to be important factors regulating the population build up of insects. A wide variety of organisms preys upon or parasitizes insects, including vertebrates like mammals, birds and reptiles and invertebrates like spiders, mites, other insects, nematodes, fungi, bacteria and viruses. Natural enemies respond in two ways to increased pest population density. In ***functional response***, each individual natural enemy attacks more prey as the prey density increases and in ***numerical response***, the number of the natural enemy increases as the prey density rises. The functional response in which the rate of attack increases is the result of the ease with which the prey can be located when its density is high. Also, in the case of vertebrates there is a tendency for predators to temporarily switch their preference to the more abundant prey and recruit other members of the group in their hunting trips. Numerical response results when the natural enemy population increases through increased reproduction because of greater prey (food) availability. Numerical response can also be brought about when greater numbers of the predator or parasitoid migrate to an area where the prey density is high. Higher population density of insects may also favour outbreak of fungal, bacterial and viral diseases because of the ease of transmission of the pathogen due to host proximity. Most pests have a large group of natural enemies, some of which exert decisive influence on population growth. However, their effectiveness depends on a number of complex biotic interrelationships. For example, annual outbreaks of the defoliator *Hyblaea puera* occur in teak plantations in India in spite of the presence of not fewer than 44 species of parasitoids, 108 predators and 7 pathogens (see Chapter 10).

Apart from these, there are countless ways in which biotic relationships involving multitudes of plants and animals affect the life of a given insect species. The potential number and magnitude of interactions involving a large number of abiotic and biotic factors affecting a herbivore is enormous and difficult to enumerate. Imagine the web of relationships involving the 44 species of parasitoids, 108 species of predators and 7 species of pathogens of *H. puera*, with their alternative preys, and over 45 plant hosts (see e.g. Fig. 10.37 in Chapter 10). Add to this the differential impact of physical factors on the life process of these organisms, all of which can influence their impact on the life of the pest under consideration. In spite of such complexity of ecological interrelationships, we often find that many of the complex interrelationships are unimportant in the life system of a species. A few key variables or key factors usually have a major role in driving the population dynamics of an insect species. They play such an important role that they can be used to predict the population growth. Such abstraction is unavoidable and has often proved sufficient to predict the outcome. It is like using a mathematical prediction equation (e.g. the girth and height of a tree can be used to arrive at the commercial volume of a tree). Thus the study of population dynamics involves the use of some simplifying assumptions.

7.4 Principles governing population dynamics

Under natural conditions, herbivorous insects are ubiquitous components of terrestrial ecosystems and they usually remain in small numbers. Pest outbreaks are exceptions. Herbivorous insects form a link in the energy cycle of the ecosystem as one of the consumers of primary production, and in turn serve as food for secondary consumers, ultimately contributing to the cycling of

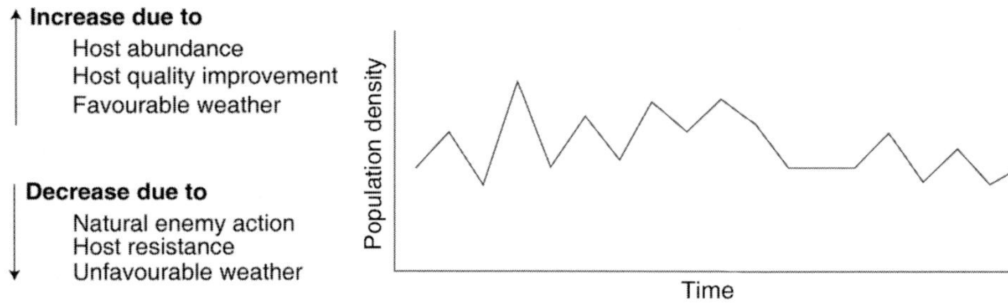

Fig. 7.3 Insect population growth under natural conditions. In natural ecosystems, the population density of a species usually remains in a dynamically steady state (i.e. fluctuates within a limited range) due to the action of opposing forces.

nutrients and energy (see Chapter 3). In the pristine forest, most insects remain in low numbers most of the time as their populations are regulated by a number of biotic and abiotic factors. This apparent constancy of numbers is the result of a dynamic equilibrium between production and destruction, i.e. increase in numbers due to reproduction, and mortality due to various factors (Fig. 7.3). Negative feedback mechanisms are involved in maintaining this stability, i.e. the dynamic equilibrium, of insect numbers (Berryman, 1986). For example, when the population of an insect species increases, populations of its parasitoids also increase. The increased parasitoid populations exert greater pressure on the host population and reduce it to a lower level. Thus an initial stimulus (increase in the pest population that causes an increase in the parasitoid population) is fed back to the pest population, causing a negative impact, in the same way as an increase in temperature which expands the bimetallic rod of a thermostat breaks the electric circuit and regulates the temperature of an oven.

The negative feedback can also be effected through the host plant when an increase in the pest population results in decreased availability of food which in turn reduces the population growth. When the negative feedback mechanism fails, uncontrolled increase of a population can occur. This can also occur when a positive feedback mechanism comes into operation. For example, an increase in the number of bark beetles boring on a pine tree can overcome the tree defences such as resin flow more effectively than fewer borers. This causes a further increase in the borer population, creating a chain reaction which constitutes a positive feedback mechanism (i.e. every increase leads to further increase), leading to uncontrolled increase in the number of bark beetles.

The mechanisms which regulate insect numbers have been the subject of intense theoretical debate since the 1930s, with two main schools of thought, one emphasising the importance of density-dependent factors (i.e. the direct or indirect negative feedback exerted by the increasing population) and the other the importance of density-independent (abiotic, like weather) factors (Clarke *et al.*, 1967; Turchin, 1995). The debate still continues, but newer approaches to the study of insect population dynamics including the application of mathematical theory has brought to light details of many forms that regulation can take, such as simple local regulation, metapopulation regulation and complex dynamics involving endogenous and exogenous factors (Cappuccino, 1995; Turchin, 1995; Berryman, 1999). There is an emerging consensus that density-dependent regulation may be very common although there are many 'counterexamples demonstrating that regulation does not always operate in all populations at all times' (Turchin, 1995, p. 36).

Berryman (1999) has listed five basic principles that govern the population dynamics of insects. The first is **exponential growth of populations** to which we have already referred (Fig. 7.1). If unchecked, it leads to an unstable, exponentially increasing population. It is unstable because the population tends to move away from the original or initial condition. The second principle is **cooperation among individuals of the same species**, which can lead to a higher rate of increase as populations become larger or denser. For example, there is increased probability of encountering a mate and reduced probability of being killed by a natural enemy in a dense population. Mathematical simulations show that under this principle, populations grow when they are above a threshold and decline when they are below it (Berryman, 1999). The third principle is **competition or struggle between individuals of a species** to obtain the resources they need to survive and reproduce (struggle for existence). Mathematical simulations demonstrate that operation of this principle leads to a logistic population growth curve (Fig. 7.2). Also, random environmental disturbances cause saw-toothed oscillations in the population curve, gradual change in the external environment causes trends and sudden changes cause shifts.

The fourth principle is **circular causality between the population and its environment**. According to this, populations can affect the properties of their environment and thus create circular causal pathways linking the populations to elements of their environment such as resources, enemies or other components. Mathematical simulations show that circular causality can induce low-frequency cycles in population dynamics, with environmental variability sustaining and amplifying these cycles. Population cycles are the result of this principle, where the adverse impact of high-density population on the environment, for example destruction of food supply, takes time to reverse, causing a delayed negative feedback effect. Circular causality can also generate extremely complex patterns in time and space. The fifth principle is **the existence of limiting factors**. It recognises that while a given population is embedded in complex webs of interaction with other biological populations and their physical environments, only one or a few of these interactions is likely to dominate the dynamics at any particular time and place. Therefore we need not know all the details of the web of interrelationships to understand and predict the dynamics of a particular population. In other words, some of the feedback loops act as limiting factors. This is a simplifying principle, although the limiting factors can change in response to changing population density and environmental conditions. Simulations show that this can lead to unpredictable population dynamics including population explosion and collapse (Berryman, 1999).

Populations governed by these five principles, that is, geometrical growth, cooperative interaction between individuals, competitive interaction between individuals, circular causality between the population and its environment and limiting factors, can display a wide array of dynamic behaviour patterns.

7.5 Types of forest insect outbreaks

What do we observe in real life situations? Chapter 10 will show that pest incidence can take many different forms, from low density infestations to very heavy outbreaks which may be regular or sporadic. Is there any consistent pattern? Unfortunately we have only qualitative knowledge of pest incidence in tropical forests. Most information on the dynamics of forest insect populations has come from studies in temperate forests. Berryman (1988) compiled detailed information on the dynamics of 27 well-known forest insect pests across the world, mostly from Europe, North America and Australia, but including one, the teak defoliator, from India. Largely based on data from temperate forest insects, Berryman (1986, 1987, 1999) has made an attempt to develop a classification system for forest insect outbreaks. A classification system tries to organize the observed patterns of population fluctuations into groups or classes according to their common characteristics. In turn, it helps us to organize the observed phenomena and probe into the cause–effect relationship. The rationale is that if we know how a pest outbreak originates, we are better able to prevent or control it.

Theoretically, population fluctuations may be caused either by endogenous factors (density-induced feedback loops) or by exogenous factors (weather, host condition etc.). However, different causes can lead to the same type of population growth behaviour (Berryman, 1999). Fig. 7.4 shows the commonly observed types of insect population growth. Two basic types have been recognized, **gradient** and **eruptive**.

Gradient population growth occurs when the environment is favourable for a particular insect. In this case, changes in the favourableness of the exogenous

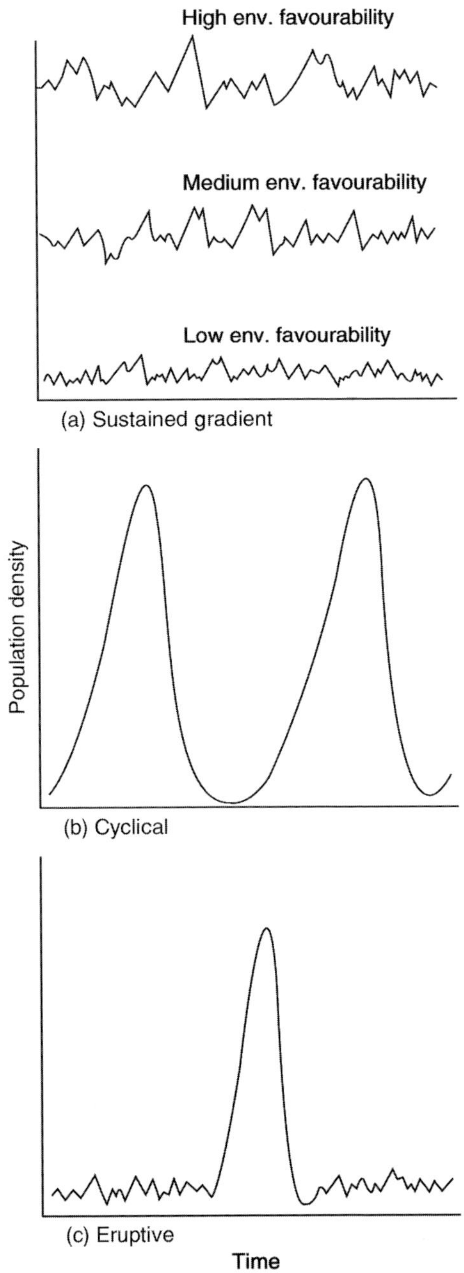

Fig. 7.4 Three common types of insect population growth. (a) Sustained gradient, (b) cyclical, (c) eruptive.

environment (e.g. food availability) leads to corresponding changes in the equilibrium density of the insect population. It is called a gradient population because the population responds in a graded manner to improvement in environmental favourableness. Within the gradient outbreak, three subclasses have been recognized, that is, sustained gradient, cyclical gradient and pulse gradient. In sustained gradients the populations may persist at a fixed level depending on the limit imposed by that environment; in cyclical gradients the populations go through regular cycles of abundance caused by delayed density-dependent feedback; and in pulse gradient the pest populations will follow a 'boom and bust' course when the environment changes from low to high favourableness and back. According to this classification, most herbivorous insects in forest environments belong to the sustained gradient type, i.e. relatively stable populations. Fast-acting, negative density-dependent feedback mechanisms regulate their populations at relatively stable levels. Outbreak of the leucaena psyllid *Heteropsylla cubana* (see Chapter 10) appears to be of the sustained gradient type where high densities are reached in the favourable exotic locations free from native natural enemies. The best studied example of a cyclical gradient is the larch budmoth in the Swiss Alps in the temperate region, in which the populations cycle violently every 9 to 10 years. During the peaks, population density is about 30 000 times the minimum density (Baltensweiler *et al.*, 1977). It is believed that these cycles are caused by delayed, negative density-dependent feedback mechanisms. A delayed negative feedback is exercised when the larch leaves (needles) produced after defoliation are fibrous, lower in nitrogen content and covered by resins. These factors lead to reduced survival of the insect and it may take several years before the normality of the foliage is restored when the insect population builds up again. We have no information about regular cyclical outbreaks in tropical forests although periodic outbreaks are common.

In eruptive population growth, populations that remain relatively stable for long periods of time erupt occasionally and irregularly, and spread over large areas, starting from *epicentres* (specially favourable locations where the outbreak begins). This is believed to be caused by positive density-dependent feedback mechanism. Generally, eruptive population growth falls into the subclasses of pulse eruption and cyclic eruption. In an eruptive outbreak, the population may spread into adjacent unfavourable habitats, unlike in gradient outbreaks. Bark beetle outbreak on pines is one example of eruptive outbreaks. In this case, larger numbers of beetles are able to overcome the resistance of hosts by reducing the defensive resin flow, leading to a positive density-dependent feedback mechanism favouring the survival and multiplication of the insects. Periodic outbreaks of the sal borer *Hoplocerambyx spinicornis* in India (see Chapter 10) is also an example of eruptive outbreaks.

As Berryman (1999) himself observes, the outbreak classification system is a human attempt to order natural phenomena. The observed spectrum of outbreaks is indeed very wide and complex and each pest may be unique, but according to Berryman (1999, p. 9) "the fact that each person is unique does not prevent the physician from practising medicine". The outbreak of *H. puera* populations, for example, is unique as it combines the characteristics of the gradient type and eruptive type. While it is a response to increased supply of food (tender leaves) during the flushing season of teak (pulse gradient type?), there are epicentres where the outbreak begins, as in an eruptive outbreak, although these epicentres are not the typical specially favourable population multiplication sites but random locations where moths are brought together, probably aided by the monsoon wind system (see Chapter 10). This example emphasises the need to study more of the tropical insects. It is typical of nature to defy neat classifications!

7.6 Causes of forest insect outbreaks

In spite of the theoretical advances, our understanding of the cause–effect relationships of insect population outbreaks is incomplete. This is particularly so in the case of tropical forest insects. One consistent trend we notice is that pest problems are more common in, but not exclusive to, plantations than in natural stands of trees. This suggests that pest problems are precipitated by environmental change, perhaps disruption of the naturally existing ecological interrelationships. On the other hand, another consistent trend is that out of several tens or even hundreds of insects associated with a particular tree species, only some become pests. In North America, for example, outbreak species represented fewer than 2% of tree-feeding Macrolepidoptera species (Nottingale and Schultz, 1987). This suggests that development of pest status has something to do with the innate biological attributes of insects, or in other words the life history strategies characteristic of the species.

Based on life history strategies, insects have been categorised to fall within a scale of *r*-*K* continuum, where *r* represents the intrinsic rate of increase and *K* the carrying capacity of the environment. At the *r* end of the scale are species selected for fast population growth, ensuring maximum food intake in a short time in an ephemeral environment, and at the *K* end are species selected for maintaining a steady population by harvesting food effectively in a crowded environment. Southwood (1977) discusses the main points, which can be summarised as follows. The *r*-strategists tend to be small, with a short generation time. They increase enormously in number starting from small beginnings (e.g. a few colonisers) in the ephemeral habitats. Their population 'booms and busts', like that of the teak defoliator *Hyblaea puera*, which 'booms' into outbreaks on newly flushed teak plantations and then collapses. At the other extreme, *K*-strategists maintain a steady population at or near the carrying capacity of the habitat. They are in equilibrium with their resources, whose renewal they do not adversely affect. The *r*-strategists are devastating pests. They become very numerous at certain times in certain places (outbreak) and may destroy their habitat. On the other hand, *K*-strategists have a minimal impact on their host plants and are not recognised as pests, except when man is sensitive to such low levels of damage or he disrupts the natural regulation and causes an increase in the pest's density. In between are the intermediate pests, which are normally held at a lower level than the carrying capacity of their habitat by the action of the natural enemies, but will occasionally erupt into outbreaks due to environmental change. The concept of *r*–*K* selection in the life history strategies of insects is an attempt to order the observed complexity of pest situations. However, there is no conclusive empirical proof for many of the traits like body size, fecundity, voltinism etc. predicted for the outbreak and non-outbreak species according to the *r*–*K* selection model.

Several authors have listed the genetic traits associated with outbreak species (see Berryman, 1999 and Cappuccino and Price, 1995). These include (1) high reproductive potential; (2) high mobility and dispersal abilities according to some authors and poor dispersal abilities according to others, (3) utilization of rare or ephemeral habitats or food supplies; (4) well-developed cooperative or aggregation behaviour; (5) reliance on food stored at immature stages, rather than adult feeding, for reproduction; (6) going through colour or phase polymorphism in response to density; (7) hibernation or overwintering in the egg stage (in the case of temperate species) and (8) having broad food preferences. According to Berryman (1999) the kind of dynamic behaviour exhibited by a particular species depends as much on the characteristics of the other organisms with which it interacts as on its own adaptive traits. The teak defoliator *Hylaea puera* satisfies most criteria attributed to *r*-strategists. Yet the absence of its outbreaks in teak plantations in Africa shows that these life history characteristics and the co-occurrence of the insect and the host tree are not sufficient to precipitate outbreaks.

Other postulated causes of outbreaks, in particular large outbreaks, include (1) dramatic changes in the physical environment, (2) qualitative changes in the host plants caused by environmental stresses and (3) changes in the genetic composition of the pest population.

It appears that much of the complexity and our difficulty in understanding the causes of a particular insect outbreak are attributable to Berryman's (1999) fifth principle governing population dynamics, i.e. limiting factors.

In the complex web of biotic and abiotic interactions in which a given species is embedded, a particular factor may act as a limiting factor under a given environmental setting but a different factor may play this role when the environmental conditions, including the density of the population, change. Thus the regulating factor may appear different at different times, creating confusion in our understanding. In fact they are different at different times. The best example for appreciating the functioning of limiting factors is to consider the limiting of growth of a crop plant by the nutrient which is in shortest supply; when the supply of that nutrient is restored, the next nutrient in shortest supply limits the growth; when that is supplied, the next, and so on. In the complex web of interactions in which a pest insect is involved, a hierarchy of feedback loops may be involved in the regulation of its population. Identifying which acts when is a problem. As Berryman (1999, p. 75) explains "some insect populations are limited by insectivorous vertebrates when their densities are low, by insect parasitoids if they escape from vertebrate limitation, by pathogen if they escape parasitoid limitation, and by food in the absence of all the above."

A multitude of interrelationships (both biotic and abiotic) exist in the natural forest, so that a different one takes over when a particular one fails, but the problems are aggravated in plantations because several of the feedback loops are severed and we do not know which are the important ones that need to be restored to maintain the equilibrium. Even when we know, it may be practically difficult to restore them under plantation conditions. However, knowledge of the type and cause of outbreak can help in its management. For example, use of chemical insecticides to suppress sustained gradient outbreaks will not be cost-effective as the insect population will quickly grow back to initial density, requiring repeated insecticide application. On the other hand, eruptive outbreaks can be prevented from spreading by controlling the epicentre populations (Berryman, 1999). Similarly, host stress-induced outbreaks can be managed by taking appropriate action to improve tree health, where feasible.

8

Some general issues in forest entomology

8.1 Introduction

Based on their ecological status, we can distinguish the forest stands as undisturbed natural forests, disturbed or degraded natural forests, and plantations. The plantations can be further categorised into those of indigenous or exotic species, and those consisting of a single species (usually called monoculture) or more than one species (usually called mixed plantation). Foresters, forest entomologists and plant ecologists have strong traditional views on the risk of pest susceptibility of these different types of natural and man-made forest stands. Speculation was unavoidable in the past because the practice of forestry could not wait for conclusions based on long-term experiments. Now that fairly adequate data have accumulated, it is possible to make a critical assessment of the hypotheses and their theoretical foundations. Three commonly held views and their underlying hypotheses are examined here. These views are (1) that natural, mixed-species tropical forests are free of pest problems (in contrast to forest plantations); (2) that plantations of exotics are at greater risk of pest damage than plantations of indigenous species and (3) mixed plantations are at lesser risk of pest damage than monocultures.

8.2 Do plantations suffer greater pest damage than natural forests? And if so, why?

That plantations suffer greater pest damage than mixed-species natural forests is a well-accepted axiom in forestry, although contrary to the conventional wisdom, tropical forests are not free of pests. Empirical data presented in Chapter 4 showed that all gradations of insect damage ranging from minor feeding with no significant impact to occasional large-scale outbreaks resulting in massive tree mortality may occur in natural tropical forests. However, the frequency and severity of pest damage is greater in plantations as summarised in Chapter 5 and described in detail in Chapter 10. Chapter 4 also showed that the most common insect outbreaks in natural forests occurred in high-density stands approaching monoculture.

A detailed analysis of the plantation effect on pest incidence in tropical tree species was made by Nair (2001a). He compared the pest incidence in natural forests and plantations of several species for which relevant published literature was available—*Eucalyptus* spp., *Gmelina arborea*, *Hevea brasiliensis*, *Swietenia macrophylla* and *Tectona grandis*, and found that all of them suffered greater pest damage in plantations. In a meta-analysis of 54 individual studies reported in the literature, Jactel *et al.* (2005) also concluded that, overall, forest monocultures are more prone to pest infestation than more diverse forests. Thus the greater pest incidence in plantations is an undisputed scientific fact.

Two main hypotheses have been proposed to explain the lower pest incidence in natural forests – the 'enemies hypothesis' and the 'resource concentration hypothesis' (Root, 1973; Carson *et al.*, 2004). Recently, Nair (unpublished) proposed a third hypothesis called the 'pest evolution hypothesis'.

8.2.1 *Enemies hypothesis*

According to the enemies hypothesis, the lower pest incidence in the mixed-species stand is due to greater action of the pests' natural enemies. This is thought to be facilitated by the diverse plant community providing (1) alternative prey or hosts on which the natural enemies can sustain themselves and build up during periods when the pest is not present in the habitat, (2) a better supply of food such as pollen, nectar and honeydew for the natural enemies that enhances their fecundity and longevity and therefore overall effectiveness and (3) greater variation in microhabitats and microclimate that provides a larger variety of shelters for natural enemies. The increased natural enemy effectiveness therefore is thought to prevent pest build-up in the natural forest.

8.2.2 *Resource concentration hypothesis*

According to the resource concentration hypothesis (Root, 1973), also called host concentration hypothesis (Carson *et al.*, 2004), monoculture favours pest build-up by providing (1) a larger absolute supply of food resources, (2) greater ease in host location due to the physical proximity of the host trees and absence of interfering non-host volatiles and (3) reduced dispersal of the pests from the host patch. Arresting the dispersal, i.e. curbing the tendency of the herbivores that arrive on a clump of host plants to leave the area, appears to be the most important factor. This 'trapping effect' of monocultures on specialized pests may largely account for the greater pest load of monocultures (Root, 1973). Reduced dispersal also ensures less exposure to the risk of mortality during dispersal.

Experimental studies in agriculture have given strong support to the resource concentration hypothesis. In a comprehensive study of the insect fauna of collard (*Brassica oleracea*) in a pure crop in comparison to the same crop surrounded by miscellaneous meadow vegetation, Root (1973) found no evidence of greater effectiveness of natural enemies in the mixed vegetation, suggesting that the host concentration hypothesis offers a better explanation. In a test of the two hypotheses in the corn–bean–squash agroecosystem, Risch (1981) also found that there were no differences in the rates of parasitism or predation of pest beetles between monocultures and polycultures. On the other hand, it was found that the pest beetles tended to emigrate more from polycultures that included a non-host plant than from host monocultures, supporting the host concentration hypothesis.

8.2.3 *Pest evolution hypothesis*

According to Nair (unpublished), pest evolution might account for the greater pest incidence in forest plantations. He argues that natural selection of the pest genotypes most adapted to the planted host and the plantation environment is the major cause. This is facilitated by the large pest populations built up in large-scale plantations, the fast turnover rate of the pest generations and the inability of plantation trees to counterevolve.

In plantations of indigenous species, all pests originate from the natural forest. Most tree species in natural forests have a large number of associated insect species, of which only some become serious plantation pests. For example, out of over 174 species of phytophagous insects associated with the teak tree *Tectona grandis* in Asia, only three, the defoliator *Hyblaea puera*, the skeletonizer *Eutectona machaeralis* and the beehole borer *Xyleutes ceramicus* are serious pests of plantations (for details see under teak in Chapter 10). The major pest *H. puera* is widely distributed across the tropics and subtropics, covering Asia-Pacific, Africa, Central America, the Caribbean and South America, but its population dynamics on teak shows differences between the major regions. It has not attacked teak plantations in Africa so far and only recently has it attacked teak plantations in Latin America (in 1995 in Costa Rica and in 1996 in Brazil), in spite of its presence on other vegetation and the long history of teak planting in these regions. *H. puera* has been recorded on at least 45 host plants but outbreaks are common only on teak and rarely on some mangrove hosts. *H. puera* is suspected to be a species-complex (CABI, 2005). These observations show that there is large variation in the biological characteristics of *H. puera* populations and that the insect which infests teak in Asia might be a teak-adapted genotype. Enormous numbers of *H. puera* moths are produced every year on teak plantations and it is logical to assume that over the more than 100 years since it was first recognized as a pest of teak plantations in India, the species has become adapted to teak through natural selection. The teak skeletonizer *E. machaeralis* also seems to be adapted to teak through natural selection. Until recently it was thought that the skeletonizer which attacks teak in India, Bangladesh, Myanmar and other counties in Asia is the same species, but Intachat (1998) showed that the teak skeletonizer present in Malaysia, Indonesia and possibly Thailand is a closely related species, *Paliga damastesalis*. The differences between the two species are very slight and it is obvious that this also represents an evolving species-complex.

Obviously, out of the many species of insects associated with a tree species in the natural forest, only some have the greater potential to adapt to the particular host species and the plantation environment and become serious plantation pests. This is shown by the spectrum of pests attacking *Eucalyptus* spp. in natural forests and plantations in Australia. Only some of the pests that occur in natural forests are found in plantations; the most notable difference is the near absence of phasmatids and the preponderance of leaf-feeding beetles (chrysomelids and scarabaeids) in plantations (Wylie and Peters, 1993; see also Nair 2001a). It is evident that species and genotypes which can better adapt to the plantation environment will be selected in the plantations.

In plantations of exotic species, new pests may originate by adaptation of indigenous insects. The number of indigenous insect species attacking the exotic *Leucaena leucocephala* in India and *Acacia mangium* in Malaysia showed an increase over time (see Chapter 10). Wylie (1992) noted that rapid expansion of eucalypt plantations in China has been accompanied by a substantial increase in the number of insect species feeding on them. The bagworm *Pteroma plagiophleps*, which has been an insignificant pest of some native species, has become a major pest of the exotic *Falcataria moluccana* in India, with expansion of plantations of the latter (see Chapter 10). Other examples of such host-adapted insects are wingless grasshoppers on pines in Africa (Schabel *et al.*, 1999); several defoliating lepidopteran caterpillars also on pine in Africa (Gibson and Jones, 1977); the myrid bug *Helopeltis* spp. on *Acacia mangium* in Indonesia, Malaysia and the Philippines and on *Eucalyptus* in India (Nair, 2000); and the noctuid *Spirama retorta* on *Acacia mangium* in Malaysia (Sajap *et al.*, 1997). These insects became serious pests of exotics over time because insects, with a shorter generation time than trees, can adapt more quickly, and the trees in plantations have no chance of developing resistance mechanisms through natural selection, unlike those in natural stands. Insects can overcome the chemical defences of exotics through adaptive evolution using population genetic mechanisms, in the same way as they develop resistance to insecticides. All these examples of newly adapted pests in exotic plantations indicate the role of pest evolution in the origin of plantation pests.

Evolution is an ongoing process which enhances the fitness of pests in plantations. This pest evolution is invisible when it does not lead to changes in the physical appearance of the pests. It has therefore gone unrecognized although it is logical to expect that genotypic variation among individual insects will result in some individuals faring better than others on a particular host species, and that large-scale and long-term monoculture of the species will lead to natural selection of the best adapted insect genotypes. Adaptive evolution must be taking place in pest insects even when it is not physically visible, as in the case of development of insecticide resistance, particularly when large populations are built up repeatedly in plantations of selected tree species within the plantation environment, which differs from the natural forest environment in many respects. While a negative selection pressure is exerted by an insecticide on individuals not possessing resistant characteristics, a plantation crop exerts a positive selection pressure on individuals better adapted to the crop. The result is the same – survival and selection of better adapted individuals, i.e. differential survival and large-scale multiplication of certain genotypes, aided by a virtually unlimited food source offered by the plantations. Indeed, formation of demes (groups of individuals of a species that show marked genetic similarity) within populations of phytophagous insects in response to isolation, variation in host quality and other stochastic events is a well-recognized phenomenon (Speight *et al.*, 1999). There is little doubt that development of pest status by an insect is an evolutionary process. Pest evolution must be the main reason for the greater pest problems of monoculture plantations compared with mixed-species natural forests. Natural forests have the advantage that the trees can also evolve defensive mechanisms by differential survival of better-adapted tree genotypes, but this cannot take place in plantations. In the tropics where a typical insect pest can complete its life cycle in less than a month and breeding may take place throughout the year, the turnover rate of pest generations, and therefore the chances of natural selection, is very high compared with that of the long-lived trees. The narrowing of the genetic base of plantation trees due to human selection and inbreeding has been recognized as a factor favouring pest susceptibility (see e.g. Gibson and Jones, 1977) but pest evolution must be playing a more crucial role.

The pest evolution hypothesis is not an alternative to the host concentration hypothesis and the enemies hypothesis, but complementary to both. Pest evolution and host concentration appear to be the more important mechanisms although all three mechanisms might be operating with varying degrees of relative importance in different situations. The biological attributes of the pest insect are also important in determining whether it attains serious pest status in a plantation in contrast to a mixed-species natural stand. For example, where the adult female of a pest is flightless, as in bagworm moths, or has limited powers of dispersal, as in the psyllid bug *Phytolyma* spp., proximity of host trees, i.e. host concentration, might be necessary for precipitating an outbreak. On the other hand, a species like the elm bark beetle may spread the tree-killing Dutch elm disease to isolated elm trees. The importance of an insect's specialiced host-finding mechanism in its successful exploitation of a monoculture vs. mixed-species stand is discussed further in Section 8.4.5.

8.3 Pest problems in plantations of indigenous vs. exotic species

8.3.1 *The issues*

A substantial percentage of forest plantations in the tropics is made up of exotic species, notably eucalypts and pines, and more recently acacias (see Chapter 1). The success of exotics in plantations has generally been attributed, apart from the adaptability of the chosen species to the site, to the absence of their native pests. While many plantations of exotic species continue to be free of major pests, there is a fear that catastrophic outbreaks of pests may occur suddenly as in the case of leucaena psyllid and pine aphids (see Chapter 10). As mentioned earlier, it is generally believed that exotics are more prone to pest outbreaks. Some typical expressions of opinion include the following.

> The world-wide distribution of forest trees is being continuously changed as exotic species are used more and more in plantation forestry . . . We should expect trouble from insects in these exotic plantations (Berryman, 1986, p. 249)

> The [indigenous] species is adapted to the environment and already filling an ecological niche. This may render it less susceptible to serious damage from diseases and pests since controlling agents (predators, viruses, climatic factors) are already present . . . As a rule, where a native species meets the need, there is no reason to choose an alternative. Indeed, for reasons of conservation, if the choice lies between two species of comparable growth and quality, one of which is native and one exotic, . . . the native species is to be preferred. (Evans, 1992, p. 103–4)

> Some important biological advantages are present with indigenous species . . . They deserve more attention: It is possible to predict their

> performance in plantations based on their performance in natural stands; the species fills an existing ecological niche – it may therefore be less susceptible to diseases and pests, since the natural enemies are already present . . . (Appanah and Weinland, 1993, p. 28)

It can be seen from the above that two main reasons are given for the presumed lesser pest damage of indigenous species – (1) they have developed resistance or tolerance against the local pests through coevolution, and (2) natural enemies of the pests are present to keep them under check.

An exception where exotics were considered to be at lesser risk from pests is the following.

> . . . the argument that establishing a species outside its natural habitat (i.e. as an exotic) increases its susceptibility to pests has not been proven . . . Growing a species as an exotic may actually release that species from its natural pests and thus improve its health and performance. (Zobel *et al.*, 1987, pp. 160–161)

Alternatively, it can be argued that the risk of pest outbreaks is associated with monocultures, irrespective of whether a species is indigenous or exotic. The question has become important in the context of the ongoing, rapid expansion of exotic plantations in the tropics, particularly large-scale industrial plantations aimed at production of pulpwood for medium-density fibreboard. The issue was examined in detail by Nair (2001a) and the following account is mainly based on that evaluation. He made detailed case studies of nine tree species commonly planted as exotics in the tropics. For each species, the pest problems in three situations were examined and compared; (1) in natural forests in countries where the species is indigenous, (2) in plantations in countries where the species is indigenous (native plantations) and (3) in plantations in countries where the species is exotic (exotic plantations). The species chosen were *Acacia mangium*, *Eucalyptus* spp., *Falcataria moluccana*, *Gmelina arborea*, *Hevea brasiliensis*, *Leucaena leucocephala*, *Pinus caribaea*, *Swietenia macrophylla* and *Tectona grandis*. The results are described below, following a brief consideration of the definition of exotics.

8.3.2 *Defining the exotic*

The term 'exotic' is generally used in relation to a country, to indicate a species introduced from outside, in contrast to 'indigenous' or 'native' species that grow naturally within the country. Since the political boundary of a country is the unit of area, a species is considered indigenous even when it occurs only in some parts of the country. Thus teak is indigenous to India, Myanmar, Thailand

and Laos, although it does not occur in all parts of these countries. This definition is not scientifically rigorous, particularly when the natural distribution of a species is limited to small parts of a big country. For example, *Acacia mangium, Falcataria moluccana* and *Eucalyptus deglupta* occur naturally in very small pockets in the eastern islands of Indonesia, and to say that they are indigenous to Indonesia is misleading as they do not form part of the natural vegetation for most of the country. For practical purposes, an exotic species is defined here as an introduced species that does not occur naturally over a large part of a country.

8.3.3 *Empirical findings*

When an exotic species is grown in monoculture, it becomes difficult to distinguish between the 'monoculture effect' and the 'exotic effect' contributing to pest problems. Analysis of the pest problems in the three habitats, that is the natural forest, native plantations and exotic plantations facilitated the segregation of monoculture and exotic effects. Comparison of the pest problems of native plantations with those of natural forests gave a measure of the monoculture effect, and comparison of the problems of exotic plantations with those of native plantations gave a measure of the exotic effect. A summary of the results from the case studies (for details see Nair (2001a)) is presented in Table 8.1.

In all the five cases for which data are available, monoculture practice itself led to greater pest damage. The species were *Eucalyptus*, *Gmelina arborea*, *Hevea brasiliensis*, *Swietenia macrophylla* and *Tectona grandis*. Data for exotic effect on pest susceptibility are available for eight species. Five of them (*Acacia mangium*, *Eucalyptus*, *Gmelina arborea*, *Hevea brasiliensis* and *Tectona grandis*) suffered lesser damage in exotic locations and two (*Leucaena leucocephala* and *Pinus caribaea*) suffered greater damage. One species (*Swietenia macrophylla*) suffered equal damage in some exotic places and greater damage in others. This shows that pest susceptibility is not exclusively determined by the exotic or indigenous status of a tree species.

It is also interesting to look at the number of insect species associated with native and exotic plantations (Table 8.2). The number of species found in exotic plantations was greater for four species, less for three and equal for one.

In summary, the empirical data shows that *neither the intensity of pest damage nor the number of insects associated with a tree species is determined by its exotic status.* While plantations are at greater risk of pest attack than natural forests, plantations of exotics are at no greater risk than plantations of indigenous tree species. They are in fact at lesser risk initially. Exotic status is only one among the many determinants of pest incidence.

Table 8.1. *Segregation of the monoculture effect*[a] *and exotic effect in pest susceptibility of tropical forest plantation species*[b]

Tree species	Monoculture effect	Exotic effect
Acacia mangium	No data	Lesser damage
Eucalyptus spp.	Greater damage	Lesser damage
Falcataria moluccana	No data	No data
Gmelina arborea	Greater damage	Lesser damage
Hevea brasiliensis	Greater damage	Lesser damage
Leucaena leucocephala	No data	Greater damage
Pinus caribaea	No data	Greater damage
Swietenia macrophylla	Greater damage	Equal damage in some places, greater in others
Tectona grandis	Greater damage	Lesser damage

[a]Monoculture effect indicates whether monoculture plantations in regions where the species is indigenous suffer greater or lesser pest damage compared to natural stands. Exotic effect indicates whether monoculture plantations in regions where the species is exotic suffer greater or lesser pest damage compared to monoculture plantations in regions where the species is indigenous.

[b]Data from Nair (2001a)

8.3.4 *Theoretical explanations*

When an exotic tree species is introduced into a new environment, it comes without its associated insect pests. Pests may originate from indigenous or exotic sources through the following mechanisms.

(*a*) *From indigenous sources*

1. Generalist feeders

This category accounts for most of the insects associated with exotics in a new location. Many insects are polyphagous and their host selection mechanism permits acceptance of a wide variety of plants. Probably they arrive on a host plant by random exploratory movements and accept it when they come in contact with it, based on some general criteria which may include absence of deterrents rather than presence of specific attractants. Thus a number of indigenous insects colonize an exotic. Examples of generalist feeders are root-feeding cutworms and whitegrubs; stem-boring hepialids and cossids; and leaf-feeding grasshoppers and caterpillars of noctuid, geometrid and lymantriid moths. Generally they are incidental feeders and therefore only minor pests, although some species like root-feeding termites on eucalypts and trunk-dwelling termites on teak in Indonesia have become serious pests of exotics.

Table 8.2. *Comparison between the numbers of insect species associated with native and exotic tree plantations*[a]

Tree species	Score[b] for number of insect species in		Whether exotic plantation has greater or lesser no. of associated insect species
	Native plantations	Exotic plantations	
Acacia mangium	1	8	Greater
Eucalyptus spp.	11[c]	40	Greater
Falcataria moluccana	-	5	-
Gmelina arborea	10	2	Lesser
Hevea brasiliensis	6	3	Lesser
Leucaena leucocephala	1	4	Greater
Pinus caribaea	3	3	Equal
Swietenia macrophylla	1	2	Greater
Tectona grandis	23	2	Lesser

[a]Data from Nair (2001a)

[b]Scores are used instead of actual numbers as the number of associated insects is only approximate. One score is assigned to one to ten species. Thus, for example, score ten indicates 91–100 species and score 40 indicates 390–400 species

[c]Excluding those in the temperate region

2. Newly adapted insects

As mentioned earlier (Section 8.2.3) some indigenous insects adapt and become serious pests of exotic tree species over time. Examples are the bagworm *Pteroma plagiophleps* on *Falcataria moluccana* in India, wingless grasshoppers on pines in Africa, the myrid bug *Helopeltis* spp. on *Acacia mangium* in Southeast Asia and on *Eucalyptus* in India, the noctuid *Spirama retorta* on *Acacia mangium* in Malaysia etc. They become adapted in a short period because of their shorter generation time than trees, and trees in plantations, unlike those in natural stands, have no chance of developing resistance mechanisms through natural selection.

3. Specialized insects preadapted to closely related plant species

The examples of *Hypsipyla robusta* on mahogany and the shoot moths *Dioryctria* spp. and *Petrova* spp. on pines in Southeast Asia (see Chapter 10) show that an introduced tree species may encounter insects already adapted to closely related tree species in the location of introduction. This leads to quick attack of the exotic by these specialized oligophagous insects because the same or a closely related host selection mechanism developed over evolutionary time may operate. This results in serious pest problem as soon as the exotic tree is introduced.

(*b*) *From exotic sources*

In this case, well-adapted pests are introduced unintentionally from the native habitat of the exotic tree. Examples are the psyllid *Heteropsylla cubana* on *Leucaena leucocephala*; the beetles *Phoracantha* and *Gonipterus* on eucalypts; and the aphids *Cinara cupressi, Pinus pini* and *Eulachnus rileyi* on pines (see Chapter 10). These introduced pests can cause havoc, as in the case of the leucaena psyllid in Southeast Asia because they come without the natural enemies that often keep them in check in the pest's native habitat. However, the initial outburst may be tempered in the course of time as the native generalist natural enemies catch up with the pest.

Among the exotic tree species examined by Nair (2001a), the number of associated insect species ranges from about 20–400 (Table 8.2). This number is determined by several factors; distance from the native habitat, the extent and diversity of the geographical area of introduction, the time elapsed since introduction and the chemical characteristics of the tree species.

The major factors that determine the risk of pest incidence on exotics are the following.

1. Presence of other closely related tree species in the location of introduction.

Closely related species, particularly of the same genus, may harbour preadapted insect pests. In some cases, plants of closely related genera may serve the same purpose (e.g. *Toona* and *Swietenia*). Similar phytochemical profile is the deciding factor.

2. Extent of area occupied by the exotic plantations

The risk of pest problems increases with an increase in the extent of planted area, for the following reasons: (1) greater numbers of indigenous insects from diverse habitats come into contact and interact with the exotic species and adapt to it; (2) the greater the area of planting, the greater is the chance of mismatched planting sites which lead to plant stress. This could promote the outbreak of some pests like bark beetles which build up on stressed trees and then spread; (3) greater habitat heterogeneity increases the chances of matching with the habitat requirement of invading exotic pests and (4) a larger planted area provides a larger receptacle for randomly dispersing preadapted exotic pests.

3. Genetic base of the introduced stock

A narrow genetic base increases the risk of pest outbreaks. The risk increases over time, due to inbreeding.

4. Distance between location of introduction and the native habitat of the tree species

The longer the distance, the less the risk of pest problems as shown by the example of teak in Asia, Africa and Latin America.

5. Existence of serious pests in the native habitat

This is important in two ways. Their absence indicates that the tree species has innate resistance to most insects and therefore indigenous insects in the new location are unlikely to adapt to it easily and acquire pest status (e.g. *Hevea brasiliensis*). Secondly, the existence of serious pests in the native habitat indicates the chance of their unintentional introduction through one or other means.

6. Time elapsed since introduction

The risk of pest outbreak increases with time due to adaptation of indigenous insects and the greater likelihood of invasion by exotic pests.

7. Chemical profile of the exotic species

Some species are less prone to pest attack due to the presence of toxic or deterrent chemicals.

8. Innate biological attributes of the insects associated with the tree species

Populations of some insect species characteristically display outbreak dynamics while others display non-outbreak dynamics (r- and K-adapted insects, see Chapter 7).

As pointed out earlier, the two main reasons postulated for the presumed lower pest risk of native plantations are resistance of trees to indigenous pests developed through coevolution and increased natural enemy action. Both are not fully valid. The first is valid to the extent that an indigenous tree species will not be wiped out by a pest because it has evolutionarily outlived such an eventuality. However, this is of little value in the plantation system of tree management because economic damage can still occur, as shown by the many examples covered in Chapter 10. The second is valid in some cases, but not in all. Although natural enemies constitute an important factor regulating the population increase of many insects, and decisively so in some, empirical observations show that pest outbreaks occur in spite of their presence, sometimes even in natural forest stands. This shows that outbreaks occur due to other reasons as well. The theoretical principles of population dynamics discussed in Chapter 7 show the possibility of complex patterns of outbreak behaviour through the interplay of endogenous and exogenous factors. While natural enemies do regulate pest population build up in some cases and in some situations, in many cases the exact causes of population outbreak remain unknown.

The theoretical considerations support the empirical findings that the risk of pest damage in plantations is not exclusively or even predominantly dependent on the exotic or indigenous status of a tree species. It depends on the interplay of a number of factors mentioned above.

8.4 Pest problems in monocultures vs. mixed plantations

As indicated in the introduction, there is a traditional view that pest problems can be reduced by raising mixed-species plantations instead of monocultures. It is argued that there is a relationship between diversity and stability and that the more diverse an ecosystem, the more stable it is. This assumption has not been subjected to adequate empirical verification. In Chapter 4 we saw that mixed natural stands are not always free from pest problems. The available evidence for and against the claim and the theoretical backing are examined here.

8.4.1 *Refining the hypothesis*

First, let us take a closer look at the hypothesis itself. We are in fact dealing with many hypotheses here. The overriding hypothesis is that there is a relationship between diversity and stability such that a more diverse ecosystem is more stable. This has led to the hypothesis that natural mixed tropical forest which has a high diversity of tree species is stable and is free from pest outbreaks. This concept has been further extended to mixed forest plantations. So the hypothesis under consideration here is that mixed forest plantations suffer lesser pest damage than pure plantations of the same species. The simplifying assumptions do not end here. What do we mean by a mixed forest plantation? Natural mixed forests in the tropics are mixtures of many species. More than 100 tree species per hectare is the norm (see Chapter 1). But most artificial mixtures tried in plantations consist of only two tree species. This is shown by the FAO documentation of mixed plantation trials across the world, covering many countries in the tropics and subtropics and involving many tree species (FAO, 1992). In theory mixtures can take many different forms because there are several variables. These include the number of tree species in the mixture, canopy layers (single, double or multi-layered), percentage composition of the different tree species, spatial arrangement (mixing within the planting line which is often called intimate mixture, line mixture, block mixture etc.), age of the tree species and choice of tree species. The most common mixed plantation is a mixture of two species, in equal proportion, planted in intimate mixture or line mixture, forming a single canopy layer. The choice of tree species in the mixture varies; it can be a combination of any two species. So, more specifically, the hypothesis under consideration is that a mixed plantation consisting of any two or more species in intimate mixture, forming a single canopy layer, suffers less pest damage than a single species plantation.

8.4.2 *Direct evidence from pure and mixed plantations of trees*

Though a large number of casual or incidental observations are available, systematic, well-planned observations on pest incidence in pure versus mixed tree plantations are rare. Available data from the tropics are summarised in Table 8.3. Excluded are several papers in which only casual observations have been made or essential details are missing. In these studies, plantations of selected species have been raised in monocultures or in mixture with other tree species and the pest incidence compared. The other tree species (one or more) constituted various percentages of the total number of stems in the plantation, as shown in the table. It may be seen that the response of pests to mixed planting was variable; the severity of their incidence was either the same as in monoculture, lower, higher or variable. In general, we can only conclude that the response of pests to mixed planting was variable. A typical example is the shoot borer of mahogany. Suharti *et al.* (1995) reported that in Indonesia, when mahogany was planted in mixture with the neem tree *Azadirachta indica*, shoot borer incidence in mahogany was much reduced. But Matsumoto and Kotulai (2002) found that in Malaysia, the same mixture did not prevent economic damage by the mahogany shoot borer. In another study, Matsumoto *et al.* (1997) reported that when mahogany plantations were surrounded or enclosed by *Acacia mangium* plantations, mahogany was not attacked by the shoot borer. It is obvious that factors other than mixing of species influenced the results. Overall, the data presented in Table 8.3 does not support the hypothesis that mixed plantations of trees suffer less damage than monocultures. There are probably several confounding factors which influence pest incidence.

Recently Jactel *et al.* (2005) made a meta-analysis of 54 observations of various authors who compared pest incidence between mixed species stands and single species stands. The data set comprised 17 observations from tropical, 32 from temperate and five from boreal forest regions. The analysis indicated that planting or managing a tree species as a pure stand, *on average* significantly increased the rate of insect pest damage as compared to a mixed stand. Among the 54 observations, the pure stand effect was an increase in pest damage in 39 cases and a decrease in 15. Further analysis showed that the overall effect was the same irrespective of forest region (boreal, temperate or tropical, although the magnitude of the effect was higher in boreal), insect order or feeding guild, but that there was difference between oligophagous and polyphagous pests.

Table 8.3. *Comparative pest incidence in pure versus mixed plantations*

		Pest incidence in					Comp. incidence	
Tree species	Pest	Monoculture	>75% mix	50% mix	20–25% mix	Unknown mix	in mixture	Reference
Swietenia macrophylla	*Hypsipyla robusta*					Lower	Lower	1
Do	Do			Attacked			Attacked	2
Milicia excelsa	*Phryneta leprosa*	Attacked				No attack	Lower	3
Sonneratia apetala	*Zuezera conferta*	51%			32%		Lower	4
Albizzia odoratissima	Psyllid	40%		6–8%	13%		Lower	5
Grewia tiliaefolia	Caterpillar	20%			38–74%		Higher	5
Do	Gall insect	37%		37%	37%		Same	5
Haldina cordifolia	Defoliator	39%		37–57%	45–49%		Higher	5
Pterocarpus marsupium	Gall insect				Higher		Attacked	5
Xylia xylocarpa				Higher	Lower		Variable	5
Pinus massoniana		Lower	Higher				Higher	6
Ailanthus triphysa	*Atteva fabriciella*	Attacked		Attacked			Same[a]	7
Do	*Eligma narcissus*	Attacked		Attacked			Same[a]	7
Vochysia guatemalensis	Defoliator	5%			18.9%		Higher	8
Virola koschnyi	Defoliator	13.2%			17.3%		Higher[b]	8
Dipteryx panamensis	Defoliator	20.3%			21.9%		Same	8
Vochysia ferruginea	Leaf-cutting ant	2.5%			3.1%		Same	8

[a]Seasonal incidence was studied over a period of 3 years; no difference was noted between monoculture and mixed culture with teak

[b]Although the percentage incidence was higher in mixed plantation, the severity of damage was higher in pure plantation

References: 1, Suharti *et al.* (1995); 2, Matsumoto and Kotulai (2002); 3, Gibson and Jones (1977); 4, Wazihullah *et al.* (1996); 5, Mathew (1995); 6, Chao and Li (2004); 7, Varma (1991); 8, Montagnini *et al.* (1995)

Contrary to the general trend, about half of the polyphagous pests caused more damage in the mixed stands. Although Jactel *et al.* (2005) concluded that the meta-analysis substantiated the widespread belief that forest monocultures are overall more prone to pest insect infestation than more diverse forests, we should not ignore the exceptions. It must also be noted that in their study no distinction was drawn between naturally occurring mixed forest stands and the more simplified mixed plantations. In addition, the number of observations from the tropical region, where it is natural for forests to occur as mixed-species stands, was small compared to those from the temperate region.

8.4.3 *Indirect evidence from natural forests and agricultural experiments*

Natural forests

Occasionally, in some natural forests, a particular tree species may occur at different densities, with some stands approaching a monoculture at one extreme. Pest incidence has been studied on some species in such stands. A well-studied example is the balsam fir *Abies balsamea* in Canada. It was found that as the percentage of broadleaf trees in the balsam fir stands increased, defoliation caused by the spruce budworm *Choristoneura fumiferana* decreased (Su-Qiong *et al.*, 1996). In Spain, pure stands of the oak *Quercus suber* suffered greater damage from the fruit-boring weevil *Curculio elephas* compared to stands mixed with *Q. rotundifolia*, another host of the weevil (Soria *et al.*, 1995). In Bulgaria, pure stands of the beech *Fagus orientalis* are more susceptible to geometrid defoliators than mixed beech/oak stands (Stalev, 1989). These examples, although from the temperate rather than tropical region, lend support to the hypothesis that mixed stands suffer lesser pest damage than pure stands. In the tropics also, particularly in the cooler tropics, although no strict comparison between pure and mixed stands has been made as above, many insect outbreaks, though not all, have been associated with high host density. Examples of such outbreaks include *Eulepidiotis phrygiona* on *Peltogyne gracilipes* in Brazil, bagworms on pines in Indonesia, *Ophiusa* spp. on *Palaquium* and on *Excoecaria agallocha* in Indonesia, *Hoplocerambyx* on sal in India, bark beetle on pines in Honduras and sawfly on *Manglietia conifera* in Vietnam, as described in Chapter 4. In spite of the occasional occurrence of insect outbreaks in mixed tropical forests, it is generally agreed that they are relatively free of persistent pest problems compared with natural stands dominated by a single species.

Agricultural experiments

Numerous experiments with agricultural crops support the hypothesis that mixed stands suffer less pest damage than monocultures.

Speight *et al.* (1999) have cited many such examples. Planting carrot with onion reduces attack by the carrot fly *Psila rosea* (Diptera, Psyllidae). Broccoli when mixed with beans shows substantially reduced infestation with the flea beetles *Phyllotreta* spp. (Coleoptera, Chrysomelidae). Maize intercropped with cowpea reduces incidence of stem-boring Lepidoptera by 15–25%. In a comprehensive, three-year study carried out in New York, Root (1973) clearly demonstrated that *Brassica oleracea* grown in pure stands had substantially higher (often more than double) herbivore biomass per unit weight of foliage than when the crop was surrounded by miscellaneous meadow vegetation. He also found that the higher herbivore load of the pure crop was concentrated on a few specialized insect species. In another detailed study, Risch (1981) found that in polycultures in which at least one non-host plant was mixed, the numbers of six chrysomelid beetle pests of squash or bean were significantly lower than the numbers of these beetles on host plants in monocultures. Jactel *et al.* (2005) reviewed the various studies in agroecosystems reported in the literature and concluded that pest densities were significantly lower in mixed crop than in monocultures in 60–62% of cases. Here again, although the majority of cases supported the hypothesis under test, the exceptions which constituted 38–40% of the cases cannot be ignored. In the 150 independent studies examined by Risch *et al.* (1983), in 18% of cases pests were more abundant in the more diversified system, in 9% there was no difference and in 20% the response was variable. It appears that the response depended on the crop combination.

8.4.4 *Inference from the evidences*

The overall conclusions from direct and indirect evidences can be summarised as follows.

1. There is no consistent evidence to assert that pest problems are less severe in mixed-species forest plantations than in single-species forest plantations.
2. In contrast, there is clear evidence that in naturally occurring mixed-species stands of trees the pest problems are less severe compared with natural single-species dominated stands, although there are exceptions.
3. In the agriculture system, there are many examples where the insect pest damage in mixed cultures is lower than in monocultures. However, the exceptions were as high as 38–40% of the cases examined.

The first conclusion is not unexpected because, as pointed out earlier, the application of the diversity–stability principle to a simple mixed-species tree plantation is an unjustified oversimplification. Although we do not know exactly how diversity brings about stability, the ecological interrelationships that exist in a mixed-species natural forest in which the biotic components have coevolved over a long period of time is qualitatively and quantitatively very different from what we can expect in a random artificial mixture of two or more tree species. Therefore the second conclusion of lower pest incidence in mixed-species natural stands is in agreement with the general expectation in the context of the overriding hypothesis of the relationship between diversity and stability. The difference between mixed-species forest plantations and mixed-species agricultural crops comes as a surprise. Why should mixed-species stands of forest trees behave differently from mixed-species stands of agricultural crops?

8.4.5 *The theoretical basis*

The difference between mixed-species forest plantation and mixed-species agricultural crop appears to be the effect of host spatial scale. For an insect, a tree canopy which occupies a large volume of space is comparable to a monoculture patch of an agricultural crop. A single tree canopy is made up of thousands of shoots spread over a fairly large area. A large host patch arrests the movement of a host-seeking insect more effectively than a small host patch (Miller and Strickler, 1984). Even in a mixed-species tree plantation, the sensory stimuli offered to the insect by the odour plume of a tree is high because of the higher resource volume, perhaps as intense as that offered by a patch of agricultural crop. Therefore the insect tends to remain on the tree longer than on the individual plants in a mixed agricultural crop. Host selection involves not only the insect finding and accepting a host but also its remaining on the host once it has arrived. Insect pests easily disperse away from a mixed-species agricultural crop because of low resource concentration but a host tree species in a mixed forest plantation acts more like a patch of agricultural monocrop because of higher resource concentration, and retains the insects. Therefore the difference in pest response between a mixed-species and a single-species forest stand is not as contrasting as between a mixed-species and a single-species agricultural stand.

The mechanisms proposed to explain the postulated difference in pest incidence between mixed plantation and monoculture include increased natural enemy action and difficulty in host finding in the mixed plantation, reducing pest build-up, and effect of host concentration in the monoculture, encouraging pest build-up. These hypotheses, which are more applicable to the natural forest situation were discussed in Section 8.2 above. It is obvious that natural enemy action will be effective in the mixed natural stand but its effectiveness in an artificial mixed stand will depend on crop composition. From the empirical facts, it is clear that none of the above theoretical explanations is able to accommodate all the observed facts. There are far too many exceptions to each of the generalisations we tried to formulate, whether it is a comparison of natural mixed-species stands versus natural single-species stands, mixed-species tree plantations verus single-species tree plantations or mixed-species agricultural planting versus single-species agricultural planting. According to Jactel *et al.* (2005), the exceptional instances of increased pest damage in mixed forest caused by some polyphagous pests were attributable to heteroecious pests and the contagion process. Heteroecious pests are those that have an obligate alternate host which is essential for completing the development of the insect, as in the case of adelgids which have sexual and asexual stages on different host species. The mixed forest in which both hosts occur is more favourable for pest multiplication than the single species stand. Contagion process refers to a situation where a pest builds up on a more favourable host and then spills over to a less favourable host, when both are present in a mixed forest. In this case, the less favourable host in a single-species stand is more likely to escape infestation. However, the majority of the exceptions do not fall under the above two categories. Thus the theoretical basis for the presumed freedom from pests in artificial mixtures of trees is weak.

Difficulty in host finding has been assumed to reduce pest incidence in a mixed stand. But this will again depend on the pest species. Host finding is a highly evolved behavioural mechanism in many insects which have a narrow food range. These insects have very efficient, fine-tuned host finding mechanisms, usually mediated by secondary plant chemicals characteristic of a group of plants and specialised sensory receptors in the insects. Usually, host volatiles attract these insects from a long distance through receptors in their antennae and once they land on the plant gustatory receptors trigger a sequence of host acceptance behaviour. So it is unlikely that the presence of non-host trees can confuse them. On the other hand, there are polyphagous insects in which host acceptance behaviour is more complex, involving a series of step by step, yes or no behaviour options. In such species there is a random search for hosts during which a large number of plant species will be probed. Some trees may attract the insect towards them and provide an acceptable food source but will not elicit egg laying. In this process of host selection, a mixed-species stand can hinder or delay the host finding of a polyphagous insect. Thus the response of an insect to monoculture and mixed stands will also depend on the insect's biological attributes. Serious infestation can occur either in a mixed-species stand or single-species stand, depending on the characteristics of a particular insect species. Our inability to extract a valid generalization, applicable to all cases not only the majority, on pest susceptibility of natural mixed-species stands, mixed-species plantations, monoculture etc. is not surprising because the driving force is not the stand composition, but the biology of the insect species, with stand composition modifying the severity of infestation.

9

Management of tropical forest insect pests

In all countries research in forest entomology manages to convey the impression that it produces little that is of direct use to the executive forest officer. The average entomological bulletin with its detailed life-cycle studies, its technical descriptions, its record of discarded theories and incidental experiments does not appeal to his taste. It is either rejected or digested hastily, and the core of practical results remains undetected in the voluminous fruit of the investigation. What the forest officer requires, it has been said, are not life histories, but death histories; not suggested remedies but tested remedies.

C. F. C. Beeson (1924, pp. 516–17)

9.1 Pest control, pest management and integrated pest management

'Pest control' was the term commonly used in the past for our attempts to limit the damage caused by pests. We tried to kill the pest insects using chemical or other means. In spite of initial success, we soon realized that it was not easy to kill off the insects; they reappeared when the effect of the insecticide waned or developed resistance to the chemicals. Humbled by the success of the pests, we also realized that pests need to be controlled only if they cause economic damage. The term 'pest management' was therefore coined to indicate management of the pest population to limit it to a tolerable level. The emphasis was on regulating the population size, not killing all the pest insects, which was impracticable anyway. The concept of 'integrated pest management' (IPM) emerged in the 1970s. It envisaged the use of all the available techniques in an integrated manner to reduce the economic damage caused by pests, with the least ill effects on the environment. In the strict sense, IPM aims at regulating all the pest species including insects, pathogens and weeds in a crop production system (Dent, 1991) but it is often understood in a limited sense, as integrated insect pest management. It marks a change in our attitude or philosophy, from supremacy over nature to acceptance of an ecologically compatible strategy to contain the pests.

9.2 Historical development and present status of tropical forest pest management

Traditionally, forest managers in the tropics have ignored insect pest problems. This is attributable mainly to four reasons: (1) in the mixed-species natural forests of the tropics, pest problems are only sporadic and less frequent than in the temperate forests, (2) plantations where more serious pest problems occur are fairly recent in origin, most of them having been established since the 1960s, (3) there has been little information on the economic impact of forest insect pests, except in rare cases like the borer outbreaks on *Shorea robusta* (sal) in India where large-scale tree mortality occurs (see Chapter 10) and (4) even when the economic damage inflicted by the insect pests was recognised to be serious, there was no easy and effective method of controlling the pests.

Research in tropical forest entomology is about a century old (see Chapter 2, Section 2.1) but a critical evaluation of past work will show that while a sound foundation of basic knowledge on insects associated with forest trees has been built up over time, very few practical problems have been addressed. This has been partly due to traditional preoccupation with taxonomic and life history studies. Control attempts were made only in exceptional cases like the periodic borer outbreaks in natural forests of *Shorea robusta* in India, annual defoliation of teak plantations in India and Myanmar caused by the caterpillars *Hyblaea puera* and *Eutectona machaeralis* and the chronic infestation by the bee hole borer *Xyleutes ceramicus* of teak trees in natural forests and plantations in Myanmar. As discussed in Chapter 10, the approach to control of borer attack of *Shorea* was mainly silvicultural and physical – thinning of overmature and infested trees to prevent the build up of the borer population, felling and removal of heavily infested trees and trapping and killing the borer adults using trap billets to which they were attracted. Methods suggested against the teak defoliators included anticipated prevention of outbreaks by enhancing natural enemy action by silvicultural manipulation of the vegetation composition in and around the plantations. Although this method was not practised and would not have worked, even if practised, due to the unique population dynamics of the main defoliator *H. puera* (see under teak, Chapter 10), the control approach adopted was ecological. In some cases, the approach was to abandon cultivation of the species which had a serious pest problem and choose alternative tree species. Examples of abandoned species are mahogany, *Gmelina arborea* and *Ailanthus* spp.

Thus it can be seen that the early approaches were ecologically sound silivicultural and biological measures. Sporadic attempts were also made to standardize the use of chemical insecticides but fortunately these have not led to routine practice, largely for economic reasons. However, insecticides have been used in some countries in recent years in privately owned, industrial plantations.

An unfortunate trend among many entomologists in the tropical countries has been to include control recommendations when they report pest problems, without critical evaluation and without themselves undertaking any control experiments. This has been facilitated by the loose refereeing system of some of the journals. As a consequence of these armchair prescriptions, one can find many recommendations that are ineffective, ambiguous, contradictory, impracticable, prohibitively costly, highly damaging to the environment or sometimes even foolish (Nair, 1986b). No examples are cited for obvious reasons.

The present status continues to be neglect of pest problems where the forests are managed by government or government-controlled agencies and occasional use of chemical insecticides or other available methods in industrial plantations raised by commercial enterprises. Much information on the pest management practices followed by the commercial enterprises is not publicly available.

9.3 Overview of pest management options

The principles and methods of pest management are common to agricultural and forestry pests and since they are discussed in many standard textbooks, the details will not be covered here. Dent (1991) gives comprehensive coverage of various aspects of the subject. A brief overview, with particular reference to forestry applications, is given below.

Two approaches are available for pest management – *prevention*, where the build up of pests is prevented by appropriate means and *remedial action*, where control measures are applied after the infestation has occurred. The success of preventive measures depends on our ability to identify the causes of pest build up. As has been said, preventive measures are like replacing the worn-out washer of a water tap to stop the leakage, while remedial measures are like collecting the dripping water and pouring it away continuously. The first is removing the cause; the second is treating the symptom. When we cannot identify the cause, only remedial action is possible.

9.3.1 *Preventive measures*

Preventive measures aim to keep pest populations at low densities and not allow them to develop into outbreaks. They rely on an understanding of the causes of pest build-up. As discussed in Chapter 7, a large number of interacting factors are involved in determining the population size of a pest and it is often difficult to identify which factor is responsible for precipitating large-scale build up. Preventive measures are possible in some cases where the causes of population build up are known. These measures usually consist of silvicultural interventions aimed at tree health improvement in order to 'tune up' the tree's innate defence mechanisms. They are effective where pest build-up is caused by poor tree health. Thus, as discussed in Chapter 10, preventing injury to trees by lopping can prevent infestation by the teak trunk borer *Alcterogystia cadambae* and prompt removal of overmature trees and regular thinning of stands of *Shorea robusta*, as well as trapping and killing of moderately high populations of adult beetles, can prevent outbreak of the sal borer *Hoplocerambyx spinicornis*. Similarly, improvement of tree health and removal of dead and unhealthy trees in a pine stand can prevent pine bark beetle outbreak. Prompt removal of tree-felling refuse from a plantation site can prevent the build-up of pests like bark beetles which infest and breed on freshly felled trees and eventually attack healthy standing trees. In the case of teak defoliator outbreaks, as discussed in Chapter 10, destroying the early epicentre populations during the pre-monsoon period can prevent at least part of the subsequent large-scale outbreaks. Silvicultural practices such as retention of plant species that support alternative hosts of pest insects, as discussed under teak in Chapter 10, or raising mixed-species plantations, as discussed in Chapter 8, can also reduce pest build-up by enhancing natural enemy action. In the case of pests introduced from other countries, quarantine measures, where potential pests are intercepted at the ports of entry of commodities such as wood or planting material, is also a preventive measure.

Use of pest-resistant trees can also be considered a preventive measure. Resistance refers to the genetic capability of trees to prevent, restrict or withstand pest infestation. There are not many instances of trees showing useful resistance to pests. When present, tree resistance to insects is usually polygenic. It may also be based on physical factors such as resin system characteristics. Conventional breeding for resistance is constrained by the long reproductive cycle of trees.

For pests of stored timber, preventive measures include immersing the logs in water and debarking newly felled logs to prevent some groups of borers from laying eggs beneath the bark.

9.3.2 *Remedial measures*

Remedial measures aim to reduce the pest population level by killing the insects by one means or other. A large variety of remedial measures has been developed and tried against insects.

Insecticides

Historically, the most common and effective means of killing insects has been the use of chemical poisons, commonly called insecticides. Insecticides are used either prophylactically or remedially. Prophylactic use involves application of the insecticide before the insects appear, as in the case of control of root-feeding termites of eucalypt saplings, where the insecticide is mixed with the soil to kill the termites that might attempt to penetrate to the tap root (see under *Eucalyptus* in Chapter 10). Other examples are insecticidal treatment of nursery soil to control ants and whitegrubs or mixing of insecticide with seeds while in storage. In remedial application, insecticides are applied to the insects and the trees after the infestation is noticed. Although inorganic poisons such as lead arsenate, calcium arsenate and sulphur were used in the early days, organochlorines have been used extensively since World War II, when DDT became popular because of its effectiveness against mosquito vectors of malaria. In the United States alone, 5 billion kilograms of insecticides were used from 1945–1970. Most major outbreaks of forest insects in North America were sprayed with DDT until it was withdrawn from the US market in 1973 (Berryman, 1986). In 1968 alone, 20 000 kg of DDT was used in the US forests to control defoliating insects. Organochlorine insecticides were also used in fairly large quantities to control bark beetle and termites (Berryman, 1986). In the developing countries of the tropics, use of organochlorines was continued for a longer time.

Other classes of insecticides that are less persistent in the environment such as organophosphates, carbamates, synthetic pyrethroids, chitin inhibitors, botanicals (like nicotine, rotenone, pyrethrin and neem products) and insect growth regulators have since been developed. Aerial application of insecticides has continued into the 1970s and 80s in many industrialised countries although over a much reduced area and with less persistent insecticides. The US Department of Agriculture guidelines for 1980 (USDA, 1980) contained recommendations for use of the following insecticides against various forest pests – acephate, aldrin, azin-phosmethyl, cacodylic acid, carbaryl, carbophenothion, chlordane, chlorpyrifos, diazinon, dieldrin, diflubenzuron, dimethoate, disulphoton, ethyl dibromide, fenitrothion, methoxyclor, methyl bromide, sulfuryl fluoride and trichorfon. The turmoil created in the USA and the entire world with the publication of Rachel Carson's (1962) book entitled 'Silent Spring' in which she vividly described the adverse impact of indiscriminate spraying of insecticides from the air over the vast stretches of forest is now part of history. Although bordering on poetic exaggeration at times, her criticism of the excessive use of pesticides, particularly in the forests, and the consequent disruption of ecological processes leading to aggravation of pest problems, resurgence of secondary pests and the accumulation of toxic residues in the human food chain hastened the development of IPM practices. IPM involves the use of various methods such as biological control, habitat management, plant varieties resistant to pests, cultural practices and selective pesticides in a harmonious manner, as appropriate to each pest situation. It aims at reducing the pest population below economic injury level, and not at complete 'control'.

The main advantages of pesticides are: (1) dramatic effectiveness by killing the insects in a short period of time, (2) broad spectrum of effectiveness and (3) commercial availability. The main disadvantages are: (1) unintended effect on non-target organisms, particularly parasitoids, predators and pollinators, (2) development of resistance by pests and (3) the temporary nature of the effect, necessitating repeated applications. Some problems like long persistence and bioconcentration in the human food chain have been overcome by the development of newer, more easily degradable pesticides. The drift of pesticides in the environment has also been reduced by improvements in application technology. Yet substantial portions of insecticides applied over the forest canopy find their way into other components of the ecosystem through drift, rain washing, leaching, etc. Its effect on non-target natural enemies is of serious concern in the forest environment where many potential pests are kept in check by their natural enemies. Application methods are still primitive in the developing countries of the tropics and entail large wastage as well as contamination of the environment. Experience in the industrialised countries has shown that unanticipated pest problems can arise as a result of widespread application of broad-spectrum insecticides as some potential pests are released from the influence of their natural enemies when these are destroyed by the insecticides.

Biological control with predators and parasitoids

All insects have natural enemies. These may be vertebrate predators (birds, bats, reptiles etc.), insect predators, insect parasitoids, nematode and protozoan parasites or pathogenic micro-organisms like fungi, bacteria and viruses. They play an important role in the natural regulation of insect numbers as discussed in Chapter 7 and have been employed for artificial suppression of pest populations. Biological control is generally considered the most appropriate method for management of forest pests. The relative freedom of mixed tropical forests from pest outbreaks is generally attributed to the 'checks and balances' exerted by natural enemies in the complex natural community. This inference rests essentially on circumstantial evidence and it is difficult to obtain direct proof. While the qualitative relationship between the insect pests and their various natural enemies has often been fully elucidated, the quantitative effects remain largely unknown. Based on theoretical considerations it is assumed, however, that in natural communities like forests, natural enemies do play a significant role in preventing the population of pests from attaining damaging levels, that is, natural biological control. We may recognise its value only when we disrupt it, just as we seldom recognise the value of good health until we lose it. The increased pest problems experienced in plantations is generally attributed to the disruption of natural enemy action. The attainment of pest status by some species when accidentally introduced into new geographical regions devoid of their natural enemies, and their suppression on introduction of the native natural enemies (classical biological control), is taken as proof of the effectiveness of natural enemies. When natural enemies are managed to control a pest, we call it biological control; in the strict sense it is applied or artificial biological control.

The literal meaning of biological control can be extended to include any technique of human intervention employing biological means. The use of naturally occurring genetically resistant trees, transgenic trees or even spray application of commercially formulated bacterial or **baculovirus preparations** are all methods which make use of biological means of intervention. So is silvicultural manipulation. However, as commonly used, biological control means use of artificially introduced or augmented natural enemies, usually insect predators and parasitoids, for suppression of pest populations.

Three methods of biological control are generally recognised: (1) introduction (introducing a natural enemy to a location where it did not previously exist), (2) conservation (conserving the existing natural enemies by habitat management) and (3) augmentation (inundative or inoculative release of mass-multiplied natural enemies). Different groups of natural enemies play different roles in regulating insect pest populations. Vertebrate predators, and some arthropod predators and parasitoids, seem to be capable of regulating their prey at low densities (Berryman, 1986). On the other hand, pathogenic organisms seem to be more important in suppressing pests after they have reached high densities. Most arthropod predators and parasitoids will act between these extremes of pest densities. Thus each natural enemy group may exert its influence in different situations. Unfortunately, in tropical forestry, blind faith has often been placed in the effectiveness of biological control. It is instructive to examine in some detail a case study of biological control from India in order to appreciate this point.

In the well-studied example of two leaf-feeding caterpillars of teak, *Hyblaea puera* (Hyblaeidae) and *Eutectona machaeralis* (Pyralidae), a very complex web of interrelationships exists among the two pests and their natural enemies.

At least 40 insect parasitoids have been recorded from *H. puera* and 60 from *E. machaeralis* (discussed in detail under teak in Chapter 10). Several of them are common to both the caterpillars and each may also attack several other caterpillar hosts. The resulting food web is very complex, particularly in the natural forest with a multitude of plant species, each supporting a variety of caterpillars. About 213 plant species indirectly support parasitoids of either of the above two teak pests by harbouring their alternative hosts. In addition to these insect parasitoids, a large number of predators including insects, spiders and birds also attack the two pests. Based on these considerations, a package of biological control practices (including silvicultural interventions) was formulated as early as in 1936 to control the two pests in teak plantations. The recommended actions included the following: (1) subdivide the planting area into small blocks of 8–16 ha, leaving strips of pre-existing natural forest in between, to serve as reserves for natural enemies; (2) improve these reserves by promoting desirable plant species and removing undesirable ones. (Desirable plants are those that support the alternative hosts of the parasitoids, and undesirable plants are those that serve as alternative hosts for the teak defoliators themselves.); (3) within the teak plantation itself, encourage the natural growth of desirable plant species as an understorey and discourage the undesirable and (4) introduce natural enemies of the teak defoliators where they are deficient.

It appeared that the above scheme was ideal. It was in agreement with the concept of IPM, although the recommendations were formulated long before the formalized IPM concept emerged in the 1970s. For a long time, Indian forest entomologists have strongly and often aggressively advocated this package of practices. However, the method was not adopted by the forest managers in practice. They ignored it for three reasons: (1) they did not recognize the need for control, (2) they were not convinced of the effectiveness of the suggested method and (3) the method was difficult to implement. Unfortunately, the entomologists failed to recognise the real needs of the forest manager and continued to advocate the method and find fault with the forest manager. The good work in the 1930s leading to the formulation of the package of biological control recommendations was not followed by additional research on the teak defoliators until much later. As discussed in detail in the pest profile for *H. puera*, under teak in Chapter 10, a fresh look at the problem was initiated in the 1980s (Nair, 1986a) and it was demonstrated (Nair *et al.*, 1985) that defoliation by *H. puera* resulted in loss of about 44% of the potential volume increment of the trees. It was also shown that, of the two pests, *E. machaeralis* did not cause any significant growth loss under Kerala conditions. Therefore, in Kerala, control is needed only against *H. puera* outbreaks which occur in the early part of the growth season. Research on the population dynamics of *H. puera* further indicated that this long-advocated package of biological control involving silvicultural manipulations could not succeed against *H. puera* because its outbreak populations are highly aggregated and mobile. The effect of a resident population of parasitoids on millions of larvae that build up suddenly from immigrant moths will be insignificant (see Chapter 10 for details). In retrospect, it was good that the forest managers did not practise the recommended biological control method.

The above case study shows that biological control may not always work. Some parasitoids do indeed exert some control over local populations, under certain conditions, but population outbreaks appear to be triggered by the plentiful food supply during the flushing period of teak as well as the monsoon wind system which aids the immigration of moths (see Chapter 10). Natural enemies become unimportant under such circumstances. Migration also serves as a mechanism of natural enemy evasion (Nair, 1987a). In fact *H. puera* outbreaks occur in natural forests as well, in spite of the presence of a large complement of natural enemies. The theoretical principles of population dynamics discussed in Chapter 7 show the possibility of outbreaks being caused by the interplay of several endogenous and exogenous factors. Unfortunately, the well-entrenched concept of 'balance of nature' and the successful examples of applied biological control have overemphasized the importance of parasitoids as regulators of pest populations. While they do regulate population outbreaks in some cases and under some circumstances, we must recognise that biological control is not a panacea. This case study of teak defoliator control also emphasises the need to field-test the recommendations before advocating them, to safeguard the entomologists' credibility.

Biological control with microbial agents

Fungi

Several species of fungi are entomopathogenic. Spores of entomopathogenic fungi germinate on the insect cuticle and penetrate into the body. In contrast, other pathogens like bacteria and viruses infect through the gut wall and therefore need to be ingested by the insect. Growth of the fungal hyphae inside the body eventually causes the death of the insect, whereupon the hyphae penetrate to the exterior and produce infective conidia or spores. Two species of fungi have shown potential for applied biological control of tropical forest insects. These are *Beauveria bassiana* (white muscardine fungus) and *Metarhizium anisopliae* (green muscardine fungus) of the class Deuteromycetes ('imperfect fungi'). The occurrence of these two species has been reported in a variety of insects, and laboratory trials have shown their potential for practical use as mentioned in Chapter 10.

M.anisopliae has a wide host range, covering species of Coleoptera, Lepidoptera, Diptera, Orthoptera, Hemiptera and Hymenoptera. In general, successful infestation by entomopathogenic fungi requires high atmospheric humidity, perhaps for spore germination, viability and sporulation after the host is dead. Because of this limitation, successful field control has been achieved only under some circumstances. Their potential needs to be further explored and conditions for successful use standardized. There is also scope for isolating more virulent strains. *Beauveria*, which can be mass-produced on artificial nutrient media, has shown potential for control of soil insects like whitegrubs in forest nurseries in China and India (Speight and Wylie, 2001). *Metarhizium* has been found effective against the pine shoot-boring moth *Rhyacionia frustrana* in Cuba. It may also have potential against root-feeding termites (see under *Eucalyptus*, Chapter 10).

Bacteria

Many species of bacteria infect insects but only a few cause serious disease. Of these, some like *Serratia marscecens*, which can cause significant mortality of *Hyblaea puera* pupae, as mentioned under teak in Chapter 10, are also pathogenic to man, and therefore not safe for insect control. *Bacillus thuringiensis*, usually abbreviated to Bt, first recognized as a disease agent in silkworm, has emerged as the most promising bacterium for control of lepidopteran and some coleopteran pests. The related *B. sphaericus* is pathogenic to mosquito larvae and *B. popilliae* to scarabaeid beetles. Different strains of *B. thuringiensis* have been isolated with different levels of pathogenicity to various insects. *Bacillus thuringiensis* var. *kurstaki* has been found the most pathogenic to lepidopteran larvae.

The incidence of Bt infection in natural populations of insects is not high enough to cause acceptable levels of mortality and therefore living Bt, unlike other natural enemies, is not effective for standard biological control practices. Living Bt is slow to act and is also killed by sunlight. For these reasons, most common formulations of Bt contain the toxin produced by Bt. It is used for control of insects in the same way as chemical pesticides are used. Bt toxin, however, is not harmful to man. It consists of proteins, called delta-endotoxins, present in large crystals in mature, sporulating cells of the bacterium. After consumption by the insect, the proteinaceous crystals break down in the high pH medium of the larval gut, releasing the delta-endotoxins which are further broken down to toxic protein molecules by the digestive enzymes. The toxins cause paralysis of the gut and mouth, lysis of the gut epithelial cells etc., leading to death of the host.

Bt has been mass-produced in fermenters and commercially formulated like chemical insecticides. The formulations do not usually contain living bacteria and therefore their application does not strictly conform to standard biological control. Bt formulations fall under the category of a bioinsecticide rather than a biological control agent. Commercial formulations of Bt have been marketed under a variety of trade names – Delfin, Dipel, Biolep, Bioasp, Biobit, Lepidocide, Thuricide etc. Bt has been used on a large scale, by aerial spraying, for control of forest-defoliating Lepidoptera in many developed countries since the 1960s. Annual worldwide usage has been estimated at over 2.3×10^6 kg and it has been found effective against several temperate forest pests such as the Douglas fir tussock moth *Orgyia psuedopstugata*, spruce budworms *Choristoneura* spp., pine caterpillar *Dendrolimus punctatus*, larch budmoth *Zeiraphera diniana*, gypsy moth *Lymantria dispar* and fall webworm *Hyphantria cunea* (Strauss *et al.*, 1991).

There are some disadvantages with the use of Bt. It is effective only when ingested and therefore sap suckers are not affected. It is pathogenic to silkworm and therefore cannot be used in areas where sericulture is practised because of the risk of contamination. Most importantly, some agricultural pests like the diamond-back moth *Plutella xylostella* have shown resistance to Bt (McGaughery, 1994), suggesting that other insects may also develop resistance.

Bt has been used in tropical forestry in a limited way. Seed orchards or other high value teak plantations in Thailand have been aerially sprayed with Bt and ground application has been made against the same insect in commercial teak plantations in India (see under teak, Chapter 10). It has also been used against the defoliating caterpillar *Theirenteina arnobia* in *Eucalyptus* plantations in Brazil, either alone or in combination with the pyrethroid deltamethrin, by aerial spraying (Zanuncio *et al.*, 1992). However, economic considerations have prevented its wider use against forest pests in the tropics.

Viruses

There are at least seven groups of viruses known to cause diseases in insects but only one group (Baculoviridae) is considered safe for applied use against them (WHO, 1973; Entwistle and Evans, 1985). Others (e.g. Poxviridae, Picornaviridae) have varying degrees of similarity in physical and chemical characteristics to viruses found in vertebrate animals. Baculoviruses comprise a large group of DNA viruses unique to invertebrate animals.

Natural outbreaks of virus diseases are common in many forest insects, particularly when the population density reaches high levels. They cause the sudden collapse of population outbreaks as in the case of the teak defoliator *Hyblaea puera* (see Chapter 10). Such disease epizootics are usually caused by baculoviruses. The disease is characterized by liquefaction of the body contents followed by rupture of the body wall. Dead caterpillars usually hang head downwards, by their prolegs.

The biology of many baculoviruses has been studied in great detail (Granados and Federici, 1986) and a wealth of information is available on their structure, disease development, transmission characteristics and ecology. It is beyond the scope of this book to cover the details. In the majority of Baculoviridae, the rod-shaped virions (their structure made up of the DNA-protein core and envelopes) are occluded within a crystalline protein coat. In one subgroup called the Nuclear Polyhedrosis Viruses or Nucleopolyhedroviruses (NPVs), several virions are embedded in the protein matrix to form polyhedron-shaped inclusion bodies (PIBs) which accumulate in the nucleus of the infected insect cells. The PIBs (also called POBs or polyhedral occlusion bodies) may range in size from 0.5 to 15 μm. In another subgroup, the virions are embedded singly in protein and they are known as granulosis viruses (GVs). In a third subgroup no inclusion bodies are formed. The most common baculoviruses are NPVs.

Baculoviruses are usually very host specific. They have no direct impact on other organisms including non-target insects. When a PIB is ingested by a susceptible insect host, the polyhedra dissolve in the mid-gut releasing the virions. The virions pass through the mid-gut and enter the insect tissues where they multiply in the nucleus of the cells and form PIBs, killing the insect in the process. A dead larva may contain up to 10^9 PIBs. The PIBs can persist in the soil and are passed on to the next generation of insects when consumed through the contaminated leaf. Many NPVs are also transmitted transovum (vertical transmission).

Baculovirus diseases have been recorded in most lepidopteran pests of agriculture and forestry. As a biological control agent, baculovirus has the advantage that it is very host-specific and does not cause any harm to non-target organisms. It is fairly quick-acting and particularly effective against early instars. The fairly stable PIBs from dead insects have been isolated and formulated as effective insecticides for many pests of agriculture and forestry and used like a chemical insecticide by spraying on to the foliage. Suitable formulations have been developed and registered for use against a wide range of lepidopteran pests. In developing the formulations and application methods, several variables that influence the effectiveness of the baculovirus such as conditions related to the host insect, the pathogen, the host tree, the physical environment and spray technology are taken into consideration (for example, see under *Hyblaea puera* on teak, in Chapter 10). Baculovirus insecticides are now routinely used in aerial spraying against many forestry pests in the developed countries. Registered formulations are available for use against European pine sawfly, spruce budworm, Douglas-fir tussock moth and gypsy moth.

Baculovirus insecticides are comparatively costly as baculoviruses can be mass-produced only on their specific hosts. In the tropics, although baculoviruses have been used for control of some agricultural pests, in forestry only the NPV of the teak defoliator, *Hyblaea puera* has been formulated and standardised for field use (see Chapter 10 for details). Cost of the product is the major limiting factor in its widespread field use.

Use of transgenic trees

An emerging method of pest management is use of transgenic or genetically modified trees which possess genes conferring insect resistance. Recent advances in biotechnology have made it possible to transfer desirable genes across species. The desirable genes can come from a variety of sources, including plants, insect pathogens or even insects themselves (Strauss *et al.*, 1991). Examples are the toxin gene from the bacterium *Bacillus thuringienis* (Bt), proteinase inhibitor genes from other plant species, chitinase genes, baculovirus genes etc. The potential for manipulation of gene expression is enormous; for example, a gene may be configured to be expressed only after insect attack has begun.

Following the success of Bt as an effective microbial insecticide against a large number of lepidopteran pests of agricultural and forestry crops, most work on genetic engineering of insect resistance in trees has concentrated on the use of the genes for the toxic Bt protein crystals called delta-endotoxins (cry). Several transgenic agricultural crops containing Bt endotoxin genes are now commercially cultivated. Transgenic cotton has led the list, with about a million ha planted in the USA in 1996. It is also now common in tropical countries. At least 33 species of transformed forest trees containing genes for various traits have so far been produced including poplars, eucalypts, *Casuarina glauca*, pines and larches (Frankenhuyzen and Beardmore, 2004). High levels of mortality have been produced under laboratory conditions for lepidopteran pests on transgenic poplar, white spruce and loblolly pine, and for leaf beetles on eucalypt (Frankenhuyzen and Beardmore, 2004). Effective resistance to natural infestations was obtained in the field with transgenic poplar in USA and China. Obviously, it will take many years before the use of insect-resistant transgenic trees percolates to the tropics.

Unfortunately, some insect pests have shown ability to develop resistance to Bt toxins (Tabashnik, *et al.*, 2003). Research is in progress to circumvent development of Bt toxin resistance and to find new sources of genes for tree resistance.

At present there is an ongoing worldwide debate on the risks and benefit of transgenic trees. The advantages are many. Bt toxins are not toxic to humans and only those species that ingest the plant material will be exposed to the toxins. Because of the high specificity of the toxins, only target species would be harmed. Internally produced toxins of transgenic trees can reach the concealed internal feeders such as shoot borers and bark beetles which are difficult to control by external insecticide sprays. The main risks are the potential for development of resistance by insects and concern whether the transgene would spread into the wild population. These issues are discussed in detail by Strauss *et al.* (1991), Frankenhuyzen and Beardemore (2004) and Velkov *et al.* (2005).

Semiochemicals (behaviour-inducing chemicals)

Chemical communication plays an important role in the life of an insect as the insect depends on it for host finding and a number of other interactions with the biotic environment, including the insect's own population. Chemical substances emitted by an individual to induce behavioural responses in other individuals of its own species are called *pheromones*. Thus there are sex pheromones specific to each species which attract members of the opposite sex for mating, and aggregation pheromones which attract other members of the population irrespective of sex for specific purposes. Substances which cause a behavioural response in individuals of another species are called *kairomones* when they benefit the receiving individuals (e.g. attraction of parasitoids) or *allomones* when they benefit the emitting individuals (e.g. a repellent which keeps away a natural enemy). A wide variety of such chemicals is produced by insects for various purposes and attempts have been made to use them for management of pests. Thus sex pheromones have been used for trapping insects and for disruption of mating. Sex pheromones of several lepidopteran forest pests of North America and Europe have been identified, synthesised and used for population management. Aggregation pheromones have also been used successfully for trapping bark beetles. Forest insects for which semiochemicals have been tested or used operationally for population suppression in developed countries of the West include the western pine-shoot borer, *Eucosma sonomana*; the gypsy moth, *Lymantria dispar* and several species of bark beetles (Berryman, 1986).

Physical methods

A variety of physical or mechanical methods has been employed for killing forest pests. The most widely practised is use of light traps to capture and kill insects. Host tree billets have been used as a trap to collect large numbers of the sal tree borer *Hoplocerambyx spinicornis* (see Chapter 10). Other examples include inserting a metallic wire probe into the tunnels of large borers such as hepialids on saplings and cossids on older trees. Scrapping the infested bark of teak trees has been recommended in Thailand to kill the early instar larvae of the bee hole borer (see under teak, in Chapter 10). Cutting and removal of infested trees to prevent the spread of infestation is also a physical method.

9.4 Unique features of forest pest management

From the pest management point of view, it is useful to classify the forest pest problems on the basis of the growth stage of the trees. Thus we have insects affecting (1) seeds, (2) nurseries, (3) young plantations, (4) older plantations and natural forests and (5) stored timber. The problems of managing pests of seeds, nurseries and young plantations are similar to those of agricultural pest management, but there are some unique features associated with older plantations and natural forests. These are examined below.

Economics of pest management

Any measure to prevent or control insect damage would involve cost. If the value of the damage prevented is not greater than the cost incurred, it is not worthwhile to prevent the damage. It is true that economic analysis is seldom carried out even in agriculture before undertaking pest control operations. For example, when a farmer sprays an insecticide to control a pest affecting his vegetable crop, he does not do so after carrying out a cost–benefit analysis. He makes an intuitive judgment of profitability based on past experience and simple calculations. He will apply control measures only if the cost of control is less than the value of the increased yield expected due to the damage prevented. He can assess the benefit easily, based on the prevailing market price of the produce. But the situation in forestry is quite different. For tree crops where timber is the harvested produce, the benefit can be realised only a long period after the protective treatment is given. For example, in traditional plantations of teak, the timber is harvested only 50–60 years after planting. We cannot therefore work out the economics of pest control in forests without the help of an economist, because the value of the returns received after 50 years, or even 10 years, cannot be compared straightaway with the value of the money spent today for control. Therefore the economist usually calculates the 'Net Present Value' of the returns to be received in future, employing the principle of discounting. Several uncertainties are involved in such calculations, and the methods are subject to debate among leading economists. Obviously, the silvicultural rotation age fixed for various common forest plantation tree species in the tropics is not based on economic analysis, and it is doubtful whether conventional cost–benefit analysis would support raising plantations of species with a rotation age of, say, 60 years at all (Nautiyal, 1988). A case study of teak defoliators given by Nair and Sudheendrakumar (1992) will illustrate the kind of problems encountered in an economic analysis of forest pest control.

The caterpillars *Hyblaea puera* and *Euectona machaeralis* are well-known pests of teak in India, as discussed earlier. The former feeds on young foliage during the early part of the growth season and the latter on older foliage during the fag end of the season. Both may cause complete and extensive defoliation, sometimes more than once during the season. The severity of damage may vary from place to place at a given time, but most plantations suffer at least one severe defoliation per year. The economic damage caused by these pests has been the subject of speculation and debate since the 1920s. The early literature has been reviewed by Nair (1986a). Estimates of loss varying from 6.6–65% of the potential volume growth were reported earlier but because of too-liberal assumptions, no reliable conclusions could be drawn. For example, in one of the estimates, Mackenzie (1921) assumed that one complete defoliation caused loss of one month's growth. A subsequent study extending over a 5-year period, using more realistic methods (Nair *et al.*, 1985) showed that under plantation conditions at Nilambur in Kerala, India, naturally occurring defoliation resulted in loss of 44% of the potential volume growth in four to eight-year-old teak plantations. The study also showed that all the loss was attributable to the defoliation caused by *H. puera*, the impact of *E. machaeralis* which feeds on older foliage being negligible, at least under Kerala conditions. It led to the conclusion that in Kerala no control measures are necessary against *E. machaeralis*. It was estimated that the protected trees put forth an annual wood volume increment of 6.7 m^3/ha compared to 3.7 m^3/ha of unprotected trees – a gain of 3 m^3/ha per year.

If we apply this to the entire rotation period in a plantation at site quality II, it can be shown (Nair *et al.*, 1985) that the protected trees would be ready for harvest in 26 years instead of 60 years, as in 26 years they would have accrued as much volume as unprotected trees would accrue in 60 years (Fig. 9.1). This is an enormous gain, if accomplished. However, such a projection is not realistic, as crowding-related limiting factors will retard the growth as soon as the normal increment is exceeded. This is because the silivicultural thinning schedules have been worked out on the basis of the normal growth trend (Nair *et al.*, 1985). In theory, it is possible to work out new thinning schedules and fertilizer and other inputs to enhance the rate of growth. In order to make a realistic

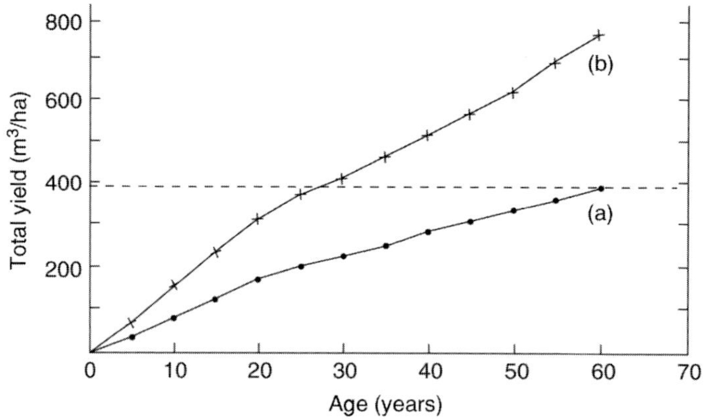

Fig. 9.1 Effect of protection against defoliator on tree growth. Volume growth of a teak stand of Site Quality II is shown. Curve (a) shows the usual growth in commercial volume over 60 years, and curve (b) shows an artificial trend that may be expected under ideal conditions of growth when defoliator damage is prevented. In theory, a defoliator-protected teak plantation can attain the volume growth of a 60-year-old unprotected plantation in just 26 years (see dotted line). From Nair and Sudheendrakumar (1992).

prediction of the increased volume production as a result of protection from defoliator, we need a stand-growth model for teak plantations. An economic analysis cannot be undertaken without data on the increased volume production, not only at the end of the rotation but also during the intermediate thinnings. Unfortunately, this must await development of stand-growth models for teak plantations, which is beyond the domain of entomologists. Once this is accomplished, we can calculate the economic gain using currently available econometric methods, assuming that it is possible to prevent defoliation completely. In practice, complete prevention of defoliation may not be feasible. In addition, cost will be incurred to control the insect. Thus we need information on the cost of control and the level of control obtainable. Since the cost will be incurred throughout the rotation period, but the returns realized only after long intervals, suitable methods need to be used for the cost–benefit analysis. If insecticides are used, it will entail environmental cost. At present, methods for control of the teak defoliator are still under development (see Chapter 10). Thus what might appear at first as a simple problem of working out the economics of pest control in teak plantations turns out to be a complicated and challenging problem when we get into the details. The question is not only whether economic damage is caused but also whether it can be economically prevented.

This example shows that economic analysis of a forest pest situation is a difficult task although it can be accomplished. It has not been attempted in most cases, due to paucity of relevant data. As observed by Schabel and Madoffe (2001), although increasing sophistication in pest management has been attained in other parts of the world, for most of the tropics forestry has low priority compared to programmes to promote food security.

Difficulty in reaching the tall canopy

Application of control agents to the tree canopy, whether chemicals or biological agents, requires the use of aircraft, as most ground-operated sprayers cannot reach the required heights. This poses more difficulties than in agriculture, particularly in the deveoping countries of the tropics.

Environmental impact

Compared to agricultural crops, forests and forest plantations occupy much larger areas. Application of insecticides in such large areas, often interspersed with agricultural fields and human settlements, can cause adverse environmental impact due to the residual effect of insecticides and unintended exposure of beneficial organisms. It is difficult to prevent drift of insecticide when it is sprayed from great heights. Therefore, unlike in agriculture, greater caution is necessary in the use of insecticides in forests and forest plantations, and as far as possible non-insecticidal methods should be preferred, if control is economically justified.

9.5 Constraints to forest pest management in the tropics

While the unique features described above are common to all forest pest management problems, whether in tropical or temperate countries, there are severe social, economic and policy constraints in the tropics (Nair, 1986b, 1991, 2000). Most of these are the direct or indirect effect of the poor economic development of the tropical countries, whether in Asia, Africa or Latin America. The important constraints are the following.

1. Small number of forest entomologists compared to the large number of pest problems.

The total number of forest entomologists in the tropics was estimated at less than 50 in 1972 (Gray, 1972). In 1995, a world directory published by IUFRO (Skilling and Batzer, 1995) listed 121 tropical forest entomologists. The number is still small. For example, Indonesia, with over 100 million ha of area under forest, had only about 40 researchers in forest protection (including entomologists and pathologists) in the year 2000, with less than half of them possessing a Ph.D. degree (Nair, 2000). India, with about 64 million ha of area under forests had less than 20 forest entomologists in 1990, compared with several hundred agricultural entomologists. The increase since then has only been marginal.

2. Lack of adequate training of entomologists in applied research and in the principles and techniques of pest management.

Many forest entomologists in tropical countries (e.g. India) have come from the pure science stream where there is a traditional emphasis on taxonomy and natural history, unlike in the agricultural stream. In the absence of subsequent training, and due to isolation, research has tended to be of an academic nature. This is typical of most tropical countries.

3. General absence of demand from practising forest managers, largely due to lack of information on the economic impact of damage.
4. Lack of adequate organizational and infrastructural facilities for entomological field work.
5. Inapplicability of sophisticated pest management methods employed in developed countries, such as ultra-low volume aerial application of insecticides, aerial release of biocontrol agents, computer-based pest prediction and alert systems, due to technological and economic constraints.
6. Centralisation of research effort in a few government-controlled research centres which imposes physical limitations to field-oriented research and hinders the scope for diversification of research approaches.
7. Poor research management and lack of incentives to researchers for carrying out applied research.
8. In most tropical countries forests are predominantly government-owned, and forest pest management has low priority relative to the more pressing agricultural pest problems. Also agriculture, being a private enterprise, generates more social demand.

Given the constraints to successful forest pest management in the tropics discussed above, a number of suggestions can be made to improve the situation, although many constraints cannot be easily removed as they are strongly linked to overall socio-economic development of the countries.

Nair (1991) stressed the need to make policy level changes in research management. In most developing countries of the tropics, research management is not given adequate attention, and as a result problem-solving research gets low priority. Most researchers are interested in publishing papers in journals which earn them professional recognition. Problem-solving research calls for imaginative research management by administrators to ensure due recognition and rewards to scientists engaged in planning and implementing applied research and extension, which may not produce papers in journals but produces results in the field. The organisational set-up of most universities and research institutions in the tropical countries promotes individualized, piecemeal research, while team effort involving scientists from more than one discipline is often necessary to develop pest management recommendations and to test them in the field.

Although many forest managers understand and appreciate the concept and goal of IPM, when it comes to practising it there are two major stumbling blocks: (1) it is not easy to translate the IPM concept into a set of actions in a given pest situation and (2) there is no evidence of the effectiveness of the suggested course of action. Unlike in insecticide trials, we do not normally test the effectiveness of IPM; we simply advocate it (Nair, 1991). There are no simple answers to overcoming these obstacles. Translating the concept of IPM into implementable action plans, and providing evidence of effectiveness involves complex and drawn out procedures and calls for dedicated effort over several years, particularly in forestry situations, as we saw in the case of teak defoliator control. It must be recognised that the responsibility of the entomologist does not cease with making a recommendation for pest management. It must be part of his responsibility to demonstrate its effectiveness. The forest manager's usual reluctance to practise an entomologist's recommendation is at least partly due to doubts on its effectiveness. Beeson (1924) has stated this problem very succinctly, as quoted at the beginning of this chapter.

Pest management research in the tropics could also be improved by international cooperation. IUFRO has been doing exemplary service through its Working Party on 'Protection of Forest in the Tropics' under the Subject Group 'Forest Health', by providing opportunities for participation of scientists from the developing tropical countries in international meetings to facilitate exchange of information among scientists of the tropical countries. IUFRO meetings have also served as a window into the world for many developing country entomologists. IUFRO's Special Programme for Developing Countries (SPDC) has been particularly useful. Bilateral programmes, promoted by some of the developed countries, have also helped in a limited way to improve the capability of tropical country scientists by training and facilitating participation in collaborative research. However, with rare exceptions, bilateral collaborative research programmes tend to be dictated by the professional and sometimes political interests of the sponsoring developed countries rather than the real needs of the participating developing tropical countries. The large research effort on leucaena psyllid, an exotic pest on an exotic plant, is an example of this skewed priority. Unless the research managers in the developing tropical countries set their priorities right, the scientists could easily be led to highly sophisticated but less relevant areas of research because donors have their own interests and priorities.

9.6 Guidelines for the practice of forest pest management in the tropics

Specific control methods for various pests of the commonly planted tree species of the tropics are discussed in Chapter 10, whenever information is available. They will not be repeated here but an overview of the general approaches is given below.

9.6.1 Seeds

Seeds of many forest trees are attacked and damaged by insects at three stages – while on the trees, when fallen on the ground and while in storage. Generally the damage is not serious in the first two stages. The problems of protection of stored forest tree seeds are similar to those of agricultural seeds and merit no special discussion. In general, systematic storage in closed containers is sufficient to prevent damage. Insecticide may be mixed with the stored seeds in exceptional cases where a problem is noticed.

9.6.2 Nurseries

Forest nurseries usually consist of 12 m × 1.2 m raised soil beds, prepared in forest areas close to the planting sites. Seeds are sown in these beds where the seedlings are maintained for varying periods, depending on the species. For some species like *Eucalyptus*, young seedlings are pricked out and transplanted into soil-filled polythene bags in which they are maintained for several months before field planting. Ants which carry away newly-sown small seeds (such as that of *Eucalyptus*), whitegrubs and termites that feed on the roots, cutworms which cut off the stems and caterpillars which feed on the leaves are the common pests of nurseries as discussed in Chapter 5. These pest problems are similar to those encountered in agriculture. The experience gained from agriculture has formed the basis for control measures. Cultural practices and use of insecticidal chemicals are the two approaches generally adopted for nursery pest management. Cultural practices involve cleaning of the nursery site of weeds and woody debris, and soil working. Soil working facilitates destruction of whitegrubs, cutworms and nests of subterranean termites. Removal of wooden debris helps to reduce feeding sites for termites and weeding helps to remove feeding sites for whitegrubs and cutworms. Good cultural practices in the nursery, such as optimal irrigation, fertilization and weeding, also help to reduce pest problems by keeping the plants vigorous. Bad practices, like retaining the plants in impervious plastic containers for too long, often result in root coiling and encourage subsequent pest susceptibility.

In addition to cultural practices, insecticides are commonly used for control of nursery pests when needed. Usually, dust formulation of a suitable insecticide such as carabaryl is sprinkled on the top of the nursery bed and mixed with the top layer of soil for protection from ants. In areas prone to whitegrub damage, a suitable insecticide is incorporated into the top layer of the nursery bed soil, prior to sowing of seeds, as a prophylactic measure. BHC and other persistent organochlorines such as aldrin, heptachlor or chlordane have been used in the past, but these are now being replaced by the more easily biodegradable organophosphates like chlorpyrifos or systemic insecticides such as phorate or carbofuran. This treatment is also effective against cutworms. A general purpose contact insecticide such as quinalphos is usually sprayed when caterpillar feeding is noticed on the foliage. Apart from such instances, insecticides are rarely used in forest nurseries in the tropics, although practice varies widely between countries, depending on the local needs and availability of insecticides. In the tropics, it has been customary to raise makeshift nurseries near the planting site. This practice is now being increasingly replaced with centralized nurseries where facilities for irrigation, supervision including pest management etc. can be more easily organized. However, centralization of the nursery increases the risk of pest build up due to continuous availability of suitable host plants over large areas.

9.6.3 *Young plantations*

In many respects, the problems of controlling pests of young plantations are also similar to those of controlling pests of agricultural crops. The plantations are usually raised in land cleared of other vegetation, so that natural enemies which are supported by alternative host insects thriving on other vegetation are usually absent or deficient. Because of the short height of young plantations, conventional application of insecticide from the ground is feasible as in agricultural crops. Insecticides are therefore often used for managing pests of young plantations.

One of the most common uses of pesticides in the tropics is for the control of root-feeding termites attacking young transplants of trees, particularly *Eucalyptus* spp. Others such as casuarinas, pines and poplars are also susceptible. The treatment details are discussed under *Eucalyptus*.

Hepialid sapling borers have been effectively managed by spot application of an insecticide taking advantage of the behavioural characteristics of the larva, as mentioned under *Sahyadrassus*, under teak in Chapter 10. A baculovirus preparation has been standardized for controlling the teak defoliator affecting young plantations. Details are given in Chapter 10. There are no effective treatments for some pests of young plantations such as the mahogany shoot borers. Locally available, non-persistent chemical insecticides, neem-based products or Bt may be used when needed against open-feeding caterpillars, when no specific control methods are available.

9.6.4 *Older plantations and natural forests*

Older plantations and natural forests are considered together because of similarities with respect to pest control options, in spite of many ecological differences. Both are characterised by the relative constancy of the biotic environment compared to the agricultural situation, where violent changes in the plant community take place annually. Generally, the relative constancy facilitates the operation of several natural control factors, most importantly the parasitoids and predators. In these situations, use of insecticides often aggravates the pest problems by interfering with the action of natural enemies. Accumulation of toxic chemicals in the environment, development of insecticide resistance by target pests and outbreaks of secondary pests are other adverse consequences. Use of chemical pesticides is therefore not a suitable option for control of pests of older plantations and natural forests, except as part of an IPM programme. Suitable IPM programmes are yet to be developed for most pest problems of older plantations and natural forests in the tropics. In general, the following guidelines, suggested by Nair (1994) for India, are useful for pest management practices in tropical forestry.

When faced with an insect pest problem in a managed natural forest or older plantation in the tropics, ask the following questions and take the suggested steps.

1. Is it economically worthwhile to control this pest?

Many growers are pesticide addicted. When an insect is found, usually the immediate response is to spray with insecticide. Pause and think. Is it causing any economic damage? When a farmer sprays insecticide to control pests in his vegetable crop, he uses his judgement. He will spray pesticide only if the value of the expected crop yield is much greater than the cost of the pesticide. He can assess the benefit, based on the prevailing market value of the produce. The forestry situation is different. Timber is the product usually harvested. Its value is realized only after a long period of growth of the tree. Leaf-feeding insects may retard the growth of the tree significantly only if the damage is extensive and often-repeated and unlike the case with the evergreen conifers of the temperate region, against which insecticides have been used in the past, defoliation does not usually kill tropical trees. The value of the returns received after, say, 30 years, cannot be compared straightaway with the value of the money spent on applying insecticides today. Therefore, to judge the profitability, an economist usually calculates the Net Present Value of the returns to be received in the future, employing the principles of discounting. However, you can make an informed judgment of profitability in consultation with a specialist who has more detailed knowledge of the nature of the damage and probability of repeated pest attack.

If you decide that adopting control measures is likely to be economically worthwhile, ask the next question.

2. Is it possible to prevent or control outbreak by adopting suitable silvicultural measures?

Some pest problems can be prevented by suitable silvicultural measures. Examples are the sapling borer *Sahyadrassus malabaricus*, the teak trunk borer *Alcterogystia cadambae*, and the sal borer *Hoplocerambyx spinicornis* discussed in Chapter 10. Wherever possible, follow silivicultural measures. These could also include avoiding planting in unsuitable areas. For example, *Acacia nilotica* growing on poor dry soil is believed to be prone to damage by the root borer *Celosterna scabrator* (see Chapter 10). If silvicultural measures are not applicable, go to the next question.

3. Are varieties or provenances resistant to this pest available?

If yes, use them for future planting and harvest the present crop as soon as economic return is expected. An example is an indigenous provenance of *Eucalyptus deglupta*, resistant to the varicose borer *Agrilus sexsignata* in the Philippines. Practise suitable control measures until the existing plantation is harvested. Go to the next question.

4. Can this pest be suppressed by natural enemies like parasitoids and predators?

If yes, use them. If no, go to the next question.

5. Can this pest be controlled by other specific biocontrol agents like baculovirus?

If yes, and if the technology is available, use it. If not, go to the next question. (At present, for most forest pests, the technology for such control measures is only now being worked out; but the guidelines are intended for future use also.)

6. Can the pest be controlled by other less specific biocontrol agents?

Commercial preparations of the bacterial insecticide *Bacillus thuringiensis* (Bt) which is effective against a wide range of caterpillar pests are now available. Prefer an asporogenous preparation of Bt. Bt can kill honey bees, but asporogenous preparations will not perpetuate the organism in the environment and will therefore be less harmful to bees. Since there is evidence of development of resistance to Bt by some pests, Bt may be used as part of an IPM programme, including use of insecticidal chemicals. Therefore ask the next question.

7. What are the most suitable chemical pesticides and their methods of application, consistent with least harm to non-target species, particularly honey bees?

In choosing an insecticide for use in the forest, care should be taken to choose one which is not only effective against the pest but also meets certain other criteria. The most important is the safety of non-target organisms. Newer insecticides like the chitin inhibitor diflubenzuron act by inhibiting chitin synthesis and is therefore safe to several other groups of organisms. Also choose those chemicals which are comparatively less toxic to honey bees. Data on toxicity rating for honey bees is available for most pesticides. Honey bees and other bee pollinators are important components of the forest ecosystem. Their protection is important since forest plantations cover much larger contiguous areas compared with agriculture.

Conventional spraying equipments are suitable only for young plantations. There is a need to develop suitable machinery for applying pesticides to tall trees, not only for chemicals but also for biopesticides. Use low-volume or

Box 9.1 Protection of stored timber

The methods used for protection will depend on the kind of wood and the purpose for which it is to be used. Apart from the use of resistant timber species to avoid pest problems, two broad categories of protection methods are available – (1) physical methods and (2) chemical methods. Each of these has its merits and demerits and it is often possible to combine some of the methods.

Use of pest-resistant timber

Pest problems can be avoided if pest-resistant timbers are used. However, there is no timber which is absolutely resistant to insect pests, although the heartwood of some tree species is practically totally resistant. Timbers are generally grouped into three classes, durable (average life > 10 years), moderately durable (average life 5–10 years) and susceptible (average life < 5 years). Susceptibility to fungal decay is also taken into consideration in this rating. In India, for example, out of 157 timbers tested, 46 fell in the durable class, 35 in the moderately durable class and 76 in the susceptible class. Some examples of durable timbers are *Albizzia odoratissima*, *Cedrus deodara*, *Dalbergia latifolia*, *Gmelina arborea*, *Hopea parviflora*, *Shorea robusta*, *Tectona grandis* and *Xylia xylocarpa*. Some examples of perishable timbers are *Alstonia scholaris*, *Bombax ceiba*, *Dillenia pentagyna* and *Vateria indica*. It must be noted that the sapwood portion of even the most durable timber is susceptible to borers and termites. Pest problems can be avoided by using the heartwood of durable timbers for such uses as doors and windows, furniture etc. The limitations of this method are shortage of such timbers, high cost, wastage of the sapwood and the unsuitability of most heartwood for some end uses like photograph frames.

Wood protection methods

Pest problems begin in the forest. To prevent or control pest problems effectively, it is essential to know when and how the pest problems originate. Pest problems begin in the forest as soon as, or sometimes even before, the trees are felled. These pests may continue to cause damage while the log is in storage. In addition, other insects subsequently invade during storage, processing or while the manufactured product is in use. Various groups of insects have become adapted to attack the wood at different stages, as indicated earlier. The preventive or control operations must begin as soon as the tree is felled, depending on the timber species and the end use requirement. All infestations originate from pre-existing populations of the pest. Since most borers are winged as adults, infestation starts with the landing of adult insects on the wood, and their egg laying.

Prevention is better than cure. It is difficult to control wood-destroying insects once they are established within the wood. Therefore preventive or prophylactic methods are essential for effective protection. Subjecting the infested wood to fumigation or pressure impregnation of chemicals is somewhat effective but not foolproof. So prevention is not only better than cure, cure is often not feasible.

Physical methods

The following physical methods are generally recommended to reduce or prevent insect infestation.

Felling during safe period. In the cooler tropics, felling and conversion of trees in the winter season reduces damage from most insects because insect borers are generally inactive during the winter season. However, the dried timber may be attacked by bostrichid borers if appropriate precautions are not taken.

Girdling of standing trees. This method reduces borer damage, possibly due to depletion of the starch content of the timber. This method, however, will not prevent termite attack.

Debarking. Debarking is useful against cerambycid borers as they lay eggs on logs with bark. The debarked timber is then given other treatments to prevent attack by other groups of borers.

Quick conversion. This prevents attack by most pinhole borers (scolytines and platypodines) because of quick drying.

Ponding. Immersion of timber in fresh water prevents borer attack because water acts as a barrier. It is also believed that leaching of carbohydrates and other unidentified constituents confer resistance, but there is a dearth of data to prove this. This method is most commonly practised by plywood factories. It is a foolproof and environmentally safe method for all kinds of timber.

Sanitation in storage yards and mills. Many insect borers breed and multiply in wood debris heaped as refuse on factory premises. Since such breeding sites serve as a source of infestation, sanitation can reduce pest attack though not prevent it altogether.

Painting of finished products. In manufactured goods where painting is permissible and feasible, it acts as a physical barrier by masking the wood.

In general, the physical methods of protection can be used as complementary to other methods, but are not fully effective by themselves. Immersion of logs in water is an exception, as noted above. We need more data on the effect of ponding of logs on the subsequent susceptibility of converted timber.

Chemical methods

A large number of chemicals and treatment methods are available for protection of wood from insects. Only brief details are given here. The chemicals to be used and method of treatment vary for different situations.

Oil-type chemicals such as coal tar, with or without fuel oil, or petroleum are suitable for treatment of timber for exterior use like railway sleepers. These chemicals have high toxicity and permanence, but are not clean to handle. Synthetic pyrethroids like permethrin are also effective but must be dissolved in an organic solvent for use. They are clean to use and are effective for long periods. Water soluble type of preservatives include the leachable boric acid–borax composition and the non-leachable ASCU and copper–chrome–boric composition.

Some timbers and the sapwood of all timbers are easily treatable, i.e. the preservatives penetrate into the wood easily. Other timbers are refractory to treatment. Various methods have therefore been developed for treatment of timber.

Surface-application. In this method, chemicals are applied by brushing, spraying or dipping for a short period. This method is useful for treating logs after debarking at site.

Soaking. In this method, the timber is submerged in the treatment solution and heated to 90 °C and then cooled. This ensures death of the insects in infested wood.

Sap-displacement method. This is used for treatment of fresh poles by dipping the lower end in a treatment solution or by the Boucherie process.

Diffusion process. In this method, the wood is dipped in the preservative solution and then closely stacked under cover to permit diffusion of chemicals into the interior. This is very effective for treatment of timber with high permeability, like rubber wood (*Hevea brasiliensis*).

Pressure process. Several processes and equipments are available for delivering the preservative to the interior of the wood under pressure. They are very effective and can be employed with any type of preservative and most timbers. Pressure process is particularly useful for treatment of railway sleepers, electric transmission poles, shuttering plywood etc. The cost of treatment is higher because of the necessity for initial investment on equipment.

Choice of the type of chemical treatment is governed by the species of timber, its sapwood content and end use. Generally, for use indoors a water-soluble type of preservative would suffice; for use outdoors pressure treatment with ASCU or creosote will be required for protection against termites. Each case will have to be considered individually. In choosing the chemical, due attention must be paid to the toxicity of the chemicals to humans. A toxic chemical can be used for treating the railway sleeper or fence post, but not tea chests or packing cases for fruits. Among the wood preservatives, boric acid–borax is the only chemical composition not toxic to humans.

ultra-low volume sprayers if available, in order to avoid wastage of insecticides and undue contamination of the environment. In some cases, application of chemicals can be limited to small areas where the pest activity is concentrated. Spot application methods could be used for such pests, as in the case of termites attacking the root of saplings or the hepialid and cossid borers attacking stems. After choosing the apropriate chemicals, go to the next question.

8. What other methods can be used in addition to chemicals and how best can they be integrated into an IPM programme?

A suitable package of IPM practices must be developed depending on the decisions made in response to the above questions. Use of more than one suitable insecticide by rotation is suggested in order to avoid the development of resistance. Other methods must also be integrated into the total pest management programme. Each pest situation requires an intelligent integration of several methods in order to ensure long-term protection.

These guidelines are only a theoretical framework within which practical field operations must be developed based on hands-on experience with the problem.

10

Insect pests in plantations: Case studies

The general aspects of pest problems of plantations were covered in Chapter 5. In this chapter, detailed case studies are given for selected, representative plantation tree species of the tropics. For each tree species dealt with, a brief tree profile followed by an overview of the pest problems is given, summarising the highlights of available information. As pointed out in Chapter 2, generally a large number of insect species is associated with each tree species, but not all of them cause serious damage. To merit categorization as a pest, an insect must be capable of causing economic damage, but relevant information is often not available. This is a handicap in listing the pests. The course followed here is to list the insects most commonly reported to cause damage to a tree species, irrespective of whether the damage is economically significant or not, so that a broad view of the actual and potential problems is obtained. This shortlisting is arbitrary, based on the available literature and the author's judgement, rather than on strict criteria. Following this listing and summary statements for each tree species dealt with, a pest profile of the major pests is given, outlining each pest's biology, ecology, impact and control.

10.1 *Acacia* species (Fabaceae: Mimosoideae)

Tropical acacias can be grouped broadly into true-leaf acacias and phyllode acacias. The phyllode acacias have their leaf modified into a thick, waxy phyllode, an adaptation to reduce water loss in dry and hot conditions. Among the nearly 1300 species in the genus *Acacia*, the majority (about 940 species) are phyllode acacias (Turnbull *et al.*, 1998), mostly distributed in the Australian region. Many phyllode acacias of Australian origin have been introduced into Asia, Africa and Latin America. *A. decurrens*, *A. dealbata* and *A. melanoxylon* were introduced into Tamil Nadu, India, in the 1840s as fast-growing trees to supply fuel for the army and were subsequently used as shade trees in tea estates (Turnbull *et al.*, 1998). Extensive plantations of the Australian black wattle *A. meanrsii* have since the 1860s been raised in many subtropical countries and the cooler regions of tropical countries – South Africa, East Africa, Indonesia, Sri Lanka, India, and Brazil for extracting tannin from its bark. Since the 1930s, *A. auriculiformis* has been planted in many countries of Asia, including Malaysia, Thailand, India and China, as an ornamental, for fuel and to revegetate denuded land. In the recent past *A. mangium*, a fast-growing phyllode acacia from Australia, has emerged as a major plantation species in Malaysia and Indonesia for production of pulpwood for manufacture of medium density fibreboard. By the year 2000, Indonesia alone had an area of about half a million ha under this species (see Chapter 1). Some true-leaf acacias like *A. nilotica* and *A. senegal* have also been planted widely in the tropics for various purposes. Pests of the most commonly planted acacias are discussed below.

10.1.1 *Acacia auriculiformis*

Tree profile

Acacia auriculiformis A. Cunn. ex Benth., a species native to Australia, Papua New Guinea and parts of Indonesia, is a phyllode acacia, valued for its fast growth, nitrogen fixing ability and tolerance of harsh environmental conditions such as acid, alkaline, saline or waterlogged soil and moderate drought. A medium-sized multipurpose tree, it has been planted widely in the tropics since the 1960s, particularly in Asia, for rehabilitation of degraded land, erosion control, fuel production and as an ornamental tree along roadsides. Extensive plantations exist in India, Indonesia, Thailand, Malaysia, Sri Lanka and China. It is also planted, to a lesser extent, in several countries in Africa and Latin America.

Overview of pests

In North Queensland, Australia, where *A. auriculiformis* is native, a bug *Mictis profana* (Hemiptera: Coreidae) was reported to attack the shoot tips of one-year-old plants, causing shoot dieback (Wylie *et al.*, 1998).

No major pests have been reported from exotic plantations. Minor pests include the following. A scolytine beetle, *Hypothenemus dimorphus*, infests the shoots of seedlings and young transplants in Malaysia. The small beetle makes a longitudinal tunnel in the centre of the shoot or twig, in which the immature stages live communally. Infestation leads to death of the seedlings (Browne, 1968). Another scolytine, *Xylosandrus compactus*, has been reported infesting seedlings in Java, Sumatra, Kalimantan and Sulawesi in Indonesia (Intari and

Santoso, 1990; Natawiria, 1990). The bostrichid beetle *Sinoxylon anale* and an unidentified species of the same genus have been reported from Thailand; they attacked small stems and branches of about 25% of saplings in an experimental plot, causing girdling and occasional breakage at the point of attack (Hutacharern and Choldumrongkal, 1989). The polyphagous borers *Sahyadrassus malabaricus* (Lepidoptera: Hepialidae) (see pest profile under teak) and *Zeuzera coffeae* (Lepidoptera: Cossidae) have been recorded, respectively, in India and Thailand (Nair *et al.*, 1996c; Wylie *et al.*, 1998), on saplings, but are of minor importance. The stem borers *Xystrocera festiva* (see pest profile under *Falcataria moluccana*) and *X. globosa* (Coleoptera: Cerambycidae) attack older trees in Indonesia, although the incidence is rare (Nair and Sumardi, 2000). Instances of minor leaf feeding by polyphagous insects are on record, but no serious defoliation has been encountered.

10.1.2 Acacia mangium

Tree profile

Acacia mangium Willd. is a phyllode acacia which is naturally distributed in coastal lowlands in some parts of northern Queensland in Australia, the western province of Papua New Guinea, and some parts of Irian Jaya and Moluccas in Indonesia. It is now a widely planted exotic in the humid tropical lowlands, particularly in Southeast Asia. The plantation history is short. It was introduced into Malaysia in 1966 and became a major plantation species. Its fast growth, tolerance of poor soils, ability to compete with tall grasses like *Imperata cylindrica* in grasslands and suitability for production of pulp for paper and medium density fibreboard has made it a preferred species for industrial plantations. By the year 2000, about 86% of the plantation forests in Peninsular Malaysia (52 000 ha) and 50% in Sabah (56 000 ha) consisted of *A. mangium* (Chey and Intachat, 2000). Indonesia had over half a million ha of *A. mangium* plantations by the year 2000 (Cossalter and Nair, 2000). Plantations have also been raised in several other countries in Asia, including Bangladesh, China, India, Laos, the Philippines, Sri Lanka, Thailand and Vietnam. Smaller plantations have also been raised in several countries in Africa and Latin America.

Overview of pests

Hutacharern (1993) and Wylie *et al.* (1998) have summarised the information on pests of *A. mangium*. Apart from some native insects associated with the tree in Australia, about 75 species of insects have been found on *A. mangium* planted as an exotic, most of them in Asia. Most of these records represent incidental feeding by polyphagous insects and in general *A. mangium* plantations are remarkably free of serious pests. Hutacharern (1993) listed about 27 species of greater importance and Wylie *et al.* (1998) about 16, for Southeast Asia. In Table 10.1, about two dozen important insects associated with *A. mangium* are listed. In the absence of suitable criteria to rate the importance of pests, the perceptions of authors are likely to vary. The plantation history of *A. mangium*, both in its native range and as an exotic, is comparatively short, with most expansion of planted area taking place during the past 15 years. Although there are no serious pests at present, some appear to be developing.

In plantations in Queensland, Australia where *A. mangium* is native, the following insects occur (Elliott *et al.*, 1998). Chrysomelid and buprestid beetles, and moth larvae feed on the leaves. Larvae of the moth *Ochrogaster lunifer* (Lepidoptera: Thaumetopoeidae), commonly known as the 'processionary caterpillar', cause occasional defoliation by gregarious feeding at night. A tortricid caterpillar, *Cryptophlebia* sp., tunnels into the terminal shoot and causes dieback; almost all trees were attacked in a year-old seed orchard (Wylie *et al.*, 1998). A cerambycid beetle, *Penthea pardalis* causes serious damage, the adults shredding the bark of stems and branches and the larvae tunnelling into the wood.

In exotic plantations, common leaf-feeding insects include several species of bagworm, some hairy caterpillars and curculionid beetles. Generally, they occur in small numbers. In nurseries and young plants, apart from generalist feeders like jassid bugs, cutworms and other caterpillars, the following insects have been noted. The termite, *Coptotermes curvignathus* killed 10–50% of field-planted saplings in Central Sumatra, Indonesia during the first year (Wylie *et al.*, 1998). The same species caused damage to 4 to 8-year-old trees in Malaysia (Kirton *et al.*, 1999); Chey and Intachat (2000) rated it as the most important pest of *A. mangium* in Malaysia. The scolytine beetle *Xylosandrus compactus* bores into the shoot of seedlings in nurseries in Malaysia and causes large-scale mortality. In Malaysia, Indonesia and the Philippines, some species of the tea mosquito bug *Helopeltis* cause damage to young plants by sucking sap from the shoot tip and causing shoot dieback, possibly as a result of injection of toxic saliva. In 6 to 18-month-old plantations in North and Central Sumatra, Indonesia such damage resulted in 'bushing' and retardation of growth (Wylie *et al.*, 1998). Also in Indonesia, caterpillars of an unidentified moth, tentatively called 'Plusia', feed on the foliage of young saplings (Nair and Sumardi, 2000) and the grasshopper *Valanga nigricornis*, which consumes leaves and terminal shoots, has shown a tendency to build-up in large numbers in nurseries and young plantations. At Perak in Peninsular Malaysia, an outbreak of a noctuid *Spirama retorta* occurred in 1992 in an 800 ha plantation (one-year-old) (Sajap *et al.*, 1997). The number of infested trees increased from 30% in May to 60% in June, declining to 55% in July and 20% in August. In the infested trees, about 20–30% of the foliage was lost.

Table 10.1. *Important insects causing damage to* Acacia mangium

Category	Species name, order and family	Countries/Regions of occurrence	Remarks
Root feeding	Termites (several species) (Isoptera: Rhinotermitidae and Termitidae)	Most countries	On saplings
Leaf feeding	*Ochrogaster lunifer* (Lepidoptera: Thaumetopoidae)	Australia	
	Cryptophlebia sp. (Lepidoptera: Tortricidae)	Australia	
	Archips micaceana (Lepidoptera: Tortricidae)	Thailand	On seedlings
	Pteroma plagiophleps (Lepidoptera: Psychidae)	India, Indonesia, Malaysia, Philippines, Thailand	
	Dasychira mendosa (Lepidoptera: Lymantriidae)	India, Malaysia, Thailand	
	Eurema spp. (Lepidoptera: Pieridae)	Malaysia, Philippines, Vietnam	
	Unidentified 'caterpillar Plusia' (Lepidoptera: Noctuidae)	Indonesia	On seedlings
	Spirama retorta (Lepidoptera: Noctuidae)	Malaysia	On saplings
	Spodoptera litura (Lepidoptera: Noctuidae)	Malaysia	
	Hypomeces squamosus (Coleoptera: Curculionidae)	India, Indonesia, Malaysia, Thailand	
	Valanga nigricornis (Orthoptera: Acrididae)	Indonesia, Malaysia	
	Leaf-cutter ants (Hymenoptera: Formicidae)	Costa Rica	
Sap sucking	*Helopeltis* spp. (Hemiptera: Myridae)	Indonesia, Malaysia, Philippines	
	Acizzia sp. (Hemiptera: Psyllidae)	Philippines	
Stem/branch boring	*Penthea pardalis* (Coleoptera: Cerambycidae)	Australia	

Table 10.1. (cont.)

Category	Species name, order and family	Countries/Regions of occurrence	Remarks
	Agrilus fisheri (Coleoptera: Buprestidae)	Philippines	
	Xylosandrus (=Xyleborus) compactus (Coleoptera: Curculionidae: Scolytinae)	Indonesia, Malaysia	On seedlings/ branches
	Sinoxylon anale (Coleoptera: Bostrichidae)	Thailand	On branches
	Xystrocera spp. (Coleoptera: Cerambycidae)	Indonesia, Malaysia	
	Onciderus saga (Coleoptera: Cerambycidae)	Brazil	On branches
	Zeuzera coffeae (Lepidoptera: Cossidae)	Thailand	
	Xyleutes sp. (Lepidoptera: Cossidae)	Philippines	

Although exotic plantations of *A. mangium* are at present, by and large, free of major pests, the situation needs watching because of the short plantation history of the species. Two insects are of particular concern. The first is *Helopeltis*, some species of which are well-known pests of economic importance in plantations of tea and cashew. As noted above, it is already recognized as a pest of young *A. mangium* in Indonesia, Malaysia and the Philippines. The other is the noctuid *Spirama retorta* (also reported from *A. mearnsii* in China), (Haojie *et al.*, 1998b) which has shown a tendency for outbreak on *A. mangium* in Malaysia. None of the native pests from Australia have found their way to exotic plantations of *A. mangium* so far, but if they do, the risk of outbreak development is very high.

10.1.3 *Acacia mearnsii* (common name: black wattle)

Tree profile

Acacia mearnsii De Wild., known as black wattle, is native to southeastern Australia. It is a small tree with bipinnate leaves having a large number of leaflets. Adapted to subtropical and temperate lowlands and tropical highlands, it is cultivated widely throughout the world for extracting high quality tannin from its bark. The tree is also used for erosion control and soil improvement. Principal growing areas include Brazil with over 200 000 ha of plantation, South Africa with 160 000 ha, East Africa (Zimbabwe, Kenya, Tanzania,

Rwanda and Burundi) with 30 000 ha, India with 20 000 ha and Indonesia with 15 000 ha (CABI, 2005). It is also grown widely in subtropical parts of China. For tannin production, it is usually grown in monoculture and managed on 8–10 years rotation.

Overview of pests

In Australia, where the species is native, a large number of insects are associated with *A. mearnsii*. The leaf-eating chrysomelid beetle *Acacicola orphana* (Erichson), commonly known as fire-blight beetle, is a serious pest in southern Australia and Tasmania. Damage caused by this insect gives the plantation a brown, scorched appearance (Elliott *et al.*, 1998). The risk of *A. orphana* damage is one of the major reasons why larger areas have not been planted with *A. mearnsii* in southeastern Australia (Floyd *et al.*, 1998). Other pests include a scarabaeid beetle, *Xylotrupes gideon*, known as elephant beetle, which feeds on the bark of branches of young trees, and larva of a cossid moth, *Endoxyla liturata*, which bores into the wood and several species of sap sucking bugs (Elliott *et al.*, 1998; Floyd *et al.*, 1998).

Over 200 species of insects have been recorded on exotic plantations of *A. mearnsii*, some causing economic damage. Whitegrubs, termites, grasshoppers and cutworms cause problems in the nursery. The bagworm *Chaliopsis* (*Kotochalia*) *junodi* (Lepidoptera: Psychidae) is the most serious pest in South Africa. It causes defoliation which affects the growth of the tree. Atkinson and Laborde (1996) estimate that it infests 12 000–20 000 ha of wattle plantations annually in South Africa; data for 42 years from 1953 to 1994 showed an average infestation of 25% of the planted area. Chemical insecticides have been used for its control and 800 to 4000 ha have been sprayed annually. Trees up to two years old are also attacked by a myrid bug *Lygidolon laevigatum*, which causes leaf lesion, leaf drop and forking. It is estimated to affect about 2000 ha annually in South Africa and insecticides have been used for its control (Atkinson and Laborde, 1996). This insect is also rated as a serious pest in young plantations in Zimbabwe. In Kenya, the lymantriids *Argyrostagma niobe* and *Dasychira georgiana* were reported to defoliate *A. mearnsii* (Bullock and Smith, 1968). In southern Brazil, the beetles *Oncideres* spp. (Coleoptera, Cerambycidae) cause heavy damage by girdling twigs and branches (CABI, 2005). Also in Brazil, the geometrid *Thyrinteina arnobia* sometimes causes complete defoliation of trees (Tarrago and Costa, 1990). In China, more than 70 species of insects were recorded on *A. mearnsii*, which included root-feeding termites, leaf-feeding caterpillars like *Clania* spp. (Psychidae), *Orgyia postica* (Lymantriidae), *Semiothisa* spp. (Geometridae), *Spirama retorta* (Noctuidae) and the sap-sucking scale insect *Icerya purchasi* (also recorded in India). However, many pests are kept effectively under check by parasitoids and arthropod predators (Haojie *et al.*, 2002). Haojie *et al.* (1998b) found that the average level of defoliation was only 4% in an experimental plot in China. There are no major pest problems for *A. mearnsii* in India.

10.1.4 *Acacia nilotica*

Tree profile

Acacia nilotica (L.) Willd. ex Delile is a true-leaf acacia. It is a medium-sized, thorny tree with a spreading crown, indigenous to the Indian subcontinent, tropical Africa, Saudi Arabia, Egypt and Sudan. Several subspecies are recognized, based on crown shape, pod characteristics and branch pubescence. It is a light-demanding, nitrogen-fixing tree which can withstand extremes of temperature and moisture and a wide variety of soil conditions such as alkaline and saline. Its wood is strong and durable and finds a number of rural uses – as posts, and for making carts, agricultural implements etc. The leaves and pods are used as fodder. The tree yields a gum that is similar, but inferior, to the gum Arabic obtained from *A. senegal*. The bark yields tannin which is used in local tanneries. The species is cultivated widely in the rural setting, as a multipurpose domestic tree, as avenue trees along road, rail lines and canals, in agroforestry systems, for afforestation of wastelands etc. In addition to countries where it is native, *A. nilotica* has been introduced and planted in many countries such as Jamaica, Australia, Israel, Iraq, Iran, Pakistan and Indonesia.

Overview of pests

About 71 species of insects have been found on *Acacia nilotica* subspecies *indica* (IFGTB, 1995). They include defoliators, bark feeders, sap suckers, seed borers and a stem and root borer. The more important pests are listed in Table 10.2.

In general, there is no major threat to *A. nilotica* plantations from pests, although in some places the stem and root borer, *Celosterna scabrator* is rated as a serious problem. This insect is discussed in detail below.

Among the defoliators, the polyphagous bagworm *Cryptothelia cramerii* can commonly be seen, although in small numbers, on *A. nilotica* trees in India. The larva covers its bag characteristically with cut thorns of the tree (Fig. 10.1). The generation is annual. Outbreaks of a smaller bagworm *Pteroma plagiophleps*, another polyphagous species, may occur occasionally. In August 1988, it caused total defoliation of about 10% of the saplings in a 20 ha plantation at Salem in Tamil Nadu, in southern India (Pillai and Gopi, 1990a). The outbreak occurred in patches and only stray incidence of the insect was noted in other areas. A pest profile of this species is given under the tree *Falcataria moluccana*. A generation is completed in 2–2.5 months. The other lepidopteran defoliators recorded

Table 10.2. *Important insects causing damage to* Acacia nilotica

Category	Species name, order and family	Countries/Regions of occurrence	Remarks
Stem and root boring	*Celosterna scabrator* (Coleoptera: Cerambycidae)	India	Saplings often killed
Leaf feeding	*Euproctis* spp. (*lunata*, *scintillans*, *subnotata*) (Lepidoptera: Lymantriidae)	India	Sporadic outbreaks
	Calliteara (Dasychira) grotei (Lepidoptera: Lymantriidae)	India	Sporadic outbreaks
	Casiala raptaria (Lepidoptera: Geometridae)	India	
	Tephrina pulinda (Lepidoptera: Geometridae)	India	Sporadic outbreaks
	Selepa celtis (Lepidoptera: Noctuidae)	India	
	Ophiusa (Achaea) janata (Lepidoptera: Noctuidae)	India	
	Pteroma plagiophleps (Lepidoptera: Psychidae)	India	Sporadic outbreaks
	Cryptopthelia (=*Eumeta*) *cramerii* (Lepidoptera: Psychidae)	India	
	Diapromorpha balteata (Coleoptera: Chrysomelidae)	India	
	Orthacris ruficornis (Orthoptera: Acrididae)	India	
Bark feeding	*Psiloptera* spp. (*fatuosa*, *cupreosplendens*) (Coleoptera: Buprestidae)	India	On thin shoots
	Indarbela quadrinotata (Lepidoptera: Indarbelidae)	India	On thin shoots
Stem boring	*Sinoxylon* spp. (*anale*, *sudanicum*) (Coleoptera: Bostrichidae)	India	On saplings
Sap sucking	*Oxyrachis tarandus* (Hemiptera: Membracidae)	India	
	Drosicha stebbingii (Hemiptera: Margarodidae)	India	
	Anomalococcus indicus (Hemiptera: Coccidae)	Bangladesh, India	
Pod and seed boring	*Bruchidius* spp. (Coleoptera: Bruchidae)	Africa, India	
	Callosobruchus maculatus (Coleoptera: Bruchidae)	Africa	
	Careydon serratus (Coleoptera: Bruchidae)	India	
	Cryptophlebia illepida (Lepidoptera: Tortricidae)	India	

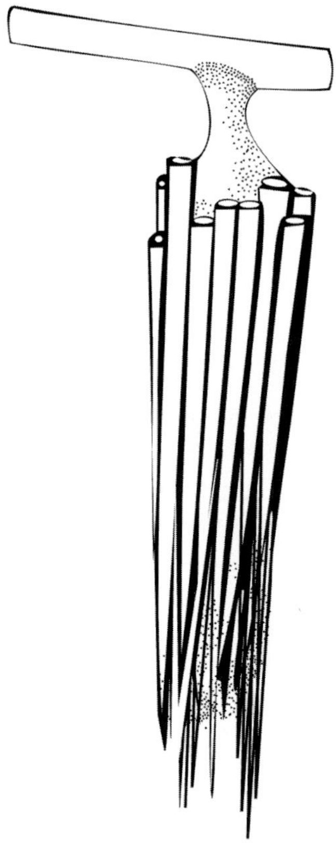

Fig. 10.1 The bagworm *Cryptothelia crameri* on *Acacia nilotica*. The larva cuts off thorns of the tree and attaches them to the bag. After Beeson (1941).

(Table 10.2) are also polyphagous. The noctuid *Selepa celtis* commonly occurs, often gregariously, in nurseries and young plantations in India in the rainy season. Outbreaks of the geometrid, *Tephrina pulinda* as well as the lymantriids *Euproctis* spp. have been noted occasionally in Tamil Nadu, India (Pillai and Gopi, 1990b; Subramonian and Krishnamurthy, 2002). Among the sap suckers, none is recognized as serious. The scale insect *Anomalococcus indicus* has been recorded in India and Bangladesh; it has been associated with weakened trees. The bostrichid beetle *Sinoxylon* sp., often tunnels to feed in the stem of saplings, the tunnel encircling the stem and rendering it liable to breakage. Among the pod/seed feeding insects, the two bruchid beetles, *Bruchidius uberatus* and *Callosobruchus maculatus* are reported to destroy about 70% of the seed crop in natural stands of *A. nilotica* in Africa (CABI, 2005); *Bruchidius* sp. has also been reported from India. An important seed pest in India is the bruchid *Careydon serratus* which attacks pods on trees and continues the infestation in stored seeds, causing 2–60% infestation, depending on the season (IFGTB, 1995).

Pest profile

Celosterna scabrator Fabr. (Coleoptera: Cerambycidae)

This beetle is commonly known in India as babul (*Acacia nilotica*) borer, because it has been frequently recorded in plantations of this tree species. However, it is a polyphagous species, attacking saplings of other tree species as well (see below). The beetle is 25–40 mm long and dull yellowish brown (Fig. 10.2a).

Life history Details of the life history have been worked out in India by Stebbing (1914) and Beeson (1931b). It has an annual life cycle. The adults emerge from the host with the onset of monsoon, in June–July. They feed on the bark of young shoots. Eggs are deposited under bark on stems, 5–23 cm girth, within 15 cm above ground level, usually one egg per stem. The newly hatched larva bores into the stem and as it grows tunnels downwards hollowing out the main root (Fig. 10.2b). The larva ejects the frass through a hole in the stem just above the ground level and the frass accumulates at the base of the stem. The larval

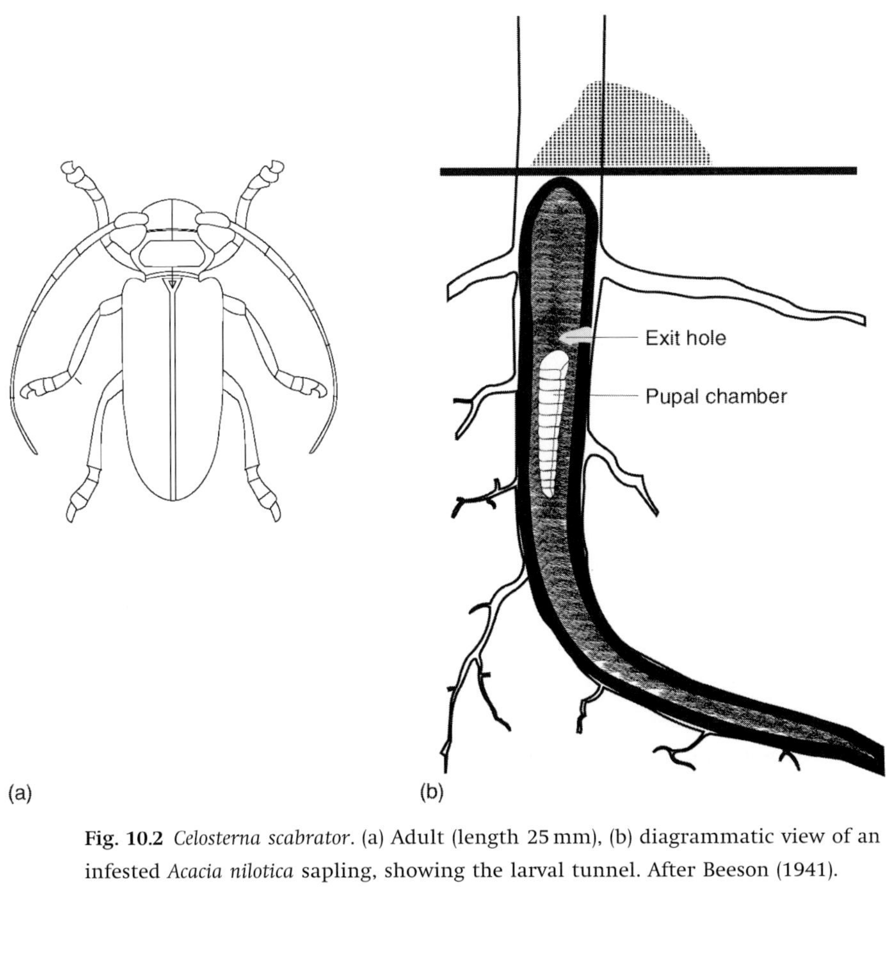

Fig. 10.2 *Celosterna scabrator*. (a) Adult (length 25 mm), (b) diagrammatic view of an infested *Acacia nilotica* sapling, showing the larval tunnel. After Beeson (1941).

period is completed in 9–10 months and pupation takes place. The beetle emerges by making a hole through the stem, just below the ground level. Beetles live for about 45 days.

Host range and geographical distribution *C. scabrator* is polyphagous; other species attacked include *Acacia catechu*, *Cassia siamea*, *Casuarina equisetifolia*, *Dipterocarpus alatus*, *Eucalyptus* spp., *Morus alba*, *Pithecolobium dulce*, *Prosopis cineraria*, *P. juliflora*, *Shorea robusta*, *Tamarix indica*, *Tectona grandis*, *Terminalia chebula* and *Zizyphus jujuba* (Beeson, 1941; Browne, 1968; Chatterjee and Singh, 1968; Gotoh, 1994; Jain, 1996). The adult beetle has also been recorded feeding on the bark of coppice saplings of *Shorea robusta* (Stebbing, 1914). In 1–2-year-old eucalypt plantations in Shimoga, Karnataka, India, 1–4 beetles were recorded per plant, their feeding on the bark leading to girdling of the stem or branches, resulting in the drying up of plants or branches (Shivayogeswara *et al.*, 1988). Root-shoot infestation similar to that on *A. nilotica* has often been noticed in plantations of *Casuarina equisetifolia* in Tamil Nadu, India. Another species of *Celosterna*, *C. pollinosa sulphurea* infests *Dipterocarpus alatus* in Thailand (Hutacharern and Tubtim, 1995).

Impact The attack of the borer results in tunnelling of the tap-root and the stem immediately above ground (Fig. 10.2b), causing cessation of growth of the sapling which, in the case of weak plants, is followed by death. The tunnel may reach about 60 cm in length. Vigorous plants on fertile soil may survive the attack by healing the wound as growth proceeds (Beeson, 1941). The incidence of attack is reported to be greater in localities where *A. nilotica* is grown on unsuitable soil, as in Berar in Madhya Pradesh where up to 80% of the plants were infested (Beeson, 1931b). At two locations in Gulberga district in northern Karnataka, India, 17–25% of *A. nilotica* plants below two years old were infested in 1983 (Ralph, 1990). Of the infested plants, 20% died at one location and 75% at another location that was less fertile.

In addition to the larvae tunnelling on the stem and root, adult beetles feed on the bark of stems and branches of saplings, in irregular patches, often girdling the shoots and causing them to dry up.

Natural enemies Beeson (1941) observed that *C. scabrator* is practically free of parasitoids and predators and that bacterial disease is infrequent.

Control Both preventive and remedial measures have been recommended (Beeson, 1941). Young, stagnating crops of *A. nilotica* on poor dry soils are believed to serve as breeding centres from where the beetles spread to other areas. Therefore as a preventive measure it has been recommended that planting should be restricted to localities entirely suitable for the tree species. Also avoid planting of other species that serve as alternative hosts of *C. scabrator* (see list above).

Remedial measures suggested include: (1) Grubbing-up the attacked saplings by inserting a thick wire probe through the borer tunnel and destroying the larvae, which may be up to 45 cm below ground level. Alternatively, the attacked sapling may be cut off at ground level and a small quantity of crude oil poured into the open tunnel, using a funnel. Current larval activity is indicated by fresh heap of ejected frass at the base of the plant, newly dead stems or coppice shoots. (2) At times when the beetle population is high, they can be picked from the plants where they settle for feeding and destroyed.

Knowledge gaps *C. scabrator* damage to *A. nilotica* has been recorded only in some places, on some occasions. It is commonly associated with dry areas. Although it is believed that plant stress is a predisposing factor for outbreaks of this borer (Beeson, 1931b; Ralph, 1990), more critical studies are needed on this aspect. The best-known instances of infestation on *A. nilotica* are those at Berar in Madhya Pradesh, recorded as early as 1892 and thereafter (Stebbing, 1914; Beeson, 1931b), and at Gulberga in Karnataka, recorded in 1983 (Ralph, 1990), (both in India). Incidence on eucalypts has been reported more frequently from several States (Chatterjee and Singh, 1968; Sivaramakrishnan, 1986; Shivayogeswara *et al.*, 1988) and on *Casuarina equisetifolia* from Tamil Nadu, in India. More research is needed to understand the conditions under which infestation becomes prevalent.

10.1.5 *Acacia senegal* (common name: gum acacia)

Tree profile

Acacia senegal (L.) Willd. is a true-leaf acacia which yields gum Arabic. It is a pioneering species found in drier lowland tropical and subtropical regions, with natural distribution stretching across Arabia to India and also across Africa. Several varieties have been distinguished; *A. senegal* var *senegal* is the most commonly cultivated for gum Arabic. The tree is also used for fuel, fodder and for soil fertility improvement. It is widely planted in Sudan, Kenya, Niger, Senegal and the Indian subcontinent, at 20–40 year rotation (CABI, 2005).

Overview of pests

The species is practically free of major pests. Nursery seedlings are attacked by root-feeding whitegrubs and termites, and leaf-feeding *Myllocerus* beetles (Curculionidae). The whitefly *Acaudaleyrodes rhachipora* has been recorded as a pest in India. In the Thar Desert in Rajastan, India, the tree is attacked by the bark-feeding caterpillar *Indarbela quadrinotata* (Lepidoptera: Metarbelidae) and the sap-sucking bug *Oxyrachis tarandus* (Hemiptera: Membracidae) (Vir and Parihar, 1993). Bruchids are known to attack fruits on trees and cause substantial loss of seed in India and some countries in Africa. In Sudan, several species of grasshoppers and locusts, and lasiocampid caterpillars feed on the foliage. Also in Sudan, several beetle borers were found associated with drought stressed trees of *A. senegal*. These include the buprestids *Sternocera castanea*, *Chrysobothris* sp., *Agrilus* spp. and *Psiloptera* sp.; the elaterids *Lanelater notodonta* and *Tetralobus cavifrons* and the cerambycid *Acanthophorus confinis* (Jamal, 1994). Mechanical damage which results from removal of bark for gum tapping provides an easy entrance for these beetles.

10.2 *Agathis* species (Araucariaceae) (common name: kauri)

Tree profile

Agathis is a tropical conifer. *Agathis* spp. are large trees, above 45 m tall, and they provide highly valued, general-purpose timber. The trees are also tapped for resin. Three species are important – *A. dammara, A. borneensis* and *A. robusta*. *A. dammara* (Lambert) Rich (syn. *A. loranthifolia; A. alba*) is native to Indonesia, Philippines and Malaysia. About 66 000 ha of *A. dammara* plantations have been established in Indonesia, in the provinces of Central and East Java, particularly in mountainous areas (Perum Perhutani, 1995). Its resin, called 'copal', is an important component of varnish. *A. borneensis* Warb. is found in Malaysia, Philippines and some parts of Indonesia. Small-scale plantations have been raised in Malaysia. *Agathis robusta* (C. Moore ex F. Muell.) F.M. Bailey (syn. *A. palmerstonii* (F. Muell.) Bailey) is native to Queensland, Australia, where plantations have been raised. It has also been introduced to some countries in Asia and Africa where small-scale plantations are raised.

Overview of pests

No major pest problem has been noticed in *A. dammara* in Indonesia, in spite of the large extent of plantations (Nair and Sumardi, 2000). Some unidentified beetles have been recorded from seeds. The case with *A. borneensis* in Malaysia is similar. On the other hand, outbreaks of a coccid *Conifericoccus agathidis* (Hemiptera: Margarodidae) is common on *A. robusta* in Queensland, Australia. Nurseries and plantations of all ages up to 40 years are attacked and repeated defoliation over several years causes the death of trees. Because of this problem, planting of *Agathis* was phased out in Queensland in the 1960s (Elliot *et al.*, 1998). Another pest, a seed-feeding moth larva *Agathiphaga queenslandensis* (Lepidoptera: Agathiphagidae) infests about 10% of seeds of

A. robusta in Queensland, but is of minor economic importance. The lepidopteran caterpillars *Achaea janata* and *Orgyia australis*, as well as a thrips *Oxythrips agathidis*, have also been recorded from *A. robusta* but are of little importance.

10.3 *Ailanthus* species (Simaroubaceae)

Tree profile

Two species of *Ailanthus* are important plantation species in the tropics. They are *Ailanthus excelsa* Roxb. and *Ailanthus triphysa* (Dennst.) Alston. Both are large, fast-growing species which produce white, light timber, highly valued for manufacture of plywood, and match splints and boxes. *A. excelsa* naturally occurs throughout the tropical and subtropical regions of the Indian peninsula, especially in drier areas, and *A. triphysa* occurs in South and South-East Asia, including peninsular India, Sri Lanka, Myanmar, Malaysia, the Philippines and parts of Indonesia, Thailand, Vietnam, Cambodia, Laos, southern China, and Queensland and northern New South Wales in Australia (CABI, 2005). Plantations have been raised largely in India but also in Senegal in Africa. In Kerala, India, *A. triphysa* is also grown on homesteads to cater to the needs of the match industry.

Overview of pests

Insect pests recorded on *A. excelsa* include the defoliators *Atteva fabriciella* (Lepidoptera: Yponomeutidae) and *Eligma narcissus* (Lepidoptera: Noctuidae), both described in detail below, and the borer *Batocera rufomaculata* (Coleoptera: Cerambycidae) which damages young trees. On *A. triphysa*, more than 35 insects have been recorded but the defoliators mentioned above are the only serious pests. Little information is available on the damage caused by the other insects which include thrips, beetles, bugs and caterpillars.

Pest profile

Atteva fabriciella Swed. (Lepidoptera: Yponomeutidae)

This insect is commonly known as *Ailanthus* webworm because of the larval habit of webbing the leaves together and feeding from within. The moth is small and slender, 14 mm in length and has a wingspan of 25–30 mm. The moth is dark orange, with white spots of variable size on the forewing. The full-grown larva is greyish green and about 20 mm long.

Life history and seasonal incidence The female moths lay eggs at night, mostly on the lower surface of young tender leaves and on buds, either singly or in small groups. The larvae feed gregariously in groups of six to ten, and stick the leaves together with silken web. The dirty white web, intermingled with faecal pellets and cut portions of leaves, may cover a large area of the shoot, making the damage conspicuous (Fig. 10.3). The larval period ranges from 13–28 days. Pupation takes place in loose cocoons, usually in the common web. The total length of life cycle may vary from 21–48 days, depending on temperature and food conditions.

In central India, the insect breeds continuously on *Ailanthus excelsa*, passing through 10 generations per year. The population increases following the onset of rains in June–July and declines thereafter, reaching its lowest in the summer months (Mathur *et al.*, 1970). One or two generations are passed in fruits of the tree. Varma (1991) studied the spatial and seasonal distribution of *A. fabriciella* in a five-year-old *Ailanthus triphysa* plantation in Kerala, in southern India. He monitored the larval population in about 540 sample trees from representative line transects across a 10 ha plantation, at monthly intervals over a three-year period. The insect was most abundant from November to February, coinciding with the general flushing period of the tree, but a smaller population was present throughout the year, thriving on small quantities of new leaves that were present (Fig. 10.4). The low population period coincided with the period of heavy rains. The spatial distribution of larvae in the 40 ha plantation was

Fig. 10.3 Damage caused by *Atteva fabriciella* to *Ailanthus triphysa* shoot. Courtesy: R.V. Varma, Kerala Forest Research Institute.

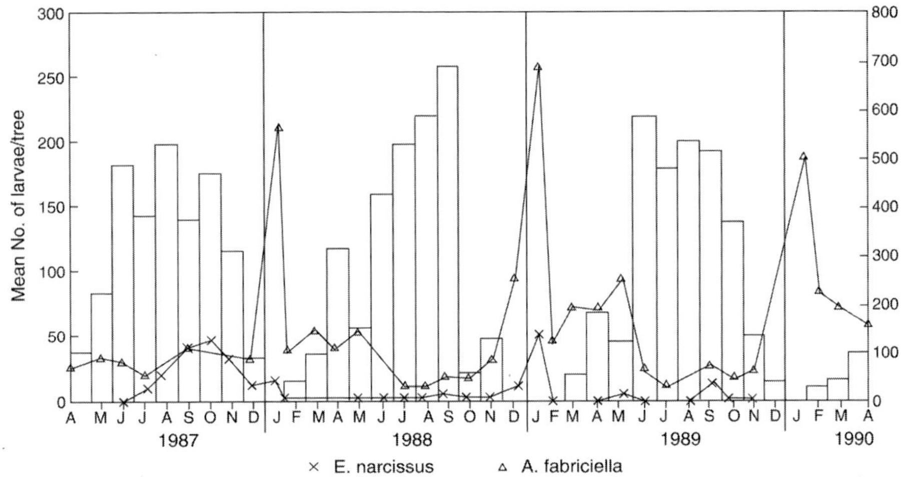

Fig. 10.4 Seasonal incidence of *Atteva fabriciella* and *Eligma narcissus* in an *Ailanthus triphysa* plantation in Kerala, India. The bars show monthly rainfall. Data from Varma (1991).

clumped, except when the population level was high when it followed a negative binomial distribution. *A. fabriciella* also attacks isolated *Ailanthus* trees grown in homesteads.

Host range and geographical distribution Other than *Ailanthus* species, *A. fabriciella* is known to attack the shrub *Boswellia serrata* (Burseraceae), *Santalum album* (Santalaceae) and *Quassia indica* (Simaroubaceae) (Beeson, 1941; Browne, 1968; Mohanadas and Varma, 1984).

Impact *Atteva fabriciella* causes three kinds of damage to the host tree – reduction in growth increment due to defoliation, forking due to damage of terminal buds and loss of seed production due to infestation of inflorescence and fruits.

Varma (1991) studied the impact of *A. fabriciella* on growth of *Ailanthus triphysa* by protecting trees from infestation by applying insecticide at monthly intervals and comparing with unprotected trees. The experiment was conducted over a period of two years in sample plots established in an infested six-year-old plantation. He found that the growth increment during the period was about 45% higher for height and 67% higher for basal area in protected over unprotected trees. When the larval population is high and the tender leaves are all eaten up, the larvae bore into and feed on the terminal bud and shoot. Damage of terminal bud causes forking of the tree. Repeated damage leads to multiple branches and bushing. Varma (1996) recorded terminal bud damage in

61% of 532 infested trees observed in a plantation. Mathur *et al.* (1970) observed that in central India, one- to two-year-old plants are sometimes killed outright by such damage, but this may be a rare event precipitated by other factors as well. Inflorescence and tender fruits are often damaged. Larvae feed on the inflorescence, wings of tender fruits and seeds. Varma (1996) found that 60% of seeds collected from six trees were bored by the insect.

Natural enemies Observations on natural enemies of *A. fabriciella* were made by Mathur *et al.* (1970) in central India and Varma (1986, 1991) in southern India. Two larval parasitoids, *Bessa remota* and *Carcelia* sp. (Diptera: Tachinidae), were recorded but the level of parasitization was low, possibly because of protection afforded by the larval web. A pupal parasitoid, *Brachymeria hime attevae* is more prevalent. It parasitized 50–80% of pupae during the winter months in central India. In southern India, it parasitized 20% of the pupae when the host population density was low and 5–10% of pupae when the host population density was high. Insect predators include three species of mantids, a carabid (*Parena nigrolineata*), a reduviid (*Panthous bimaculatus*) and unidentified ants and robberflies. An adult *P. bimaculatus* can consume two to three *A. fabriciella* larvae per day and the predator was present throughout the year in a study plot in Kerala, India. Several species of birds also feed on *A. fabriciella* larvae, particularly when the insect population is high.

Occurrence of a viral epizootic of *A. fabriciella* during monsoon months has been reported from Madhya Pradesh. A fungus, *Paecilomyces farinosus*, isolated from the *Ailanthus* pest *Eligma narcissus* in Kerala was pathogenic to *A. fabriciella* larvae in laboratory tests, but no infection was recorded under field conditions.

Control Mathur *et al.* (1970) tested and recommended some chemical insecticides for plantations up to five years old. They also suggested (without experimental proof of effectiveness) encouragement of a vegetational complex within plantations, consisting of plants which support parasitoids of *A. fabriciella* through alternative caterpillar hosts. However, no control measures are generally practised.

Knowledge gaps While the parasitoids and predators may be exerting some control against *A. fabriciella* population build-up under natural conditions, their diversity and effectiveness appear to be limited and it is not known whether any of them can be harnessed for practical pest management in plantations. Investigations on viral disease may hold promise.

Pest profile

Eligma narcissus Cramer (Lepidoptera: Noctuidae)

This colourful moth has a wingspan of 55–80 mm, with females being larger (Fig. 10.5a). The golden yellow hind wings have an incomplete black border and the brownish forewings have a black-dotted diffuse white band. The abdomen is golden yellow with black spots. The larvae are hairy and conspicuous, with yellow and black bands (Fig. 10.5b). They feed on the leaves of seedlings in nurseries and young plants up to about five years of age.

Life history and seasonal incidence The female moths lay eggs in groups on the under-surface of young leaves. The caterpillars feed on the leaves leaving

(a)

(b)

Fig. 10.5 *Eligma narcissus*. (a) Adult (wingspan 70 mm), (b) larvae feeding on *Ailanthus triphysa* foliage. Courtesy: R.V. Varma, Kerala Forest Research Institute.

only the mid-ribs. Larvae reach maturity in 22–23 days. Mature larvae congregate at the base of the saplings and pupate in cocoons attached to the stem of the host. The cocoons simulate the colour of the stem as bits of bark are attached to the surface of the cocoon. In heavy infestations, cocoons may cluster the stem for 60–80 cm height.

Varma (1986) studied seasonal incidence in a two-year-old plantation covering 40 ha in Peechi Forest Range in Kerala, India, over a two-year period. He found 1–27% of plants infested during the period September to April, with peak incidence in October to December in one year. The next year, infestation was noticed only in October and that in only 0.8% of the plants. No population was present at other periods. In a later study in a five-year-old plantation covering 10 ha in a different area, Varma (1991) recorded infestation peaks in September–October or December–January in different years (Fig. 10.4). In general observations in various plantations over a wider area across Kerala, India, he recorded *E. narcissus* infestations in most months except April–May (summer months) at different places. In general, *E. narcissus* did not have continuous generations in the same place, unlike *A. fabriciella*. Even within the same plantation, infestation was spatially discontinuous. Often the infestation appeared suddenly on a few plants. Based on these infestation characteristics, Varma (1991) suggested that infestation starts with moths arriving from outside the area.

Host range and geographical distribution *E. narcissus* feeds on several species of *Ailanthus*, viz., *A. excelsa*, *A. grandis* and *A. triphysa* in India, *A. glandulosa* in China, *A. kurzi* in Myanmar, *A. fauveliana* in Cambodia, *A. mollis* in eastern Malaysia and *A. moluccana* in Indonesia.

The species is widespread in South and Southeast Asia, with distinctive geographic races – *Eligma narcissus narcissus* in southern China and the smaller *E. narcissus indica* in India.

Impact *E. narcissus* larvae feed on young as well as mature leaves. At times of heavy infestation, 20–40 larvae may be seen feeding voraciously on each compound leaf (Varma, 1986) and the infestation is conspicuous because of the bright larval colouration and abundance. Seedlings in nurseries are often totally defoliated. While saplings up to five years old are infested, infestation of mature trees has not been observed. The reason for this escape of older trees is not known. Obviously, defoliation of *Ailanthus* by *E. narcissus* can result in loss of growth increment.

Natural enemies An unidentified sarcophagid fly *Sarcophaga* sp. was found to parasitize up to 30% of larvae/pupae in a study plot in Kerala, India

(Varma, 1986). Other parasitoids include the tachinid flies *Eutachina civiloides* and *Sturmia inconspicuella* (Mathur *et al.*, 1970). Predators include the ant *Crematogaster prelegens* feeding on eggs, and a mantid and the reduviid *Panthous bimaculatus* (which also feeds on *Atteva fabriciella*) feeding on larvae.

The fungal pathogens *Beauveria bassiana* and *Paecilomyces farinosus* have been reported from *E. narcissus* (Chatterjee and Sen-Sarma, 1968; Varma, 1986). In one plantation in Kerala, India, *P. farinosus* infestation was found in about 60% of 216 insects collected from 100 plants. In experiments where larvae were released on host leaves treated with *P. farinosus* spores, mortality ranged from 77% for late instar larvae to 90% for early instar larvae. A bacterial pathogen *Bacillus firmus* was isolated from a field population of *E. narcissus* and its pathogenicity confirmed in laboratory tests (Varma and Ali, 1986).

Control Chemical insecticides have been tested against *E. narcissus* under nursery and field conditions and recommended (Varma, 1986; Roonwal, 1990). However, in general, no insecticidal control is practised. Defoliated plants put forth new leaves in about two weeks.

Knowledge gaps Control potential of the fungal and bacterial pathogens reported from *E. narcissus* needs to be further investigated.

10.4 Bamboos (Poaceae)

Tree profile

Bamboos come under the angiosperm Class Monocotyledonae, Order Cyperales. Although they are grasses the larger, woody bamboos serve the many functions of trees and are raised in plantations. There are vast stretches of bamboo forest in South and Southeast Asia, where the majority of about 1250 world species are distributed. Bamboos occur gregariously in large patches or as under-storey in mixed forests. India alone has about 10 million ha of area under bamboos (Sharma, 1987). Bamboo culms can reach about 30 m in height and 18 cm in diameter at breast height. Bamboo has a variety of uses and is known as 'poor man's timber'. Apart from its use as a long-fibred raw material for paper pulp, bamboo culms are widely used as structural material for rural houses and a variety of other purposes in the rural setting such as scaffolding and banana support. Important plantation species are *Bambusa bambos*, *B. vulgaris* and *Dendrocalamus strictus*. *B. bambos* (syn. *B. arundinacea*) is a tall, thorny bamboo which thrives in deciduous and semi-deciduous forests and is common in India, Myanmar, Bangladesh and Sri Lanka. It flowers gregariously in 30–50 years and the clump dies after seeding. Flowering occurs in some parts of India almost every year. It is widely planted in India and used for paper pulp. *D. strictus*, also a thorny bamboo, is found in drier forests throughout India, Myanmar and Thailand and is also widely planted and used for paper pulp and board. Both are also widely used for light construction. A number of other species have similar uses and are harvested from natural stands and also planted on a small scale. *B. vulgaris*, known as 'yellow bamboo', also native to Asia, is planted very widely, pantropically, because its cuttings root readily, it is thornless and ornamental, and flowering is rare. Although most bamboos are naturally distributed in Asia, some 43 species are found in East Africa (of which 40 are distributed predominantly in Madagascar) (Kigomo, 1990) and a few also occur in South America. Many species of bamboos are confined to the subtropical and temperate regions, for example, the large genus *Phyllostachys*, with about 50 species, is concentrated in China and Japan. Many of them are cultivated on a large scale in eastern China and Japan and used as raw material for wood-based industries. Some bamboos are also cultivated for their edible tender shoots.

Overview of pests

Insect pests of bamboos have been reviewed and described by Haojie *et al.* (1998a). They gave a checklist of 70 defoliators, 117 sap suckers, 29 culm and shoot borers, 4 gall makers, 17 rhizome and root feeders and 2 seed feeders, totalling 239 species. However, a large number of them have been recorded only from the temperate regions of China and Japan. None of the pests can be rated as serious pests in the tropics, although Haojie *et al.* (1998a) have given 'major pest' status to six defoliators, eight sap suckers, nine borers, one gall maker and one seed pest. The more important tropical bamboo pests are listed in Table 10.3 and discussed briefly below.

The important pests can be grouped into three major categories – defoliators, sap suckers and shoot or culm borers. Among the defoliators, the most important is *Algedonia* (*Pyrausta*) *coclesalis*, known as the greater bamboo leaf roller. It causes occasional large-scale defoliation in bamboo plantations in northern India during the rainy season (Mathur, 1943; Singh and Bhandari, 1988; Singh, 1990). The larva rolls the leaf and feeds from within, moving on to a new leaf when the old leaf is half eaten. Normally this insect is kept under check by tachinid and ichneumonid parasitoids and carabid and mantid predators. *A. bambucivora* is a minor pest causing similar damage. Another important group of defoliators is the polyphagous grasshoppers, several species of which feed on bamboo foliage, sometimes in large numbers.

Sap feeders include mirid and coreid bugs, aphids, scale insects, whiteflies and thrips. Among them, outbreaks of the aphid *Oregma bambusae* sometimes

Table 10.3. *Important insects causing damage to living bamboos*

Category	Species name, order and family	Countries of occurrence	Remarks
Defoliators	*Algedonia (Pyrausta) coclesalis* (Lepidoptera: Pyralidae)	Bangladesh, Cambodia, China, India, Indonesia, Laos, Korea, Malaysia, Myanmar, Pakistan, Sri Lanka, Vietnam	Occasional, small-scale outbreaks
	A. bambucivora	Bangladesh, India, Pakistan	
	Grasshoppers (several species) (Orthoptera)	China, India, Pakistan	
	Pionea flavofimbriata (Lepidoptera: Pyralidae)	India, Sri Lanka	
	Pantana pluto (Lepidoptera: Lymantriidae)	China, India, Indonesia, Myanmar, Vietnam	
Sap suckers	*Oregma bambusae* (Hemiptera: Aphidae)	Bangladesh, China, India, Pakistan, Sri Lanka	Occasional, small-scale outbreaks
	Pseudoregma sp. (Hemiptera: Aphidae)	Thailand	
Shoot and culm borers	*Cyrtotrachelus dux* (Coleoptera: Curculionidae)	Bangladesh, India, Myanmar, Sri Lanka	
	C. longimanus	Bangladesh, India, Myanmar, Sri Lanka	
	Estigmene chinensis (Coleoptera: Chrysomelidae)	Bangladesh, India, Malaysia, Myanmar	
	Tetramesa gigantochloae (Hymenoptera: Eurytomidae)	Malaysia	Superficial culm wall damage
Seed feeder	*Udonga (=Ochrophara) montana* (Hemiptera: Pentatomidae)	Bangladesh, India, Myanmar	Occasional, large-scale outbreaks

occur in India. During outbreaks, the aphids cover the shoots in large numbers. Excessive loss of sap affects the vitality of the culm, often causing shrivelling and death (Chatterjee and Sebastian, 1964, 1966).

Among the shoot and culm borers, a group of weevils (Coleoptera: Curculionidae) of the genus *Cyrtotrachelus* (Fig. 10.6a,b) are the most damaging. Adults and larvae of *C. dux* and *C. longimanus* feed on the young, sprouting shoots (Haojie *et al.*, 1998a; Singh, 1990). The adult weevils, 20–40 mm long, become active at the onset of monsoon. They suck the sap of tender shoots and lay eggs on the culm. The larva bores into the culm, making a long tunnel, passing internally through several internodes and perforating each. The mature larva drops to the ground and pupates in a cocoon in the soil. The life cycle is annual. Feeding usually results in death of the culm or sometimes development of multiple shoots of little commercial value. A single larva can destroy a developing culm.

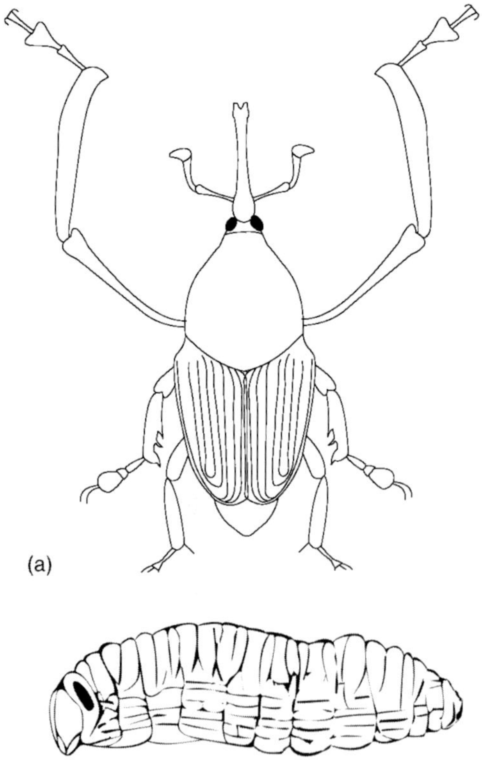

Fig. 10.6 The bamboo shoot weevil *Cyrtotrachelus* sp. (a) Adult (length 40 mm). After Beeson (1941); (b) larva. After Haojie *et al.* (1998a).

C. dichrous occurs in Thailand. It completes the egg and larval period in about 17 days but remains as a pupa in soil for 7–9 months, with the adult emerging after the first heavy rains (Attajerusit *et al.*, 2002). Another well-known borer is *Estigmene chinensis*. This chrysomelid beetle of the Subfamily Hispinae, commonly known as the hispine bamboo beetle, is 10–16 mm long, and becomes active by the onset of monsoon. The biology has been described by Beeson (1941). The female beetle lays batches of 2–4 eggs on the surface of the internode under the free part of the culm sheath and covers the eggs with chewed up fragments of leaf. The larvae feed gregariously between the outer sheath and the surface of the culm and later bore into the wall of the internode. Each internode may harbour 1–5 larvae. The larva excavates a small tunnel in the culm wall, which is enlarged into an irregular chamber in course of time. Pupation takes place inside the tunnel. The life cycle is annual and the beetle emerges with the next rainfall. The tunnelling of the larvae degrades the culm which is sometimes bent at the point of damage. Sometimes all the culms in a clump are attacked. A few species of the genus *Tetramesa*, a hymenopteran of the predominantly parasitic family Eurytomidae (superfamily Chalcidoidea), also bore into and feed on the bamboo culm. *T. gigantochloae* attacks a few species of bamboos including *Gigantochloa* spp. and *Dendrocalamus* sp. in Peninsular Malaysia (Narendran and Kovac, 1995); other species occur in temperate regions. The insect inserts eggs into the culm wall near the base of the internode during the bamboo sprouting season and the larvae make superficial galleries. Although the galleries are not deep, several of them may be found close together on the same internode, and this causes disfigurement.

Although not an economically serious pest of plantations, outbreaks of the pentatomid bug *Udonga montana* (Hemiptera: Pentatomidae) is spectacular and biologically interesting. The bugs feed on the developing seeds on the flowered culms as well as the seeds which have fallen on the ground. Eggs are laid in groups on flower heads. The newly hatched nymphs are black in colour and about 2 mm in length; the adults are ochreous yellow. Due to gregarious flowering of the bamboo which provides an abundant food supply, the bugs multiply rapidly in enormous numbers and create an outbreak situation. They assemble in large numbers on all kinds of trees, shrubs and ground vegetation, causing the branches of trees to droop and sometimes break off. Often the bugs are found in places far off from bamboo flowering sites, on other vegetation. For example, during an outbreak in 1991–92 at Wynad in Kerala, southern India, the insect was found in large numbers in about 50 ha of plantations covering teak, coffee and pepper (Mathew and Sudheendrakumar, 1992). The mass build-up occurred in May 1991 and large numbers of the bugs persisted until November 1991 at places where there was no bamboo flowering. Apparently the insects were able to survive for several months without any visible feeding although some feeding signs were seen on the tender foliage of some plants.

Aggregations of the pentatomid bug were found again in March 1992, after a build-up occurred in February 1992 on flowered bamboo some distance away. The insects completely destroyed the bamboo seed crop. Occasional outbreaks of *U. montana*, associated with bamboo flowering have also been reported from several other places in India – in 1917 at Mysore, Karnataka State; on various occasions in Madhya Pradesh, including one during 1982–83 at Chandrapur and adjoining areas etc. (Beeson, 1941; Singh, 1990). Outbreaks have also occurred in Myanmar. These outbreaks are spectacular because of the production of millions of bugs which aggregate on all kinds of vegetation and persist for several months. Apparently, the enormous population build-up during the mass flowering of bamboo declines slowly when the food supply is depleted. Some birds are also known to feed on the bugs.

In general, bamboo plantations in the tropics are practically free of pests, although as noted above some defoliators, sap suckers and culm and shoot borers are potentially capable of causing damage. It appears that normally the pests are kept in check by natural enemies.

10.5 ***Casuarina*** **species (Casuarinaceae)**

Two species of *Casuarina*, i.e. *C. equisetifolia* and *C. junghuhniana* are important plantation species in the tropics. They are relatively fast-growing, nitrogen-fixing tree species. The leaves are minute and scale-like; the jointed green branchlets (needles) perform the functions of leaves.

10.5.1 Casuarina equisetifolia

Tree profile

Casuarina equisetifolia L. occurs naturally on the tropical coastlines of Australia, on several islands in the Pacific, and in several countries in Southeast Asia, including Indonesia, Malaysia, Myanmar, the Philippines and Thailand. It has been introduced to many countries elsewhere in Asia, west and east Africa, central and south America and the Caribbean. Extensive plantations have been established in China, Cuba, India, Kenya, Puerto Rico, Thailand and Vietnam (CABI, 2005). The tree attains a height of over 30 m and is dioecious. It is salt-tolerant and grows well on sandy coastlines. It is planted widely for coastal erosion control and to act as a windbreak. The wood is hard and strong and is used for poles, scaffolding and fuel. Planted also as an ornamental, this nitrogen-fixing tree is socially well accepted in the tropics as a multipurpose tree.

Overview of pests

Over 70 species of insects have been recorded on *C. equisetifolia*, but serious pest problems have not occurred except in Taiwan, where it is an introduced species.

In the nurseries, seedlings are attacked by root-feeding termites, cutworms and leaf-feeding caterpillars, beetles, crickets and grasshoppers. In Malaysia, seedlings and young transplants are also attacked, and often killed, by a small scolytine beetle *Hypothenemus birmanus* which usually breeds on slash (Browne, 1968). The female beetle makes an irregular, longitudinal tunnel in the centre of the shoot and lays a cluster of eggs. The larvae live communally in the tunnel and the life cycle is completed in a few weeks. The more important pests of saplings and older trees are listed in Table 10.4. Several species feed on the leaves (needles). The bagworm *Cryptothelia crameri* is common in India and *Acanthopsyche reimeri* in Nigeria. In Taiwan, the polyphagous tussock moth, *Lymantria xylina*, is considered a serious pest (Chang, 1990). A large-scale outbreak of this insect was reported on trees planted as windbreak on the west coast of Peng-Hu island, and in a plantation at Taichung Harbour, and aerial spraying of insecticide was carried out to suppress this pest. Tsay *et al.* (2001) reported that the fungus *Beauveria bassiana* was effective against *L. xylina*. In India, the larva of a small moth *Eumenodora tetrachorda* (Cosmopterygidae) mine in the needles of *C. equisetifolia*, and the attacked needles turn white from the tip downwards and eventually fall off. Young plantations suffer heavy damage (Beeson, 1941). Other miners include *Labdia xylinaula* of the same family and *Metharmostis asaphaula*, a lithocolletid, which is more common in nurseries. The curculionid beetle, *Cratopus punctum* causes severe defoliation in young plantations in Mauritius. The cotton locust, *Chondracis rosea rosea* is a destructive pest in the Guangdong Province of China, where it is reported to damage 150–200 ha of trees annually (Su and Yu, 1979).

The white-spotted longhorn beetle *Anoplophora chinensis* is a serious stem borer of casuarina in Taiwan. It is estimated that between 1937 and 1946, over 70% of casuarina planted as wind-break, totalling about 4.5 million trees, were killed by this insect. The female beetle lays eggs singly under the bark of trees. The newly hatched larva first feeds on the cortex and later tunnels towards the xylem, in a circular path, causing wilting of trees. The life cycle is annual. Attack usually occurs near the ground, up to a height of about 1 m. Trees with about 10 cm diameter suffer most (Kan, 1958; Chang, 1990). The pest is also reported to attack casuarina in Hong Kong (Browne, 1968) and China (Fang *et al.*, 1997).

Table 10.4. *Important insects causing damage to* Casuarina equisetifolia

Category	Species name, order and family	Countries/Regions of occurrence	Remarks
Leaf (needle) feeders	*Cryptothelia (=Eumeta) crameri* (Lepidoptera: Psychidae)	India	
	Acanthopsyche reimeri (Lepidoptera: Psychidae)	Nigeria	
	Lymantria detersa (Lepidoptera: Lymantriidae)	India	
	L. xylina	Taiwan	Outbreaks in Taiwan
	Eumenodora tetrachorda (Lepidoptera: Cosmopterygidae)	India	Leaf miner
	Metharmostis asaphaula (Lepidoptera: Lithocolletidae)	India	Leaf miner
	Chondracis rosea rosea (Orthoptera: Acridae)	China	
Stem borers	*Anoplophora chinensis* (syn. *Melanauster chinensis)* (Coleoptera: Cerambycidae)	Taiwan, China	Serious pest in Taiwan
	Celosterna scabrator (Coleoptera: Cerambycidae)	India	On saplings
	Sahyadrassus malabaricus (Lepidoptera: Hepialidae)	India	On saplings
	Zeuzera coffeae (Lepidoptera: Cossidae)	India, Malaysia	On saplings
	Z. multistrigata	China	On saplings
	Stromatium fulvum (Coleoptera: Cerambycidae)	Egypt	
	Macrotoma palmata (Coleoptera: Cerambycidae)	Egypt	

The ambrosia beetles (Coleoptera: Curculionidae: Playpodinae) *Doliopygus chapuisi*, *D. serratus* and *Platypus hintzi* infest apparently healthy trees in the drier forests of Nigeria, making small holes in the wood and causing degrading of timber (Browne, 1968).

The other insects listed in the table cause occasional minor damage. In addition, there are other minor pests such as mealy bugs, cercopids, seed pests etc. which are not listed in the table. Although *C. equisetifolia* suffers serious damage from insects in subtropical Taiwan as noted above, in most countries insects cause little serious damage.

10.5.2 *Casuarina junghuhniana*

Tree profile

Casuarina junghuhniana Miq. (syn. *C. montana* Junghuhn ex Miq.) occurs naturally in Indonesia (East Java and some islands east of Java). It grows up to 35 m tall and 80 cm in diameter and is found mainly in the mountains (CABI, 2005). The species is dioecious. It often forms pure stands. It has been introduced to several countries in Asia and Africa and widely planted in Thailand where hybrids between *C. junghuhniana* and *C. equisetifolia* have also been raised successfully.

Overview of pests

Although occasional severe outbreaks of the lasiocampid caterpillar *Voracia casuariniphaga* in natural stands of the tree in mountain ridges and peaks in East Java, Indonesia are on record (Kalshoven, 1953), except for the common nursery pests, no serious pest problem has been reported in plantations.

10.6 ***Dalbergia*** **species (Fabaceae: Faboideae)**

Three species of *Dalbergia* are important in plantations – *D. sissoo*, *D. latifolia* and *D. cochinchinensis*, in the order of the extent of existing plantations. They occur naturally in South and Southeast Asia. They are nitrogen-fixing trees and produce valuable timber.

10.6.1 *Dalbergia cochinchinensis*

Tree profile

Dalbergia cochinchinensis Lanessan, known as Siamese rosewood, is native to Cambodia, India (in the State of Assam), Laos, Myanmar, Thailand and Vietnam and plantations have been raised in the latter two countries (CABI, 2005). The wood is used for heavy construction, furniture, domestic woodware and carvings.

Overview of pests

Eighteen species of insects have been found associated with *D. cochinchinensis* in Thailand (Hutacharern and Tubtim, 1995). These include leaf-feeding lepidopteran caterpillars and chysomelid beetles and wood-boring cerambycid beetles. Except for the wood-boring cerambycid *Aristobia horridula*, for which a pest profile is given below, none is considered a serious pest. One of the defoliators is *Plecoptera reflexa* (Lepidoptera: Noctuidae) which has assumed pest status on *D. sissoo* in India and is discussed in detail under that tree species.

Pest profile

Aristobia horridula hope (Coleoptera: Cerambycidae)

Aristobia horridula (Fig. 10.7a,b) is an emerging serious pest of *Dalbergia cochinchinensis* in Thailand. It also attacks *D. sissoo* in India. It bores into the trunk of living trees. The beetle, 27–32 mm long, is brownish, with bluish hairs on the elytra. A characteristic of the species is the presence of a dense tuft of hairs on the distal portion of the first and second antennal segments, those on the second segment being longer. The prothorax has a pair of lateral spines. The mature larva is 55–60 mm long and is creamy white.

Life history Adult beetles are active during daytime, feeding on the bark of young branches. They fly only short distances (10–20 m) at a time (Hutacharern and Panya, 1996). The female beetle makes a transverse groove on the bark of trees and lays eggs singly. The larva makes irregular, upward galleries in the sapwood initially and finally bores into the heartwood where

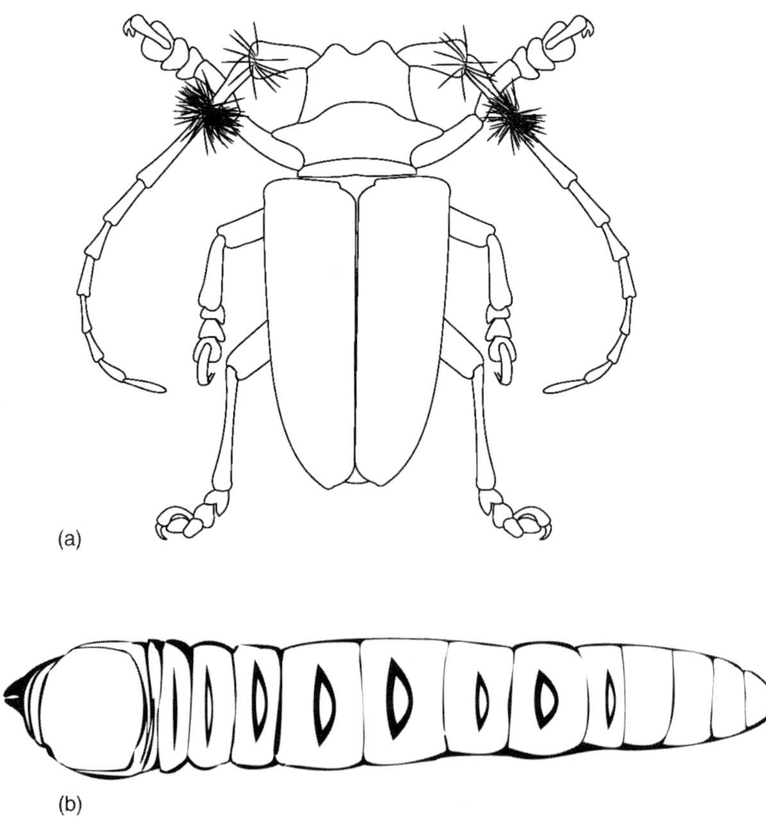

Fig. 10.7 *Aristobia horridula*. (a) Adult, (b) larva. After Mishra *et al.* (1985).

it pupates. The galleries, 50–75 cm long, are packed with frass and excreta. In *D. cochinchinensis* the feeding of young larva causes reddish resin exudation from the bark. In addition, feeding around the inner bark causes swelling of bark around the stem. In *D. sissoo* the entrance hole exhibits 'weeping symptom' (oozing of black fluid). In *D. sissoo*, the attack is restricted up to 4 m height, with maximum attack taking place at about breast height. There may be 10 to 15 entrance holes per plant but only one to seven larvae survive (Mishra *et al.*, 1985). The life cycle is annual, with most adults emerging from July to September in India and April to June in Thailand, although adult emergence was noted throughout the year in Thailand (Beeson, 1941; Hutacharern and Panya, 1996). The exit hole of the beetle is circular in outline.

Host range and geographical distribution *A. horridula* has been recorded on *Dalbergia cochinchinensis*, *D. paniculata*, *D. sissoo*, *D. volubilis*, *Pterocarpus indicus* and *P. macrocarpus* (Mishra *et al.*, 1985; Hutacharern and Panya, 1996). It is known to occur in India (in the States of West Bengal and Uttar Pradesh) and in Thailand.

Impact Borer damage degrades the timber. In Thailand, 33% of an eight-year-old *D. cochinchinensis* plantation was infested. *Pterocarpus indicus* is more susceptible, with 100% of a 10-year-old, roadside plantation being found infested. In *P. macrocarpus*, about 33% of trees in an eight-year-old plantation and 83% of trees in a 16-year-old plantation suffered damage (Hutacharern and Panya, 1996). In India, the borer was reported in epidemic form in all girth classes of a *D. sissoo* plantation in West Bengal, with the incidence ranging from 10% in one-year-old plantation to 80–90% in older plantations (Mishra *et al.*, 1985).

Control Painting the stem of trees with heavy oil or spraying chemical insecticides for 2 m above ground have been suggested for control, based on preliminary trials (Hutacharern and Panya, 1996), but more critical studies are needed.

Knowledge gaps *A. horridula* is an emerging pest of *D. sissoo* in India as well as *D. cochinchinensis* and *Pterocarpus indicus* in Thailand. The circumstances leading to pest outbreak are not fully understood.

10.6.2 *Dalbergia latifolia* (Common name: Indian rosewood)

Tree profile

Dalbergia latifolia Roxb., well-known as Indian rosewood, is a highly prized, decorative wood, used for furniture, cabinets, paneling, carvings, decorative veneers etc. The tree occurs naturally in Bangladesh, India, Indonesia, Myanmar and Nepal. Much of the commercial timber comes from natural forests;

plantations are recent. Plantations have been raised in India and Indonesia but their slow growth is a handicap.

Overview of pests

About 40 species of insects have been recorded on *D. latifolia*, mostly from India, but none is a serious pest. No serious pest has been reported from Indonesia also (Nair, 2000). Defoliators include *Plecoptera reflexa* (Lepidoptera: Noctuidae) which is a more serious pest of *D. sissoo* in India, a pest profile of which is given under *D. sissoo*.

10.6.3 Dalbergia sissoo

Tree profile

Dalbergia sissoo Roxb. ex DC., commonly known as 'shisham' in India, is a multipurpose, fairly fast-growing tree, occurring naturally in the sub-Himalayan tract in India and Pakistan. It is a gregarious pioneering species and grows well in river beds, canal banks and irrigated plantations. Plantations have been raised in India, Pakistan and Indonesia. The tree produces a good general-purpose timber used for high-class furniture, cabinet, carvings etc.

Overview of pests

About 125 species of insects have been recorded on *D. sissoo* (CABI, 2005). The important ones are listed in Table 10.5 and include defoliators, a sap sucking bug and stem borers. Beeson (1938) published a detailed account of insects attacking *D. sissoo*. The defoliator *Plecoptera reflexa* is the only pest recognized to

Table 10.5. *Important insects causing damage to* Dalbergia sissoo

Category	Species name, order and family	Countries of occurrence	Remarks
Defoliators	*Plecoptera reflexa* (Lepidoptera: Noctuidae)	India	Annual outbreaks
	Dichomeris eridantis (Lepidoptera: Gelechiidae)	India	
	Leucoptera sphenograpta (Lepidoptera: Lyonetiidae)	India	Leaf miner
	Apoderus (Strigapoderus) sissu (Coleoptera: Curculionidae)	India	
Sap sucker	*Aspidiotus orientalis* (Hemiptera: Coccidae)	India	Scale insect
Borers	*Aristobia horridula* (Coleoptera: Cerambycidae)	India	Occasional outbreaks
	Cladobrostis melitricha (Lepidoptera: Blastobasidae)	India	Twig borer

cause economic damage; a profile of this species is given below. Other notable pests include the defoliators *Dichomeris eridantis*, *Leucoptera sphenograpta* and *Apoderus sissu*. *D. eridantis* (Lepidoptera: Gelechiidae) is a minor defoliator in irrigated plantations of *D. sissoo* in northern India. In the absence of the principal defoliator, *P. reflexa*, it may cause partial defoliation during the period June – August. The greenish yellow to dark grey caterpillar, which grows up to about 25 mm in length, feeds on young and old leaves, within a shelter made by rolling a single leaf or fastening two leaves together. The life cycle usually takes 20–40 days and the insect passes through seven generations a year. The winter generation of larvae hibernate in their shelter for about five months (Beeson, 1941).

Leucoptera sphenograpta (Lepidoptera: Lyonetiidae) is a leaf miner destructive to young foliage of young trees, causing premature leaf shedding. The light green caterpillar, up to 5 mm in length, mines into the leaf in a tortuous line which expands into an irregular blotch. The spotted foliage later curls up and drops. There are nine generations in a year and the winter is passed as an inactive moth. Oviposition commences with the appearance of new foliage. *L. sphenograpta* is also a pest of poplar in India (Shujauddin and Khan, 1999).

The defoliator *Apoderus* (*Strigapoderus*) *sissu* is a golden yellow, black-spotted weevil, 5–6 mm long. It cuts the young leaf blade near the base from opposite edges up to the midrib and folds the leaf blade longitudinally along the midrib to make a compact cylindrical packet in which an egg is laid. The roll remains suspended from the petiole for a while or falls to the ground. The larva feeds on the inner rolls. There are up to eight generations per year in northern India. Sometimes this insect is abundant enough to destroy the entire flush of new leaves.

The sap-sucker *Aspidiotus orientalis* (Hemiptera: Coccidae) is a flat, circular to oblong, yellow or pale brown to dull reddish brown scale, 1.6 mm in diameter. It occasionally occurs in large numbers on *D. sissoo* in India and causes shedding of the foliage and drying of the branches. It is usually noticed first in localities unsuitable for the growth of the tree.

The stem borer *Aristobia horridula*, occasional outbreaks of which occur on *D. sissoo*, has already been described under *D. cochinchinensis*. The caterpillar, *Cladobrostis melitricha* (Lepidoptera: Blastobasidae) bores into living twigs of *D. sissoo*. The moth lays eggs on young shoots and branches of about pencil thickness. The whitish larva excavates a tunnel in the shoot, leaving only a thin outer layer. Before pupation, the larva severs the shoot except for the epidermis. The branches later break off and the attack results in heavy pruning of the crowns. The insect is only abundant when the trees are growing in open stands and are producing many weak branches from the lower bole (Beeson, 1941).

Pest profile

Plecoptera reflexa Guen. (Lepidoptera: Noctuidae)

Plecoptera reflexa Guen. (Fig. 10.8) is a well-known and serious defoliator of *D. sissoo* in India and Pakistan. The moth is greyish brown and has a wingspan of 30–35 mm. The caterpillar is a green semi-looper, turning pinkish and measuring about 25 mm long when full-grown. Outbreaks are common and regular in irrigated plantations of *D. sissoo*. The species has also been recorded as a pest of *D. latifolia*.

Life history and seasonal incidence The moths, which hide during the day in low vegetation, swarm in the early evening and lay eggs at night on young leaves, usually only one egg per leaf. Each female may lay up to 400 eggs over a period of six days. The young caterpillar feeds on the lower surface of the leaf, but later instars consume the whole leaf including petiole and the green shoot.

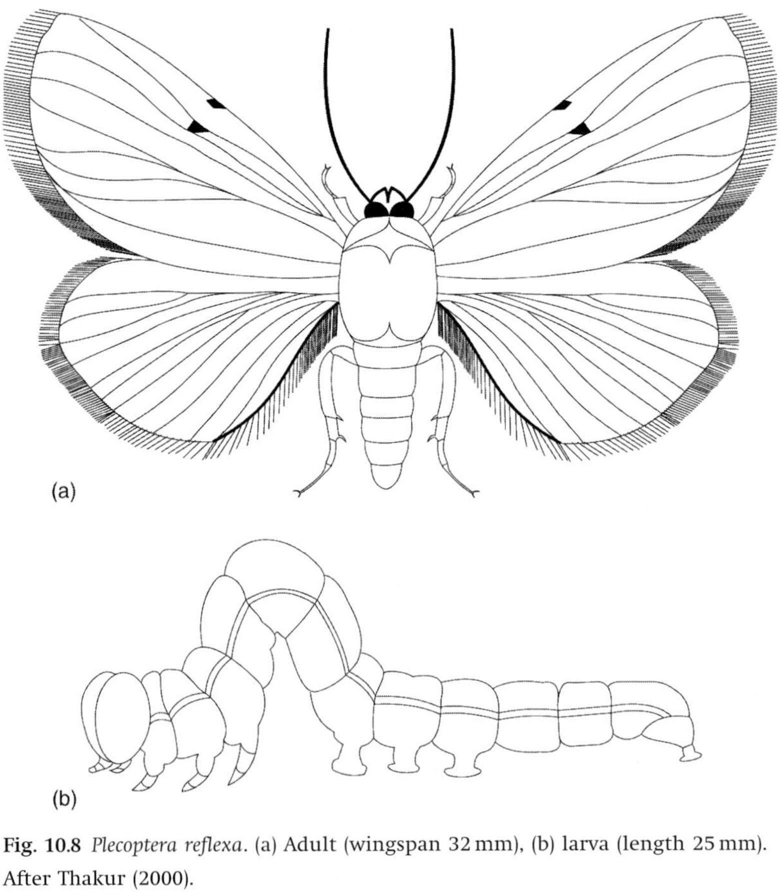

Fig. 10.8 *Plecoptera reflexa*. (a) Adult (wingspan 32 mm), (b) larva (length 25 mm). After Thakur (2000).

Young foliage is essential for the survival of the early instars. The larval period ranges from 10–20 days, depending on the temperature and the maturity of the leaf (longer on older leaves). There are five larval instars. Pupation occurs on ground, in a pupal cell of silk and particles of soil, leaf etc. Ten to 13 generations occur per year. In winter, the insect hibernates in the pupal stage for 80–100 days, usually starting in October. The first generation of moths appears in spring, but noticeable defoliation occurs only about a month later. Later generations of the insect do not cause visible defoliation and the population is believed to be kept in check by natural control factors.

Host range and Geographical distribution *P. reflexa* attacks *D. sissoo* as well as *D. latifolia*, but outbreaks are noticed only in irrigated plantations of the former in India and Pakistan. Plantations of *D. latifolia*, however, are not extensive. The pest has not been reported from Indonesia although there are over 25 000 ha of *Dalbergia* plantations (both the species together) in Java (Nair, 2000).

Impact Defoliation of *D. sissoo* by *P. reflexa* was first reported in 1899 in Changa Manga (Pakistan) and annual defoliation is common in Pakistan and northern India. Plantations about three years old and above are attacked. In repeated severe attacks, the tree remains leafless for the greater part of the growing season and epicormic branches are produced. The economic impact of this defoliation is considered serious and poor quality plantations have often been abandoned or converted (Beeson, 1941). Defoliation is severe in roadside and canal-bank plantations but not in gregarious natural stands.

Natural enemies *P. reflexa* is attacked by several parasitoids (Mathur, 1942). The eulophid *Euplectrus parvulus* is the most effective as it attacks early larval instars. The braconids *Disophrys sissoo* and *Microgaster plecopterae* also attack larvae. The tachinids *Exorista civiloides*, *E. picta* and *Podomyia setosa* attack both larvae and pupae and the chalcidid *Brachymeria nursei* attacks pupae. The recorded levels of parasistism range from 2–24% (Mathur, 1942). As usual, the level of parasistism increases towards the later part of the growth season when the pest population density is low. About a dozen species of predators, including carabids, mantids, reduviids and ants have also been recorded.

Control It is generally observed that *P. reflexa* is a serious pest only in irrigated plantations in arid regions. As the newly hatched larvae are dependent on tender foliage for establishment, trees which flush early in the season do not generally get attacked. Restricting the planting of trees to areas with an adequate moisture regime is generally recommended to ensure early flushing, so that by the time the first brood of *P. reflexa* appears, the proportion of tender leaves in the total foliage is relatively small (Beeson, 1941). Commercial preparations of the bacterial pathogen *Bacillus thuringiensis* has been shown to be effective against this insect (Kalia and Lall, 2000).

Knowledge gaps It is believed that in places with adequate subsoil water level, the trees come into flush and possess mature foliage early in the season, before the overwintering moths start laying eggs and thus escape attack as the insect lays eggs only on tender leaves. Critical field observations are necessary to establish the usefulness of this method of control. The possibility of the insects adapting to the early flushing trees cannot be ruled out.

10.7 *Eucalyptus* species (Myrtaceae)

Tree profile

Eucalyptus is a widely planted tree genus in the tropics. Most species of this genus of more than 600 species are endemic to Australia, where they occupy various ecological niches from sea level to alpine zones and from moist to dry areas (Wylie and Floyd, 2002). Three species, *E. deglupta*, *E. urophylla* and *E. pellita* occur outside Australia. *E. deglupta* is native to Papua New Guinea, eastern Indonesia and the Philippines (Mindanao Island); *E. urophylla* is native to Timor and the adjacent islands and *E. pellita* is native to Indonesia (Irian Jaya), Papua New Guinea and Australia (CABI, 2005). The adaptability of eucalypts to a variety of site conditions, their fast growth, coppicing power and suitability for paper and rayon grade pulp have made them well suited for short rotation plantations. They are usually grown on a six to eight-year rotation, with two coppicing cycles. Several species are planted in the tropics; the most common are *E. camaldulensis*, *E. globulus*, *E. grandis*, *E. robusta*, *E. saligna* and *E. tereticornis*, all of Australian origin, and *E. deglupta* and *E. urophylla* of non-Australian origin.

Although many species of eucalypts were introduced into the tropics earlier, for various purposes, large-scale planting for pulpwood production began in the 1960s in most countries. An estimate in the year 2000 put the global area under eucalypt plantations at 10 million ha (Brown and Ball, 2000), although some of it is in the subtropical zone. For example, eucalypt plantations occupy 3.1 million ha in India, 2.7 million ha in Brazil and 670,000 ha in the tropical south of China (Brown and Ball, 2000; Wylie, 1992). Exotic eucalypt plantations exist in over 80 countries and at least 20 species are grown on a commercial scale. Tropical Australia had only a few thousand ha under eucalypt plantations until recently, but the plantation area is increasing rapidly (Wylie and Floyd, 2002).

Overview of pests

Since eucalypts are the dominant tree species in Australia, there is a large variety of insects associated with them in that country. In contrast, the eucalypt insect fauna is less diverse in the exotic plantations. Consequently there is a large difference in the pests of eucalypts in Australia and other countries. It is therefore useful to discuss the pests of eucalypts separately for Australia and the rest of the tropics. As very little host species specificity is exhibited by the insects associated with the various commonly cultivated eucalypt species, pests are not discussed separately for the different species.

Altogether about 920 species of insects have been recorded from eucalypts worldwide (Wylie and Floyd, 2002), but most of them are from the temperate zone species in Australia. Although some serious pests have been recognized in the temperate zone plantations in Australia, they are not discussed here. Only those in the tropical part of Australia, mainly Queensland are considered. Strict separation into tropical Australian pests is difficult because even in Queensland some plantations are located in temperate areas. In a study of eucalypt plantations in Queensland, Wylie and Peters (1993) recorded over 100 species of insects, of which most were leaf or sap feeders that caused only minor damage. The more important pests are listed in Table 10.6. The most damaging are the cossid wood borer *Endoxyla cinerea* which tunnels into the centre of small diameter stems and the longicorn borer *Phoracantha solida* which tunnels in the sapwood and heartwood (Wylie and Peters, 1993; Lawson *et al.*, 2002). In some observation plots, the former attacked 13–16% of trees and the latter 8–47%. Among the leaf feeding insects, the scarabaeid beetle *Epholcis bilobiceps* often causes almost complete defoliation of many species such as *E. camaldulensis*, *E. grandis* and *E. robusta*, necessitating insecticidal control. Other scarabaeid beetles, chrysomelid beetles, moth caterpillars and a sawfly cause severe damage occasionally (Table 10.6). The termites, chiefly *Coptotermes* spp., infest the trunk of older plantation trees and hollow them out.

The important pests in exotic plantations of eucalypts are listed in Table 10.7. An ubiquitous problem in the tropics is the mortality of young transplants caused by subterranean termites. Because of the importance of this problem, a pest profile of termites on eucalypts is given separately below. In view of the large area under eucalypt plantations in India, China and Brazil and the detailed information available on pests, the situation in these countries is examined first.

Information on the pests of eucalypts in India has been summarized and reviewed by Mathur and Singh (1960), Sen-Sarma and Thakur (1983) and Nair *et al.* (1986b). In addition to the root feeding termites and whitegrubs, the pests recorded include some leaf feeders, sap suckers and borers, but none of them

Table 10.6. *Important insects causing damage to eucalypt plantations in Queensland, Australia*

Category	Species name, order and family	Remarks
Defoliators	*Cryptocephalus iridipennis* (Coleoptera: Chrysomelidae)	
	Monolepta australis (Coleoptera: Chrysomelidae)	
	Paropsis spp. (Coleoptera: Chrysomelidae)	
	Chrysophtharta cloelia (Coleoptera: Chrysomelidae)	
	Anoplognathus boisduvali, *A. pallidicollis* and *A. porosus* (Coleoptera: Scarabaeidae)	Occasionally severe
	Etholcis bilobiceps (Coleoptera: Scarabaeidae)	Occasional outbreaks
	Liparetrus discipennis (Coleoptera: Scarabaeidae)	
	Doratifera casta (Lepidoptera: Limacodidae)	
	Agrotera amathealis (Lepidoptera: Pyralidae)	
	Strepsicrates semicanella (Lepidoptera: Tortricidae)	
	Perga kirbyi (Hymenoptera: Pergidae)	Occasionally severe damage to young trees
Sap suckers	*Cardiaspina fiscella* and *C. maniformis* (Hemiptera: Psyllidae)	Occasional outbreaks
	Eriococcus coriaceous (Hemiptera: Coccidae)	
Stem borers/ Termites	*Endoxyla cinerea* (syn. *Xyleutes cinereus*) (Lepidoptera: Cossidae)	Giant wood moth
	Phoracantha (= *Tryphocaria*) *acanthocera* and *P. solida* (Coleoptera: Cerambycidae)	Bullseye borer
	Coptotermes spp. (Isoptera: Rhinotermitidae)	Hollows out the trunk of older trees

Data from Wylie (1992), Wylie and Peters (1993) and Wylie and Floyd (2002)

causes serious damage. About 50 species of leaf feeding insects are on record, of which over 40 are lepidopteran caterpillars belonging to the families Geometridae, Lasiocampidae, Lymantriidae, Noctuidae etc. (Sen-Sarma and Thakur, 1983); others include beetles, crickets, grasshoppers and stick insects. No outbreaks of these insects have occurred. Important pests in nurseries are cutworms and gryllids which cut off young seedlings above ground; tortricids which feed from within webbed leaves and curculionid beetles (*Myllocerus* spp.) which sporadically cause extensive defoliation. Sap sucking insects include aphids, psyllids (*Trioza* spp.) and a myrid bug. The myrid bug *Helopeltis* sp. sucks the sap from the tender terminal shoots of saplings, causing drying of shoots; about 2.5% of the saplings suffered damage in a seven-month-old,

Table 10.7. *Important insects causing damage to exotic eucalypt plantations*

Category	Species name, order and family	Countries/Regions of occurrence	Remarks
Root feeders	Termites (many species) (Isoptera)	Most countries	Kill young transplants
Defoliators	*Thyrinteina arnobia* (Lepidoptera: Geometridae)	Brazil	
	Glena unipennaria (Lepidoptera: Geometridae)	Brazil	
	Buzura sp. (Lepidoptera: Geometridae)	China, southern Africa	
	Neocleosa sp. (Lepidoptera: Geometridae)	southern Africa	
	Nystalea nyseus (Lepidoptera: Notodontidae)	Brazil	
	Sarsina violascens (Lepidoptera: Lymantriidae)	Brazil	
	Eupseudosoma aberrans and *E. involuta* (Lepidoptera: Arctiidae)	Brazil	
	Strepsicrates sp. (various species) (Lepidoptera: Tortricidae)	China, India, Malaysia, Philippines, Ghana, Nigeria	
	Carea angulata (Lepidoptera: Noctuidae)	China	
	Chalia larminati (Lepidoptera: Psychidae)	China	
	Narosa viridana (Lepidoptera: Limacodidae)	Zambia	
	Atta spp. (Hymenoptera: Attidae)	Central and South America	Leaf-cutting ants
	Acromyrmex spp. (Hymenoptera: Attidae)	Central and South America	Leaf-cutting ants
	Gonipterus spp. (Coleoptera: Curculionidae)	Brazil, Kenya	Australian origin
Sap suckers	*Helopeltis* spp. (Hemiptera: Myridae)	India, Indonesia, Congo	
	Amblypelta cocophaga (Hemiptera: Coreidae)	Solomon Islands	
	Icerya purchasi (Hemiptera: Margarodidae)	India, Malawi	Australian origin
	Ctenarytaina eucalypti (Hemiptera: Psyllidae)	Brazil, South Africa	Australian origin

Table 10.7. (cont.)

Category	Species name, order and family	Countries/Regions of occurrence	Remarks
Borers	*Agrilus sexsignatus* (Coleoptera: Buprestidae)	Philippines	
	Celosterna scabrator (Coleoptera: Cerambycidae)	India	On young trees of 5–23 cm girth
	Aristobia approximator (Coleoptera: Cerambycidae)	Vietnam	
	Phoracantha semipunctata (Coleoptera: Cerambycidae)	Brazil, Malawi, Zambia, South Africa	Australian origin
	Batocera horsefieldi (Coleoptera: Cerambycidae)	China	
	Timocratica palpalis (Lepidoptera: Stenomatidae)	Brazil	

95 ha plantation of *E. grandis* in Kerala State (Nair *et al.*, 1986b). The cerambycid borer *Celosterna scabrator* (see pest profile under *Acacia nilotica*) attacks young trees, the adult feeding on the bark of the main stem and branches and the larva tunnelling into the root. In a plantation in Karnataka State, Ralph (1985) recorded damage to 20% of two to three-year-old plants. The lepidopteran borer *Sahyadrassus malabaricus* (see pest profile under *Tectona grandis*) bores into the pith of saplings and feeds on the bark. One insect of Australian origin (where it breeds mainly on species of *Acacia*), the cottony cushion scale *Icerya purchasi* (Hemiptera: Margarodidae) has been recorded on *E. globulus* in southern India. The species was first noticed in India during 1927–28, mostly on fruit trees, and outbreaks have occurred in *Acacia mearnsii* plantations in the cooler hilly tracts. Its association with *Eucalyptus* appears to be incidental.

Information on pests of eucalypts in China was summarised by Wylie (1992). Eucalypts were first planted in China more than 100 years ago and most plantations are located in the tropical southern provinces of Guangdong, Guanxi and Hainan. Over 160 species of associated insects have been found, but most of them cause only minor damage. As in India, root-feeding termites and whitegrubs cause serious damage to newly out-planted saplings. Crickets and grasshoppers damage young plants. Important leaf feeding insects include four lepidopteran caterpillars (Table 10.7). The tortricid *Strepsicrates* sp. nr. *semicanella* attacks new leaves of very young trees, sometimes causing multiple branching. Outbreaks of the geometrid, *Buzura suppressaria* have occurred annually since 1987 in a forest farm at Wei Du, and in the worst outbreak in 1990, about

1000 ha of *E. saligna*, *E. grandis*, *E. citriodora* and *E. urophylla* were severely defoliated. During the outbreak, each tree harboured 800–1000 larvae. Although 36 species of sap sucking insects were recorded, none caused serious damage. In the Yunnan Province, the larvae and adults of a stem borer *Batocera horsefieldi* (Coleoptera: Cerambycidae) tunnel through the stem causing death of the plants (Yonghi and Kuiguang, 1996).

In Brazil, fairly large-scale planting of eucalypts was made in the 1920s and 1930s, but the programme has received a boost since the 1960s. The gigantic Jari plantation and the later Aracruz plantations which captured phenomenal increase in yield through genetic selection and clonal propagation are well known. Over 177 species of insects have been recorded on eucalypts in Brazil, including some introduced species. However, most of them are casual feeders. As in other countries, root-feeding subterranean termites attack saplings. In addition, in some areas, the termite *Coptotermes testaceus* attacks the heartwood of older trees (Nogueira and De-Souza, 1987). Numerically, the most dominant pests are the leaf-feeding lepidopteran caterpillars, of which up to a dozen species have been recognized as serious pests by different authors. Aerial application of insecticides has been made to control some of them. The most important pests are included in Table 10.7. Outbreaks of *Thyrinteina arnobia*, *Glena* sp., *Sarsina violascens* and *Eupseudosoma aberrans* have occurred in plantations of *Eucalyptus camaldulensis*, *E. grandis*, *E. saligna* and *E. tereticornis* in Mato Grosso (Filho *et al.*, 1992). *Eupseudosoma involuta* infested a 2000 ha plantation of eucalypts at Mogi-Guaco, Sao Paulo in 1970 and caused premature leaf fall and withering of the branches, necessitating aerial application of insecticide (Balut and Amante, 1971). Infestation by *Nystalea nyseus* has also required aerial application of insecticide in the Aracruz Celulose eucalypt plantations in Espirito Santo State and Bahia State (Laranjeiro, 1994).

Next in importance are the leaf-cutting ants *Atta* and *Acromyrmex* species, characteristic of Central and South America. They cut the leaves into small pieces and transport large quantities to their underground nests for cultivation of fungi. Generally, the population of ant nests is high and control measures such as pre-planting destruction of ant colonies and use of insecticidal bait bags after planting are usually practised. Some pests of Australian origin have also reached Brazil. These include the leaf-feeding curculionids *Gonipterus scutellatus*, *G. gibberus* and *G. platensis*; the sap-sucking psyllid *Ctenarytaina eucalypti* (Burckhardt *et al.*, 1999); and the stem-boring beetle *Phoracantha semipunctata*. *Phoracantha* spp. which attack mainly water-stressed trees are more troublesome in exotic locations than in Australia, in the absence of their natural enemies.

The situation in other countries is examined below. As in India, China and Brazil, several species of indigenous insects have adapted to feeding on eucalypts wherever they have been planted. The most common are the root-feeding subterranean termites (see pest profile below). In most parts of Latin America, as in Brazil, leaf-cutting ants cause substantial damage to eucalypts if control measures are not taken. Although a few species of leaf-feeding lepidopteran caterpillars have adapted to feeding on eucalypts, in many countries other than Brazil and China outbreaks have seldom occurred. The most common are leaf rollers of saplings – *Strepsicrates* sp. (Tortricidae) in the Philippines, Malaysia and Ghana, and an unidentified pyralid in Indonesia. Among sap-sucking insects, as in India, *Helopeltis* spp. (Hemiptera: Myridae) cause dieback of shoots of saplings in Indonesia and the Congo (Hardi and Intari, 1990; Diabangouaya and Gillon, 2001). In Indonesia, infestation has been reported in up to 57% of plants. In the Solomon Islands a coreid bug, *Amblypelta cocophaga*, was reported to cause severe dieback of *E. deglupta* saplings (Bigger, 1982). Among the wood borers, a buprestid beetle, *Agrilus sexsignatus*, known as 'varicose borer', caused 28–63% mortality of *E. deglupta* saplings of a Papua New Guinea provenance planted in the Philippines, while an indigenous provenance was resistant (Braza, 1987). Also, infestation by an unidentified borer (*Agrilus* sp.) killed 1000 ha of two to three-year-old *E. deglupta* plantation in East Kalimantan, Indonesia (Soepangkat, 1998). The cerambycid borers, *Oxymagis horni* and *Hastertia bougainvillea* cause damage in Papua New Guinea and *Aristobia approximator* in Vietnam. In Southeast Asia, minor damage is caused to saplings by hepialid and cossid borers. The cerambycid borers *Phoracantha* spp. and the leaf feeding curculionid beetles *Gonipterus* spp. of Australian origin have also reached Africa, as noted above for Brazil. *Phoracantha semipunctata* is already very widespread, having been reported from about 20 countries between the latitudinal range of 56°S to 47°N. Heavy infestation by *P. phoracantha* usually leads to death of trees. *Gonipterus scutellatus*, known as the *Eucalyptus* snout beetle, has emerged as an important defoliator of eucalypts, wherever it has spread, although it is more prevalent in temperate regions.

Based on the above observations, the following conclusions and generalizations can be drawn.

1. In plantations of the native eucalypts in Australia, leaf-feeding beetles, sap-feeding psyllids and wood-feeding cerambycid beetles and termites cause occasional serious damage. The area under eucalypt plantations is still small in the tropical regions of Australia, and with the expansion of plantations the pest problems are likely to increase, as in the temperate part of Australia (Nair, 2001a).
2. In all tropical countries several indigenous species of insects have adapted to feeding on the exotic eucalypt plantations but at present

serious pest outbreaks are exceptional. Such outbreaks of leaf-feeding insects have occurred in Brazil and China.

3. The most common pest problem of exotic eucalypts is mortality of saplings caused by root-feeding termites. This is not common in Australia where termite attack of older trees is a more serious problem.
4. Leaf-cutting ants are important defoliators of eucalypts in Central and South America.
5. Indigenous sap-sucking and wood-boring insects have not posed a major threat to exotic eucalypts.
6. Some eucalypt pests from Australia have found their way to exotic eucalypt plantations in Brazil and Africa (Table 10.7) as elsewhere in Europe, USA and the Mediterranean region.

Pest profile

Termites (Isoptera)

Although most species of termites characteristically feed on dead wood or other organic matter, some species feed on living eucalypts. They either hollow out the trunk of mature trees by feeding and nesting or kill young plants by feeding on the root. Termites are social insects that live in colonies, in nests that may be on trees, on dry wood or on soil with or without conspicuous mounds. Individuals belong to different castes like the queen, reproductive males and females, and the most numerous and commonly encountered workers and soldiers. Damage is caused by the food gathering worker caste. Three ecological groups of termites can be recognized: (1) drywood termites, comprising the family Kalotermitidae, which live generally on dry wood, without contact with soil; (2) dampwood termites, comprising the family Termopsidae, which are primitive termites that live in damp wood, including living trees and (3) subterranean termites comprising the five remaining families, Mastotermitidae, Hodotermitidae, Rhinotermitidae, Serritermitidae and Termitidae. The most dominant families are Rhinotermitidae and Termitidae, the last one comprising over three-quarters of all termite species. Species that cause damage to living trees of eucalypts belong to the families Mastotermitidae, Termopsidae, Rhinotermitidae and Termitidae, although *Neotermes* spp. of the family Kalotermitidae cause damage to living trees other than eucalypts (see under mahogany and teak).

Damage to mature trees Termite damage to mature trees of eucalypts, both in natural forests and plantations, is common in Australia. Termites of the families Mastotermitidae, Termopsidae and Rhinotermitidae are involved. *Coptotermes* spp. (Rhinotermitidae) are the most prevalent. *C. acinaciformis* is a very damaging species in Queensland (Elliott *et al.*, 1998). Greaves and Florence (1966) reported that 33.5% of 26–29-year-old *E. pilularis* trees examined in a plantation were infested by this species. Individuals from a *C. acinaciformis* colony within one living tree can attack other living trees in the vicinity up to a distance of about 40 m (Greaves, 1959). *C. frenchi* also infests the trunk of living eucalypt trees in Queensland. *C. testaceus* has been reported to attack living eucalypt trees in 10-year-old plantations in Brazil (Nogueira and De-Souza, 1987). In Malaysia, *C. curvignathus* destroys the heartwood of eucalypts by tunnelling upwards from the taproot (Chey, 2002). In India, *C. heimi* which infests the trunk of several other tree species has often been found infesting the root of *Eucalyptus* but there are no reports of serious damage to mature eucalypt trees. *Mastotermes darwiniensis*, the only living representative of the family Mastotermitidae, confined to Australia, is also very destructive to eucalypts and attacks living trees by tunnelling in the roots, trunk and branches (Elliott *et al.*, 1998). *Porotermes adamsoni* (Termopsidae), which is a dampwood termite, also causes serious damage to eucalypt trees in Australia, but is more prevalent in the alpine forests in New South Wales, Tasmania and Victoria (Greaves, 1959).

In addition to the above, some species of the family Termitidae, which build arboreal carton nests, enter the heartwood of trees through snags and sometimes cause death of the infested trees. Unlike in Australia, Brazil and Malaysia, termite infestation leading to mortality of older eucalypt trees is rare in most other countries, although many species feed on the dead bark of trees under the cover of conspicuous mud plaster that extend up all around the trunk. As an exception, Roonwal (1979) recorded *Postelectrotermes militaris* (Kalotermitidae) in the heartwood of living trees of *Eucalyptus* sp. in Sri Lanka.

Damage to seedlings and saplings The most prevalent pest problem of exotic eucalypt plantations in the tropics is damage caused by termites to young plants. Customarily, eucalypt seedlings are first raised in nursery beds. Three to four-month-old seedlings are then pricked out and transplanted into polythene bags, usually 12 cm by 18 cm, filled with loamy soil. These seedlings are nurtured in container beds, usually 12 m by 1.2 m, and after one to two months planted out in field pits 30 cm × 30 cm × 30 cm. At the time of planting, the polythene container is slit open and discarded, the undisturbed soil column is placed in the centre of the pit and the sides filled with soil and lightly packed. Seedlings in the primary bed and container plants are occasionally attacked by termites, but out-planted saplings suffer the most damage.

In recent years, clonal plantations of eucalypts have been raised from cuttings of selected high-yielding trees, to enhance productivity. Two-leaved, single-node shoot cuttings are rooted in mist chambers and then transferred to root-trainers (trays of conical, hard plastic containers, usually 150 ml in volume) containing various media such as soil, compost, vermiculite or their combinations. At about four months old, rooted cuttings are removed from the root-trainers and planted out in the field.

Nature of damage, seasonal incidence and impact Typically, the termites feed on the taproot of the saplings a few centimetres below the soil surface and taper it out like a sharpened pencil, severing the main root system from the stem (Figs. 10.9 and 10.10). An early symptom of attack is the flaccid, drooping appearance of the tender terminal leaf pairs while the lower leaves remain green and apparently healthy. At this stage, the plant can be pulled off the ground with very little force, because the taproot has been almost severed. In older saplings, the taproot is extensively ring-barked and the smaller roots are tapered off. Because the feeding occurs underground, usually up to 20 cm below the soil surface, early stages of attack are not recognizable and by the time the symptoms appear the plant is damaged beyond recovery.

Fig. 10.9 Vertical section through soil, showing termite tunnels leading to the taproot of a *Eucalyptus tereticornis* sapling, about eight months after field-planting. Reprinted from *Journal of Forest Ecology and Management* (Nair and Varma, 1985), with permission from Elsevier.

Fig. 10.10 Characteristic dumb-bell shaped region in the below-ground portion of a *Eucalyptus tereticornis* sapling, formed by termite feeding, prior to severance of the root system. Reprinted from *Journal of Forest Ecology and Management* (Nair and Varma, 1985), with permission from Elsevier.

In the out-planted clonal plants which have no taproot, the termites attack the stem portion of the plant in the soil, causing ring-barking. Termites also feed on the small adventitious roots (Varma, 2001).

There has been some debate in the literature as to whether termite attack on eucalypt saplings is primary or secondary. Some authors have claimed that termites will attack only those plants that are weakened by other causes such as drought or fungal infection, and others have suggested that increased plant vigour will lessen the liability of termite attack. The idea that termite attack is secondary has stemmed from the fact that termites do attack saplings that have been killed or greatly weakened by other causes and it is often difficult to distinguish between the primary and secondary causes. Based on detailed field observations, Nair and Varma (1981, 1985) recognized three kinds of situation.

Primary termite attack This is common in eucalypt plantations up to one year old in India and many other countries, and also in somewhat older plantations in Africa. In smaller saplings, the taproot is attacked a few centimetres below ground surface and tapered off, whereas in older saplings, extensive feeding occurs from the root collar downwards. In Africa, even saplings up to 15 cm diameter at base may be attacked and killed by some species of Macrotermitinae. Nair and Varma (1985) recorded unequivocal evidence of vigorously growing saplings suddenly killed by termites. Similar observations were made by Nkunika (1980) in Zambia. In general, primary attack of healthy saplings accounts for the greater part of eucalypt sapling mortality caused by termites.

Secondary termite attack Roots of saplings killed primarily by other causes and standing *in situ* are subsequently eaten up by termites. Drought is one of the primary causes of death, particularly in shallow soil near rocky outcrops; root infection by pathogenic fungus is another.

Complementary termite attack In this case, termite attack complements the action of other factors and the combined effect leads to death of the plant. Thus termite feeding on the roots may complement the action of drought, fungal infection, root coiling due to faulty bagging, mechanical injury etc. Each factor by itself may be sublethal or slow acting.

In India, peak incidence of termite attack occurs within about four months of planting out the container saplings (Fig. 10.11), although staggered deaths continue to occur almost every month during the first year, sometimes with a

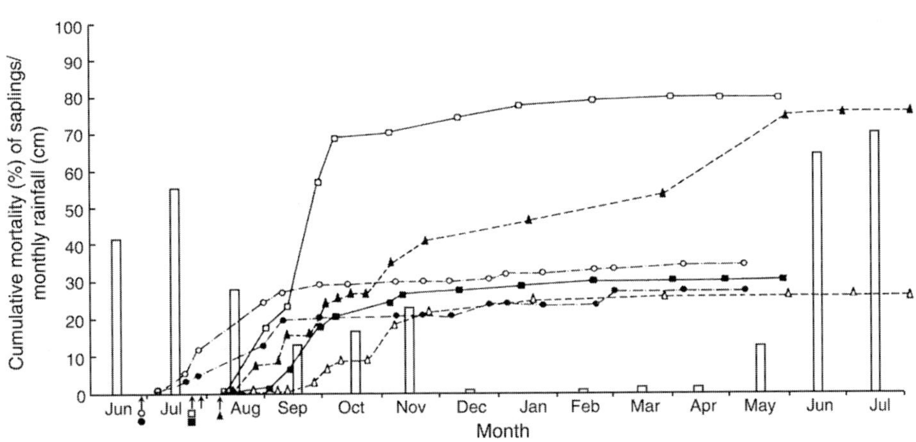

Fig. 10.11 Progress of incidence of termite attack in out-planted eucalypt saplings in Kerala, India, in various field trials in the years 1976–9. The lines show cumulative mortality of saplings in the untreated controls and in some ineffective insecticidal treatments. The bars show mean monthly rainfall for the three years. Date of planting indicated by arrow on the x-axis. Adapted from *Journal of Forest Ecology and Management* (Nair and Varma, 1985), with permission from Elsevier.

second peak during the next planting season (Nair and Varma, 1985). In African countries, where older saplings are killed also, most losses occur in the first year of planting, particularly in the first few months (Brown, 1965). Several authors have suggested that transplanting shock, by lowering the vitality of the saplings, may enhance susceptibility to termites during the first few months. On the other hand, since even healthy saplings are attacked by termites, Nair and Varma (1985) argue that other factors such as weather-related seasonal activity rhythm of termites and age-related susceptibility of the plants may be more important. Whether some environmental factor(s) predispose the plant to termite attack is not clear, although there is a general belief that plants under stress are more susceptible. Termites do attack healthy, vigorously growing plants but we cannot rule out the possibility that only some species do so, while others attack stressed plants. For example in Zambia, Nkunika (1980) found that *Macrotermes falciger* attacked the taproot of vigorous eucalypt saplings while *Odontotermes* sp. nr. *kibarensis* fed on the roots of moribund saplings. Critical field experiments with experimentally stressed plants are necessary to test the role of stress as a predisposing factor. In any case, several interacting factors may influence the incidence of attack – the species of termites present, their population density and seasonal activity rhythm; the site characteristics such as vegetation, accumulation of woody litter on ground and soil moisture content; and plant characteristics such as species, age and physiological state. Generally, there is wide variation in the incidence of termite attack between different places.

Termite attack has a serious impact on plantation success and is recognized as a major limiting factor in eucalypt plantation establishment in the tropics. In India, loss of saplings due to termite attack is generally high; losses ranging from about 20–80% in Kerala (Nair and Varma, 1981), 6–52% in Karnataka (Rajagopal, 1982), 13–31% in Uttar Pradesh (Thakur *et al.*, 1989) and 60–80% in Orissa (Patel and Sahu, 1995) have been reported. Incidence of attack is unpredictable and in most places plantation success is dependent on adequate prophylactic protection against termites. In Guangxi Province in southern China, losses up to 73% of saplings have been reported (Wylie, 1992). In Malawi, losses of 50–66% are common (Chilima, 1991). Similar levels of mortality have also been reported from many other tropical countries.

Species of termites causing damage Not all species of termites present in a locality are destructive to eucalypt saplings. In a study in central Kerala in India, Nair and Varma (1985) found that only 4 out of the 17 species of termites present in the plantation sites were injurious to eucalypts. At least 21 species of termites have been unequivocally associated with eucalypt root feeding, from various places in India (Table 10.8). All the species belong to the family Termitidae and all the four subfamilies are represented, with the majority belonging to the subfamily Macrotermitinae. The genus *Odontotermes* is the most common, accounting for 15 of the 21 species. Root-feeding termites include both mound-building and non-mound-building species. Many of the species also have other feeding habits; they may be found on sound or rotten wood, on the dead bark of standing trees or on other dry organic material, including cow-dung. Thus feeding on live

Table 10.8. *Species of root-feeding termites associated with eucalypts in India*

Family, subfamily and species	States where recorded
Termitidae	
Amitermitinae	
Eurytermes topslippensis	Kerala
Microcerotermes minor	Karnataka
Termitinae	
Pericapritermes assamensis	Kerala
P. vythirii	Kerala
Macrotermitinae	
Microtermes obesi	Kerala, Karnataka, Uttar Pradesh
Odontotermes anamallensis	Kerala
O. bellahunisensis	Kerala, Tamil Nadu
O. brunneus	Karnataka
O. ceylonicus	Kerala
O. distans	Kerala, Uttar Pradesh
O. feae	Uttar Pradesh
O. guptai	Kerala
O. gurudaspurensis	Rajastan
O. horni	Karnataka
O. malabaricus	Kerala
O. microdentatus	Uttar Pradesh
O. redemanni	Tamil Nadu
O. obesus	Kerala, Karnataka, Rajastan, Uttar Pradesh
O. roonwali	Kerala
O. wallonensis	Karnataka
Nasutitermitinae	
Trinervitermes biformis	Tamil Nadu

Data from Roonwal and Rathore (1984), Nair and Varma (1985), Thakur *et al.* (1989) and Varma (2001)

roots is not a specialized habit of these species, although many species present in a locality do not feed on roots.

Some species of *Odontotermes* are also associated with eucalypt mortality in Zambia and Zimbabwe (Nkunika, 1980; Selander and Nkunika, 1981; Mitchell, 1989). In many parts of Africa, larger species of Macrotermitinae like *Macrotermes bellicosus, M. natalensis* and *M. falciger* are mainly responsible for lethal damage to older saplings (Sands, 1962; Brown, 1965). They forage closer to the ground surface and ring-bark the root collar region. Thus, the age-related susceptibility difference between Africa and other regions is attributable to differences in termite fauna. Other species of termites associated with eucalypt root feeding in Africa include soldier-less termites of the subfamily Apicotermitinae (Termitidae), *Amitermes truncatidens, Macrotermes michaelseni, Pseudacanthotermes militaris, P. spiniger, Ancistrotermes latinotus, Allodontotermes schultzei, Microtermes* spp., *Microcerotermes* sp. nr. *parvus* and *Synacanthotermes zanzibarensis* (Nkunika, 1980; Mitchell, 1989). In Brazil, *Heterotermes tenuis* and *Cornitermes* sp. damage eucalypt saplings (Raetano *et al.*, 1997). In East Kalimantan, Indonesia, *Macrotermes malaccensis* and *Schedorhinotermes malaccensis* cause damage (Santoso and Hardi, 1991). In southern China, the root-feeding termites are *Odontotermes formosanus, O. hainanensis, Macrotermes barneyi, M. annandalai* and *Coptotermes formosanus* (Rhinotermitidae) (Wylie, 1992).

Natural enemies Termites have comparatively few natural enemies as they are better protected in the underground environment and work under the cover of mud tunnels when above ground. Indeed, termites are eaten by a wide range of vertebrates and invertebrates, but most mortality is caused only to termites in the open (Logan *et al.*, 1990). Several species of ants feed on termites, but they attack only when the termites are exposed and therefore their effectiveness is limited, with the exception of some species of burrowing doryline ants which invade the subterranean termite nests.

Many micro-organisms, including nematodes, fungi, bacteria, protozoans and viruses have been found in association with termites, but they are not significant pathogens and it is often difficult to distinguish between beneficial (symbionts or commensals) and harmful (parasites or pathogens) organisms. The mutualistic association between termites and the bacteria or protozoans in their gut or the fungi they cultivate in the fungus combs in their nest, which aid cellulose digestion, is well known. Logan *et al.* (1990) gives a detailed review of such associations; the potential biological control agents are discussed below.

Control Termite control in eucalypt plantations is one of the few success stories in tropical forest pest control. Effective and economical protection has been obtained by use of insecticides. Although the success was variable in the earlier days due to use of a variety of insecticides and application methods, the techniques have since been improved and standardized. However, the recent shift from the long-persistent organochlorine insecticides to the comparatively ecofriendly low-persistence insecticides and from polythene-bag-raised saplings to root-trainer-raised, clonally propagated, planting material with a smaller treatable soil core, have posed some difficulties. Currently, the search is on for non-chemical termite control methods. Logan *et al.* (1990) made a comprehensive review of these methods.

Chemical control Destruction of the termite colonies in the planting area by demolishing the mounds and drenching the colony with an insecticide has been suggested. This was often recommended in addition to spot treatment of the planting site. Since mound-building species are not the only ones attacking eucalypt saplings, this method alone cannot ensure safety from termite attack. Also, it is unnecessary to kill all the termites in an area, since only a small fraction of them is injurious.

For the primary nursery bed, drenching the bed with a suitable quantity of insecticide prior to sowing of seeds is recommended (KFRI, 1981).

For protection of the out-planted saplings, on the basis of early field trials carried out in some African countries employing localized application of chemical insecticides (Parry, 1959; Lowe, 1961; Sands, 1962; Wilkinson, 1962; Brown, 1965), a multitude of recommendations was made in the literature by various authors. These have been reviewed with particular reference to India (Nair and Varma, 1981), Africa (Wardell, 1987) and Africa and Indo-Malaysia (Cowie *et al.*, 1989). The most effective method is to create an insecticidal barrier in the soil core immediately surrounding the taproot of the sapling, through which the termites cannot penetrate. In addition to various chemicals and their formulations, various techniques of application have been tested to accomplish this. These include: (1) application of an insecticide, as dust or liquid, to the planting pit and mixing it with the soil; (2) mixing an insecticidal dust with the potting soil; (3) drenching the polybag soil with a liquid insecticide prior to planting out the sapling (with the treated soil core) into the field; (4) method 2 or 3 above, followed by drenching the surface soil around the sapling with an insecticide after planting it out in the field and (5) dipping the roots of the seedling into a concentrated insecticidal liquid, at the time of pricking it out into the polybag.

In a series of field experiments over a four-year period in Kerala, India, Nair and Varma (1981) tested the above methods of application, using selected insecticides and their formulations at different dosages. They found that all the above methods of application, except dipping the bare-rooted seedling at the time of pricking out into the polybag, gave satisfactory protection. Drenching the polybag soil with an insecticide emulsion, prior to planting out was the simplest and most cost-effective. Planting pit treatment is labour intensive and its success depends on adequate mixing of the insecticide with the soil, which cannot be ensured in large-scale planting operations employing unskilled labourers. In India, post-planting treatment (drenching the surface soil around the plant with insecticide) confers no additional advantage, but this is essential in Africa, where the termites attack the plants at the ground level, approaching the stem through the unprotected surface layer of soil (Selander and Nkunika, 1981). Incomplete removal of the polythene bag at the time of planting, leaving a 4-cm-wide collar of polythene around the top, to keep the treated soil projecting above ground level has also been recommended (Sands, 1962; Cowie *et al.*, 1989) but it is doubtful whether it can be accomplished without disturbing the integrity of the treated soil column.

Among the four organochlorine insecticides tested, aldrin, heptachlor and chlordane were effective, but HCH (BHC) was not. Among the effective ones, chlordane showed slight phytotoxicity. A dosage of 0.03 g (a.i.) of insecticide per container (12 cm by 18 cm) was sufficient for satisfactory protection but double the dose was recommended for routine treatment under large-scale planting operations (Nair and Varma, 1981). This is a very small quantity of insecticide, and works out at only 150 g per ha (at 2500 saplings per ha) and application is required only once in 30 years (with two coppice rotations). However, organochlorine insecticides such as aldrin and heptachlor have been phased out due to environmental concerns. Chlorpyrifos, an organophosphate, has emerged as an alternative to the conventional organochlorine soil insecticides, although a comparatively higher dosage is required (Varma and Nair, 1997). Its persistence in tropical soils is considerably lower than that of organochlorines, but this poses no difficulty as most termite attack occurs during the first year, particularly within the initial four to six months of planting out. The synthetic pyrethroids, fenvalerate and permethrin have also shown some promise as termiticides (Mauldin *et al.*, 1987), but more as wood protectants, than as soil pesticides (Varma and Nair, 1997). Controlled release formulation of carbosulfan has also been tested in some tropical countries in Africa and in Brazil and found effective (Atkinson, 1989; Selander *et al.*, 1989; Mitchell, 1989; Chilima, 1991; Resende *et al.*, 1995), but is costly.

Appropriate silvicultural practices As discussed earlier, it is not known whether plant stress is a predisposing factor for termite attack, but it is prudent to follow good silvicultural practices. It can reduce the complementary termite attack. Drought is a major stress factor and therefore it is essential to plant out the saplings at a time when the soil moisture level is adequate. It is also important to ensure healthy planting stock. Of particular importance is to ensure that the roots are not twisted and coiled inside the polythene bag. The effect of site preparation is not clear; while it is generally held that availability of alternative food sources by way of wood residues and litter in the planting site reduces the attack on saplings by attracting the termites away from the plants, others argue that this allows termite numbers to build-up at the site and increases the risk (Cowie *et al.*, 1989). The role of other vegetation in the planting site in providing alternative food sources for the termites is also not clear. In an experiment in Kerala, India, Varma (2001) found that growing a crop of the exotic leguminous plant, *Stylosanthes hamata* around the eucalypt sapling reduced the incidence of termite attack to negligible level.

Biological control Predators are not effective biological control agents as they attack only exposed individuals, as pointed out earlier.

A naturally occurring strain of *Bacillus thuringiensis* Berliner (Bt) was isolated from the termite *Bifiditermes beesoni* in Pakistan and its pathogenicity to other termite species has been demonstrated. Also some commercial strains of Bt have been shown to be pathogenic to termites, but Bt has not proved effective in field tests, probably because of its poor survival in the soil (Logan *et al.*, 1990).

Two fungi, *Metarhizium anisopliae* and *Beauveria bassiana* are reported to have potential for biological control of termites, the first being more pathogenic (Jones *et al.*, 1996). The pathogenicity of *M. anisopliae* when the conidia were applied directly to the termite or to the mound has been demonstrated (Hanel and Watson, 1983; Sajap and Jan, 1990; Milner *et al.*, 1998) although the termites have a tendency to wall-off treated areas of the mound (Milner, 2000). Varma (2001) field-tested *M. anisopliae* var. *major* by applying a conidial suspension to the root-trainer potting medium (5 ml of a suspension containing 2×10^7 conidia/ml applied to 150 ml volume of potting medium) prior to out-planting the sapling and found that it gave effective protection. In laboratory tests, the fungal conidia caused mortality of the termite *Odontotermes guptai* within seven days (Varma, 2001), but its mode of action in the field trial remains unknown. In the field, the fungus may have acted as a repellent (Varma, 2001); repellency of *M. anisopliae* conidia to *Coptotermes lacteus* has been demonstrated in the laboratory (Staples and Milner, 2000). More field trials are necessary to establish the usefulness of *M. anisopliae* as a biological control agent.

A nematode, *Steinernema carpocapsae* strain BJ2 has been reported to effect 100% parasitization of *Odontotermes formosanus* on *Eucalyptus* within 12 days (Zhu, 2002). Obviously more field trials are needed under different conditions.

Tree resistance No species of eucalypts commonly raised in plantations in the tropics is known to be absolutely resistant to termite attack. In a study in South Africa (Atkinson *et al.*, 1992) *Eucalyptus dunnii* was found to be somewhat resistant, followed in decreasing order by *E. macarthurii*, *E. smithii* and *E. viminalis*, against the termite *Macrotermes natalensis*. For example, in a typical trial, only 12% of *E. dunnii* suffered damage, compared to 65% of *E. grandis*. Differences in susceptibility were also noted between the clones of a given species. *E. grandis*, a species commonly planted in the tropics, was found to be highly susceptible, but one pure clone of *E. grandis*, and two of *E. grandis* X *E. camaldulensis*, were resistant. Hybrid clones of the resistant species *E. macarthurii*, crossed with *E. grandis*, were susceptible; but two clones of *E. macarthurii*, one probably crossed with *E. camaldulensis* and the other with an unknown species, were resistant. These findings suggest the need to explore the scope for utilization of genetic resistance.

Knowledge gaps Although fairly acceptable levels of protection from termite attack can be obtained by localized application of chemical insecticides, and the quantity of insecticide used is very small, it is desirable to search for non-chemical alternatives. While there are problems with biological control agents due to the behavioural defences of termite colonies (Logan *et al.*, 1990), the fungus *Metarhizium anisopliae* has shown some promise, possibly as a repellent (Varma, 2001). This needs to be critically investigated. Many plant products are credited in folklore as being toxic or repellent to termites, mostly in India and some African countries. Logan *et al.* (1990) gives a comprehensive list of these products and their proposed use. One such product is neem seed cake, but Varma *et al.* (1995) found it to be ineffective when mixed with potting soil. However most products and techniques have not been scientifically evaluated; some may hold promise.

10.8 *Falcataria moluccana* (= *Paraserianthes falcataria*) (Fabaceae: Faboideae)

Tree profile

Falcataria moluccana (Miq.) Barneby & J. W. Grimes, known until recently as *Paraserianthes falcataria* (L.) I. C. Nielsen, is an exceptionally fast-growing leguminous tree, native to the eastern islands of the Indonesian archipelago, Papua New Guinea and the Solomon Islands. The species was also known formerly as *Albizia falcata*, *A. falcataria* and *A. moluccana*, and classified under the family Leguminosae, subfamily Mimosoideae. It is a medium to fairly large-sized tree, with bipinnately compound leaves, and can grow up to 40 m high

(CABI, 2005). The bole is generally straight and cylindrical. The crown is narrow in dense stands but becomes umbrella shaped in the open. The tree can reach 7 m in height in one year, 15 m in three years and 30 m in 10 years, with wood volume yield of 39 to 50 m^3/ha per year on a 10-year rotation (CABI, 2005). The wood is soft and not durable, suitable for pulping, matchsticks, plywood, light-weight packing materials etc.

The species is widely planted in the humid tropics: in Bangladesh, India, Indonesia, Malaysia, the Philippines and Sri Lanka in Asia; Cameroon, Côte d'Ivoire, Malawi and Nigeria in Africa; Mexico in Central America; and Hawaii and Samoa in the Pacific (CABI, 2005). In 1990, the plantation areas included 12 000 ha in Bangladesh and 11 550 ha in Malaysia (Pandey, 1995). Indonesia where it is planted in large-scale industrial plantations for pulpwood as well as in smallholder community forestry plantations had more than 48 000 ha in 1999 (Cossalter and Nair, 2000). In different countries the tree is planted as an ornamental, in agroforestry systems, for shade, for intercropping in forest plantations, for afforestation etc. In Indonesia, it is the most preferred species in community forestry programmes.

Overview of pests

Very little information is available on pests of *F. moluccana* in native stands, except that a cerambycid stem borer *Xystrocera festiva*, attacks live trees in Indonesia (Alrasjid, 1973, cited by Nair, 2001a). This insect is also a serious pest in plantations (see pest profile below).

A large number of insects have been found in plantations of *F. moluccana*. They include root-feeding whitegrubs of saplings, leaf feeders, sap suckers, bark feeders, stem borers and seed feeders. About 40 species have been recorded in India alone (Mathew and Nair, 1985; Pillai and Gopi, 1991) but only a few cause serious damage. The more important species in Asia are listed in Table 10.9; no published pest records are available for countries in other continents.

The leaf feeders include caterpillars of several families of Lepidoptera, and chrysomelid and curculionid beetles. Most are general feeders but two have acquired pest status. First in importance is the bagworm *Pteroma plagiophleps*, for which a pest profile is given below. Second in importance is the caterpillar of the yellow butterfly *Eurema* spp. (mainly *E. blanda* and a small proportion of *E. hecabe* and others) that occasionally build-up in large numbers and feed gregariously, causing locally widespread defoliation in nurseries and young plantations in India, Indonesia, the Philippines and Malaysia (Mathew and Nair, 1985; Braza, 1990; Chey, 1996; Irianto *et al.*, 1997). In Indonesia, this insect occasionally causes severe defoliation in Java, Sumatra, Kalimantan and Sulawesi, leading to dieback of branches, but usually the infestation is transient and the damage not serious

Table 10.9. *Important insects causing damage to* Falcataria moluccana

Category	Species name, order and family	Countries of occurrence	Remarks
Leaf feeding	*Pteroma plagiophleps* (Lepidoptera: Psychidae)	India, Indonesia, Malaysia, Philippines, Thailand	Occasional outbreaks
	Eurema blanda and *E. hecabe* (Lepidoptera: Pieridae)	India, Indonesia, Philippines, Malaysia	
	Adoxophyes sp. (Lepidoptera: Tortricidae)	India, Malaysia	
	Semiothisa sp. (Lepidoptera: Geometridae)	India, Malaysia	
	Catopsilia pomona (Lepidoptera: Pieridae)	Sri Lanka	
Sap sucking	Acizzia *sp.* (Hemiptera: Psyllidae)	Philippines	
	Oxyrachis tarandus (Hemiptera: Membracidae)	India	
Bark feeding	*Indarbela quadrinotata* (Lepidoptera: Metarbelidae)	Bangladesh, India,	
	I. acutistriata	Indonesia	
	Sahyadrassus malabaricus (Lepidoptera: Hepialidae)	India	Tunnels in the pith of saplings
Stem boring	*Xystrocera festiva* (Coleoptera: Cerambycidae)	Indonesia, Malaysia	Major pest
	X. globosa	Indonesia	Minor pest
	Callimetopus sp. (Coleoptera: Cerambycidae)	Philippines	Shoot pruner on saplings
	Euwallacea fornicatus (syn. *Xyleborus fornicatus*) (Coleoptera: Curculionidae: Scolytinae)	Sri Lanka, India	On saplings
	Xylosandrus morigerus (Coleoptera; Curculionidae: Scolytinae)	Indonesia	Twig boring

(Irianto *et al.*, 1997). The bark-feeding caterpillars, *Indarbela* spp. often cause moderate damage. Although a polyphagous minor pest of several tree species, *I. quadrinotata* has been reported to build-up in damaging numbers in some plantations of *F. moluccana* in Kerala, India (Mathew, 2002). *Indarbela acutistriata* occurs in Java, Indonesia (Suharti *et al.*, 2000). Among the stem borers, *Xystrocera festiva* is a serious pest in Indonesia and Malaysia. A pest profile for this species is given below. The closely related *X. globosa* is a minor pest which sometimes occurs together with *X. festiva* in Indonesia. The scolytine twig borers (Table 10.9) are more prevalent in unhealthy plantations; *E. fornicatus*, commonly known as the shothole borer of tea, has been reported to cause mortality of apparently unhealthy, two to three-year-old saplings, in small patches of plantations (Mathew and Nair, 1985).

Pest profile

Xystrocera festiva Thomson (Coleoptera: Cerambycidae)

Xystrocera festiva Thomson (Fig. 10.12a,b), commonly known as 'albizzia borer', is a serious pest of *F. moluccana*. The beetle lays eggs on the bark of live trees and the larvae bore into the stem, causing extensive damage, often leading

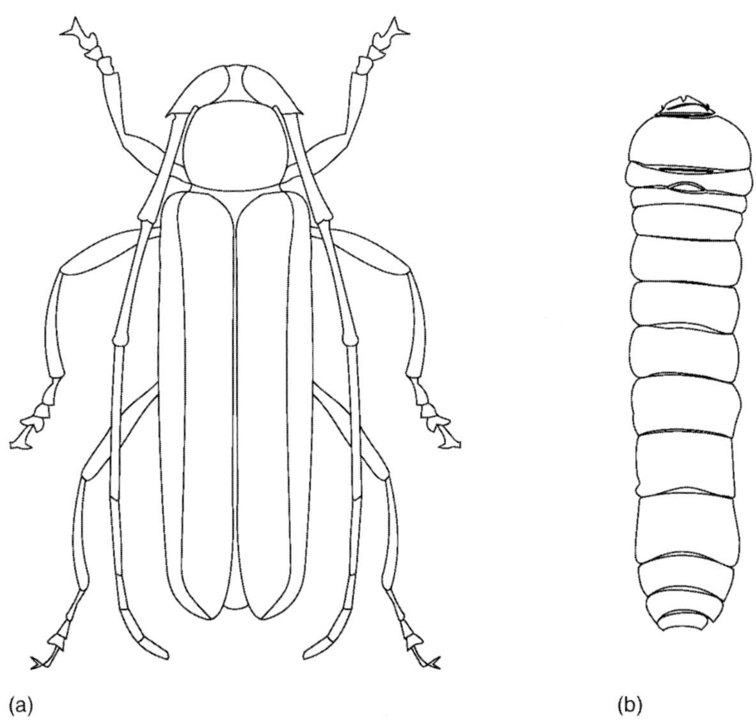

Fig. 10.12 *Xystrocera festiva*. (a) Adult (length 35 mm), (b) larva. After Abe (1983).

to death of the tree. The medium-sized beetle, 30–35 mm in length, is reddish brown in colour with dark blue-green lateral stripes on the prothorax and elytra. The larvae are yellowish green and grow up to 40–50 mm in length.

Life history and seasonal incidence Adults are nocturnal (Kasno and Husaeni, 2002) and live for only 5–10 days (Matsumoto and Irianto, 1998). According to Fauziah and Hidaka (1989), the male releases a pheromone to attract the female for mating. Eggs are deposited in clusters of over 100, in one or two batches, preferably in crevices on the stem or branch stubs, generally 3–4 m above ground (Kasno and Husaeni, 2002; Matsumoto and Irianto, 1998). Based on laboratory breeding, Matsumoto and Irianto (1998) estimated the average number of eggs laid per female as about 170. Newly hatched larvae bore into the inner bark and as the larvae grow, they feed on the outer sapwood, making irregular downward galleries, packed with frass (Kasno and Husaeni, 2002). The larvae remain gregarious. Oviposition in clusters and aggregation of larvae are unusual in cerambycids. The newly injured bark usually exudes a brownish liquid and powdery frass is expelled through crevices in the bark. The larval development is completed in about four months and each larva bores an oval tunnel upward in the sapwood in which it pupates. *X. festiva* has overlapping generations, with all developmental stages pesent at any one time. Thus new infestation takes place continuously (Kasno and Husaeni, 2002).

Impact *X. festiva* attack usually begins when the trees are two to three years old and the infestation increases with age (Suharti *et al.*, 1994). Since a large number of larvae develop on a single tree, the growing larvae create a labyrinth of tunnels on the trunk (Fig. 10.13), the bark dries and cracks, and the heavily infested trees dry up. Weakened stems are sensitive to wind, particularly during the rainy season. Even when the trees are not killed, borer attack reduces the growth rate and timber quality. Borer attacked timber is often classified as firewood and sold at much lower prices. In a field study in East Java, Indonesia, Notoatmodjo (1963) estimated that the yield loss due to this borer was about 12% if the trees were harvested in the fourth year and 74% if harvested after the eighth year. *X. festiva* is also recognized as a major pest of *F. moluccana* in Sabah, Malaysia, but not in India.

Host range and geographical distribution *X. festiva* has also been recorded on *Acacia auriculiformis*, *A. mangium*, *A. nilotica*, *Albizia chinensis*, *A. lebbek*, *A. stipulata*, *Archidendron jiringa*, *Enterolobium cyclocarpum*, *Pithecellobium dulce* and *Samanea saman*, all belonging to the family Fabaceae (Abe, 1983; Suharti *et al.*, 1994; Hardi, *et al.*, 1996; Matsumoto and Irianto, 1998). However, it has not become a serious

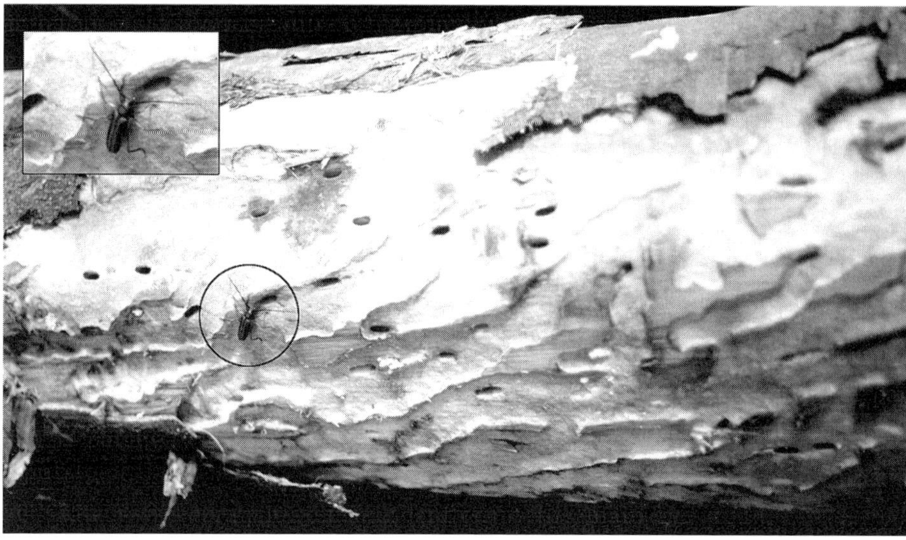

Fig. 10.13 A freshly felled log of *Falcataria moluccana* infested by *Xystrocera festiva* in Java, Indonesia. Inset shows the beetle.

pest of any of these trees, although it is commonly noticed in species of *Albizia* grown as shade trees in tea estates.

X. festiva is present in Indonesia, Malaysia and Myanmar. A related species, *X. globosa*, which is smaller in size and has a blue-green stripe on the middle of the elytra instead of on the side, occurs in several countries in the Oriental, Australasian, Palaearctic and Neotropical regions. Its hosts include many species of *Acacia, Albizia* and other leguminous genera. It is believed to mainly attack trees in poor health (Browne, 1968). A small population of this species has also been found associated with *F. moluccana* in Indonesia (Nair, 2000).

Natural enemies Some natural enemies of *X. festiva* have been reported from Java, Indonesia and Sabah, Malaysia. These include an encyrtid egg parasitoid, *Anagyrus* sp. (vide infra), an unidentified tachinid and birds.

Control Kasno and Husaeni (2002) have reviewed the control practices and options in Indonesia, where both private and government owned plantations exist. The most commonly practised control measure is removal of infested trees through silvicultual thinning. As noted earlier, *X. festiva* infestation begins when the trees are two years old, and intensifies as age advances. In Government plantations, a thinning strategy has been introduced for borer control. Thinning is carried out in the 3rd, 4th, 5th and 6th years and the infested trees are prioritized for cutting. The plantation is then clear cut when eight years old. This practice has significantly reduced the incidence of borer attack, but 4–10% of trees are still attacked (Kasno and Husaeni, 2002). Such systematic thinning is not carried out in privately owned smallholder plantations and consequently the borer damage is greater. Smallholder farmers generally harvest the trees when they are 4–6 years old, before the borer incidence intensifies.

Mechanical destruction of infestation is possible if the infested portion of the stem is debarked before the larvae are ready for pupation; the larvae then readily fall off from exposed galleries. Good control was achieved in an experiment in East Java where this method was practised by regular inspection at three-monthly intervals (Matsumoto, 1994).

Some biological control trials have also been carried out. At Ngancar in East Java, inundative release of 5000 adults of the encyrtid parasitoid *Anagyrus* sp. in the centre of a 19 ha plantation compartment gave promising results – all egg clusters introduced to the stem of trees after the parasitoid release were found parasitized when observed after two weeks (Kasno and Husaeni, 2002).

Kasno and Husaeni (2002) proposed an integrated control strategy which involves the following steps: (1) carry out a three-monthly inspection of plantations and locate the infested trees. When infested trees are found, debark the attacked portion of the stem, if accessible, or cut and remove the tree if the infested portion is inaccessible; (2) carry out thinning operations at the 3rd, 4th, 5th and 6th years, removing the infested trees preferentially. During these thinning operations, collect adult beetles from infested stems, place them in wooden boxes to mate and lay eggs, collect the newly laid egg clusters and place them on exposed stem surfaces in the plantation to attract parasitoids. This will augment the population of the parasitoid, *Anagyrus* sp.

Knowledge gaps *X. festiva* infestation of *F. moluccana* can be effectively controlled by early detection of infested trees and mechanical destruction of the insect by debarking the affected portion of stems or removal of the tree. Therefore the constraint to control is not a knowledge gap, but a gap in knowledge extension and training. Systematic inspection is the first step for control.

However, more knowledge is necessary on the potential use of pheromone as well as light traps for trapping and killing the beetles. The existence of a sex pheromone has been suspected, as noted earlier. Further studies are needed for confirming this and prospecting it for practical use. Although some of the earlier studies did not indicate the attraction of beetles to light, Kasno and Husaeni (2002) found that adults are preferentially attracted to green, followed by blue light, and the greatest numbers of beetles are trapped around midnight. Obviously further studies are needed. A constraint to inundative release of the encyrtid parasitoid, *Anagyrus* sp. is that at present it can be cultured only on egg clusters of *Xystrocera* spp. Research is needed to mass-produce this parasitoid on alternative hosts.

Pest profile

Pteroma plagiophleps Hampson (Lepidoptera: Psychidae)

Pteroma plagiophleps Hampson (Fig. 10.14a,b) is a small bagworm. It is a serious pest of *F. moluccana* in India and Indonesia. The male moth is brownish, with a wingspan of 14–16 mm and is an active flier. The adult female, however, is wingless and highly degenerate, confined to the pupal bag. It has a sclerotized posterior part but the rest of the body is virtually a bag of developing eggs. The larva is known as a bagworm as it constructs a bag around itself. The bag is made of silk, with pieces of leaf or bark material stuck on the outer surface. The larva usually remains concealed within the bag, with the head and thorax projecting out while feeding. The full-grown larva is 9–10 mm long.

Life history and seasonal incidence The life history of *P. plagiophleps* on *F. moluccana* was studied in Kerala, India by Nair and Mathew (1988). The male moth flies to the bag harbouring the adult larviform female and copulates with

Fig. 10.14 The bagworm *Pteroma plagiophleps*. (a) Adult male (wingspan 15 mm). The adult female is wingless. (b) Larva taken out of the bag.

it by inserting the abdominal tip into the bag while remaining in a suspended position in flight. The fertilized eggs develop synchronously within the female body cavity. When the eggs are ready for hatching, the body wall ruptures and the posterior part of the abdomen, which by then has become shrunken, falls to the ground, permitting the neonate larvae to hang on silken threads and disperse. Each female produces 110–200 larvae. Dispersal is aided by wind. The newly hatched larva that lands on a host leaflet immediately starts constructing a bag around itself and completes the work within an hour. It then starts feeding on the leaflets. Starting usually from the under-surface of the leaf blade, the larva consumes the epidermal layer and the mesophyll tissues containing the chloroplasts, leaving the single layer of epidermis on the other surface. Generally, feeding is patchy, with some portions of each leaflet left unfed (Fig. 10.15).

Older larvae usually migrate to the branch stems and often to the main trunk, and feed on the live surface layers of bark, leaving feeding scars on the stem (Fig. 10.16). Bark feeding is common when the infestation is heavy. Larvae resting or feeding on stems, with their conical bags held upright, resemble

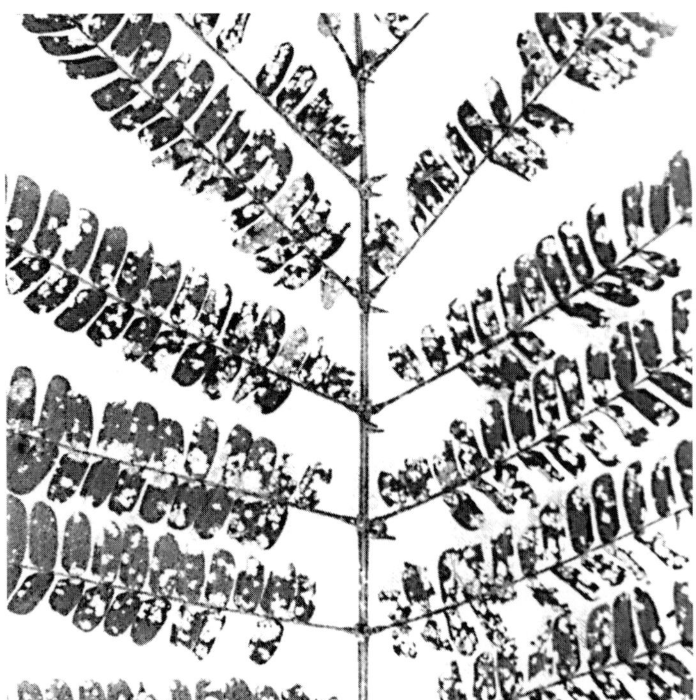

Fig. 10.15 *Falcataria moluccana* leaf showing feeding damage by the bagworm *Pteroma plagiophleps*.

Fig. 10.16 Larvae of the bagworm *Pteroma plagiophleps* feeding on the live bark of the branches of the tree *Falcataria moluccana*. With the bag held upright, the larvae resemble thorns.

thorns. When reared on *F. moluccana* saplings in outdoor cages, the larval period lasted 49–66 days, the females taking longer than the males. Prior to pupation, the larva attaches the bag, with a thick silk thread, to a twig and closes the mouth of the bag: the cocoons thus hang on the branches (Fig. 10.17). The male cocoon is comparatively short, with a truncated posterior end, while the female cocoon is longer and has a tapering posterior end. The pupal period of the male is about 14 days. The male moth has atrophied mouthparts and lives for about four days in laboratory cages. Thus the duration of the total life cycle from egg to adult is about two to two and a half months.

In Kerala, India, bagworms were present in *F. moluccana* plantations throughout the year, and up to five generations per year have been recognized (Nair and Mathew, 1988). However, over a three-year study period, outbreaks leading to heavy defoliation occurred only once or twice a year. In one year, outbreaks occurred in April and June, causing total defoliation in some patches, but the population collapsed in August, probably due to incidence of a fungal disease. Outbreak populations were usually of similar age although overlapping of developmental stages was common at other times.

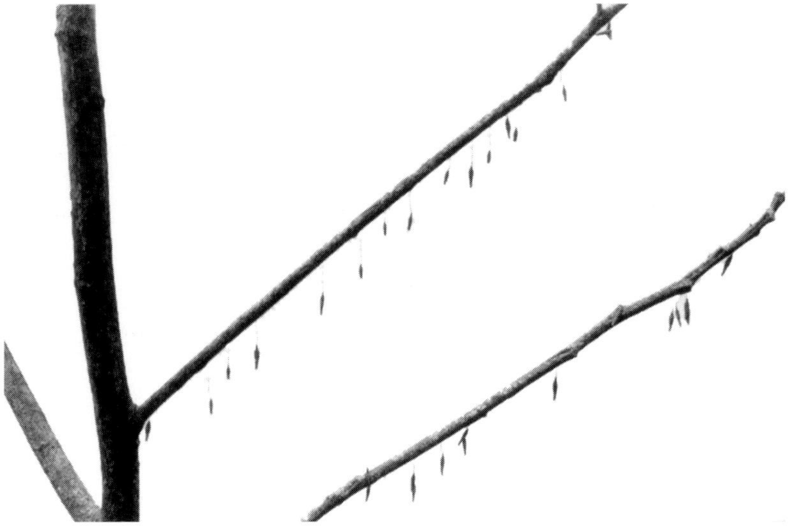

Fig. 10.17 Cocoons of the bagworm *Pteroma plagiophleps* hanging from the branches of the tree *Falcataria moluccana*.

Nature of damage and impact In heavy infestation, each compound leaf may harbour hundreds of bagworms and their feeding causes the whole leaf to dry up. The dried leaves remain on the tree for some time, giving a scorched appearance to the tree.

Within a plantation, bagworm infestation is often patchy. Nair and Mathew (1988) studied the infestation characteristics in a three-year-old plantation of *F. moluccana* in Kerala, India. In this plantation, about 10% of the trees were not infested while 51% had a low level of infestation, 19% medium, 11% high and 8% a very high level of infestation. A clumped distribution of infestation was evident (Fig. 10.18). Within the 20 ha plantation, there were two epicentres of highest infestation from where the intensity decreased gradually towards the periphery. Trees in these two patches were totally defoliated. The defoliation was the result of feeding by two generations of larvae, the second infestation being mainly centred around the first. This is to be expected as the females are wingless. Defoliation had a serious impact. Among the totally defoliated trees, some were killed outright, others were killed above the lower half of the main trunk and in still others only some of the small, top branches died. A survey of 5% of trees in the plantation, carried out two and a half years after the infestation was first noticed, showed that 22% of the trees were totally dead, 7% suffered damage to three-quarters of the main trunk, 5% each suffered damage to half and one-quarter of the trunk and 65% were healthy. The healthy trees may have suffered growth loss. Most tree mortality was centered in the two epicentres of outbreak.

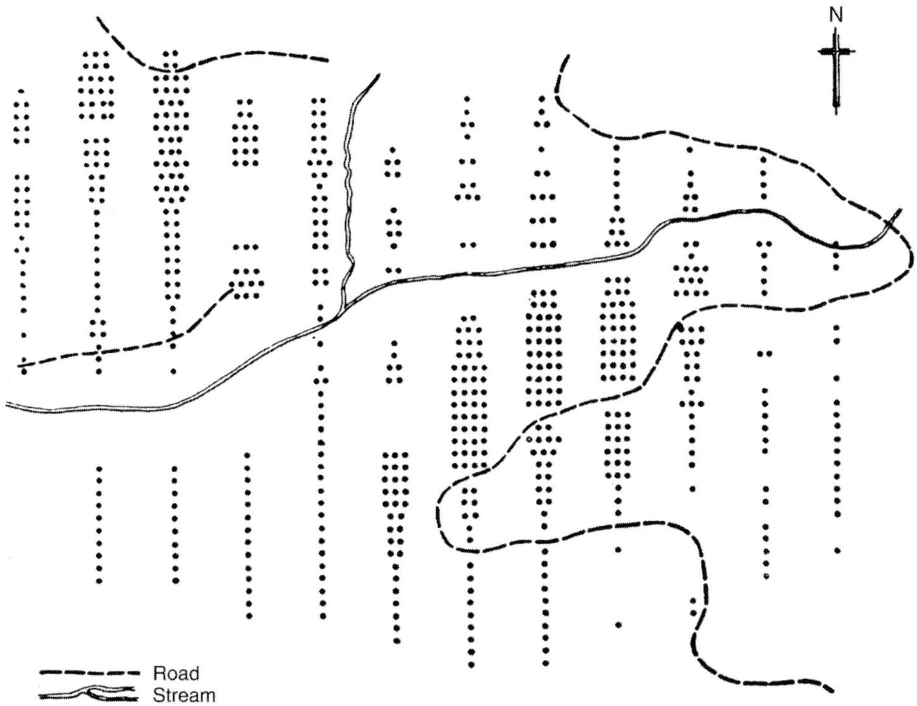

Fig. 10.18 Spatial distribution of infestation by the bagworm *Pteroma plagiophleps* in a three-year-old, 20 ha plantation of *Falcataria moluccana* at Vazhachal in Kerala, India. The intensity of infestation is indicated by the number of dots in a row, four dots representing very high intensity with total defoliation, and blank representing no infestation. Each data point presents the median score of four trees in the row; every 20th row in the plantation was scored. Note the clumped distribution of infestation. From Nair and Mathew (1992).

Incidence of bagworm attack in *F. moluccana* plantations in Kerala was erratic. Of several plantations, only some were infested. In Indonesia also *P. plagiophleps* is a sporadic pest, with severe defoliation occurring in some endemic patches within *F. moluccana* plantations in Sumatra. A five-year-old plantation in South Sumatra had a severe chronic attack from 1994 to 1997 (Zulfiyah, 1998).

Host range and geographical distribution *P. plagiophleps* has a wide host range, covering several unrelated families. Nair and Mathew (1992) listed 17 host plants under the families Arecaceae, Cannaceae, Euphorbiaceae, Fabaceae (Caesalpinioideae and Faboideae), Lamiaceae, Lauraceae, Myrtaceae, Punicaceae, Salicaceae, Theaceae and Ulmaceae. Apart from *F. moluccana*, hosts of importance to forestry include *Acacia auriculiformis*, *A. mangium*, *A. nilotica*, *Delonix regia*, *Syzygium cuminii*, *Populus deltoides*, *Tectona grandis* and *Trema orientalis*, with sporadic outbreaks occurring on *A. nilotica* and *D. regia*. Santhakumaran *et al.* (1995) reported it on the mangrove *Rhizophora mucronata* (Rhizophoraceae). The outbreak of an undetermined species of *Pteroma* reported in natural stands of the pine *Pinus merkusii*, subjected to resin tapping and growing on comparatively poor soil in North Sumatra, Indonesia, is also probably attributable to this species (Nair and Sumardi, 2000).

The circumstances under which *P. plagiophleps* develops pest status are not well understood. In India, this bagworm has been known for a long time as an insignificant pest of the tamarind tree *Tamarindus indica* and the pomegranate *Punica granatum*. Outbreaks appeared for the first time in 1977 on *F. moluccana* and later on *Delonix regia* (Nair and Mathew, 1992) and *Acacia nilotica* (Pillai and Gopi, 1990a). Nair and Mathew (1992) recognized three types of infestation: (1) sparse infestation, with very low numbers of insects, noticed on the majority of the recorded hosts, (2) dense infestation of isolated, individual plants of some species, leaving other plants of the same species in the vicinity unaffected and (3) heavy outbreak, affecting a large number of trees in patches, as in *F. falcataria* and *A. nilotica*. The reasons for these different types of infestations are not understood. There are indications that host stress is a predisposing factor for *P. plagiophleps* outbreak. *Eucalyptus tereticornis* is not normally attacked but an outbreak was noticed on trees growing in sulphur dioxide polluted premises (Nair and Mathew, 1992). In multiple choice outdoor cage experiments, saplings of *Tamarindus indica* and *Punica granatum* were the most susceptible hosts and survival was poor on *F. moluccana*. The insect failed to develop on *Delonix regia* saplings, even when the parental stock originated from naturally infested *D. regia*. Much remains to be learnt about the factors controlling host selection in this insect, but it appears that plants under stress are more susceptible to attack.

The known distribution of *P. plagiophleps* includes Bangladesh, India, Indonesia, Malaysia, the Philippines and Thailand.

Natural enemies Natural enemies appear to play a decisive role in regulating the populations of *P. plagiophleps* larvae during some periods. Often large populations of larvae were found dead inside their bags, sometimes as early instars and sometimes as late instars. Most deaths were suspected but not proven to have been caused by microorganisms (Nair and Mathew, 1988). Fungal pathogens have been recorded in other bagworms like *Crematopsyche pendula*, *Acanthopsyche junodi* and *Thyridopteryx ephemeraeformis* (Sankaran, 1970; Berisford and Tsao, 1975).

On some occasions the bagworms were also heavily parasitized. A total of 18 species of parasitoids were recorded, all hymenopterans. They included species of Braconidae, Chalcididae, Eulophidae, Eurytomidae, and Ichneumonidae

(Nair and Mathew, 1988). The ichneumonids *Goryphus* sp. and *Acropimpla* sp. nr. *leucotoma*, and the chalcid *Brachymeria plutellae* were the most common. The parasitoids emerged through cleanly cut holes in the bag, after pupation of the host. The rate of parasitism, mainly attributable to the above three parasitoids, was as high as 25–38% on some occasions. Unidentified syrphid larvae were sometimes observed within larval bags along with dead bagworm larvae but their predatory role is not proven.

Control The larval bag affords protection to the insect against direct deposition of insecticidal sprays. Varma *et al.* (1989) tested nine commercial insecticides, applied to leaf, in laboratory experiments, and found 0.05% a.i. quinalphos and methyl parathion the most effective.

Natural regulation by parasitoids and unknown diseases appears to play a role and may be responsible for the absence of widespread outbreaks. Also, the part played by tree health is not known.

Knowledge gaps An interesting characteristic of *P. plagiophleps* is the occurrence of population outbreaks only on some host species in spite of its wide host range. Even on the same host species, outbreaks occur only on some plants and some plantations. As pointed out earlier, there are indications that poor tree health is a predisposing factor for outbreaks. Much remains to be learnt about the factors controlling the host plant acceptance and the influence of host quality on the population dynamics of *P. plagiophleps*.

10.9 ***Gmelina arborea*** (Lamiaceae)

Tree profile

Gmelina arborea Roxb. is indigenous to Asia and occurs in India, Bangladesh, Pakistan, Myanmar, Sri Lanka, Thailand, Laos, Cambodia, Vietnam, and the Yunnan and Guangxi provinces of China (CABI, 2005). It occurs mostly in deciduous and moist-deciduous forests, but sometimes also in evergreen forests, and usually below 1200 m latitude. *G. arborea* is a fairly fast growing tree which produces a lightweight, creamy-white timber suitable for construction and carving, as well as for production of good quality pulp. It is often grown on short rotations of 15–20 years. It is a pioneering species and prefers full sunlight, although it can withstand partial shade (CABI, 2005). Plantations are raised from potted seedlings or 7 to 10-month-old stumps.

In Asia, *G. arborea* plantations have been raised both within its natural distribution range and outside, in India, Peninsular and East Malaysia, the Philippines and Indonesia. It has also been introduced into many countries worldwide. Large-scale plantations exist in some countries in Africa such as Nigeria, Sierra Leone and Malawi as well as in Brazil in Latin America (CABI, 2005). Some of the available planted area figures for the year 1990 are: Nigeria 91 000 ha; Sierra Leone 4000 ha; Bangladesh 6000 ha and Malaysia 11 000 ha (Pandey, 1995). In 1999, India had about 148 000 ha under *G. arborea* plantations (FSI, 2000), the largest for the species, and Indonesia had about 48 000 ha (Cossalter and Nair, 2000).

Overview of pests

Pests in native and exotic plantations are considered separately.

Pests in native plantations

A large number of insects have been recorded in native plantations of *G. arborea*; 101 species in India and at least 20 in Thailand (Mathur and Singh, 1961; Mathew, 1986; Hutacharern, 1990). Most are casual or occasional feeders, but some are serious pests. The more important species are listed in Table 10.10.

The defoliators include chrysomelid beetles and lepidopteran caterpillars. The chrysomelid, *Craspedonta leayana* (see pest profile below) is a serious pest. Defoliation caused by this insect has become a constraint to expansion of plantations, particularly in northeast India, Myanmar and Thailand (Garthwaite, 1939; Beeson, 1941). Largely due to this pest, *G. arborea* has been dropped from the planting list by forest departments in many countries where the tree is indigenous. Other leaf feeding insects (Table 10.10) are polyphagous and cause only minor damage. The nettle grub *Parasa lepida* feeds gregariously at first on the leaf surface but holes are made by the later stages during a total larval period of about two months in Maharashtra, India (Meshram and Garg, 2000).

Among sap feeding insects, the tingid *Tingis beesoni* which attacks young trees in native plantations causes dieback of shoots and is recognized as a serious pest (see pest profile below). The bark feeders are not major pests.

Among the wood borers, *Xyleutes ceramicus*, primarily a pest of teak (see pest profile under teak), has been recorded as causing occasional damage to *G. arborea* in Myanmar and Thailand. The sapwood borer, *Glena indiana*, which occurs in India, Myanmar and Thailand, is reported to have ruined some plantations in northeast Thailand. Its attack begins in year-old saplings and continues in the following years, often causing death of the trees at 8–10 years (Hutacharern, 1990). A small weevil, *Alcidodes ludificator*, about 10 mm in length, bores into young green shoots of saplings and lays eggs. The larva bores down the centre of the shoot, making small holes through the bark at intervals for the ejection of frass and excrement (Beeson, 1941). The tunnelling of the shoot causes the plant to die back. The life cycle is annual. The insect occurs as a minor pest in India,

Table 10.10. *Important insects causing damage to* Gmelina arborea

Category	Species name, order and family	Countries of occurrence	Remarks
Native plantations			
Defoliators	*Craspedonta* (= *Calopepla*) *leayana* (Coleoptera: Chrysomelidae)	India, Myanmar, Bangladesh, Thailand	Major pest
	C. mouhoti (Coleoptera: Chrysomelidae)	Thailand	
	Prioptera spp. (3 species) (Coleoptera: Chrysomelidae)	Thailand	
	Epiplema fulvilinea (Lepidoptera: Epiplemidae)	India	
	Hapalia (= *Prionea*) *aureolalis* (Lepidoptera: Pyralidae)	Thailand	
	Eupterote undata (Lepidoptera: Eupterotidae)	India	
	Parasa lepida (Lepidoptera: Limacodidae)	India	
Sap sucker	*Tingis beesoni* (Hemiptera: Tingidae)	India, Myanmar, Thailand	Causes shoot dieback in saplings
Bark feeders	*Acalolepta cervina* (syn. *Dihammus cervinus*) (Coleoptera: Cerambycidae)	India	
	Indarbela quadrinotata (Lepidoptera: Indarbelidae)	India	
	Sahyadrassus malabaricus (Lepidoptera: Hepialidae)	India	
Stem borers	*Alcidodes ludificator* (syn. *Alcides gmelinae*) (Coleoptera: Curculionidae)	India, Myanmar, Thailand	On green shoots of saplings
	Euwallacia (= *Xyleborus*) *fornicatus* (Coleoptera: Curculionidae: Scolytinae)	India	Shothole borer of saplings

Table 10.10. (cont.)

Category	Species name, order and family	Countries of occurrence	Remarks
	Xyleutes ceramicus (Lepidoptera: Cossidae)	Thailand, Myanmar	
	Glena indiana (Coleoptera: Cerambycidae)	India, Myanmar, Thailand	
	Acalolepta rusticator (Coleoptera: Cerambycidae)	India, Malaysia	
Exotic plantations			
Root feeders	*Coptotermes curvignathus* (Isoptera: Rhinotermitidae)	Malaysia	On saplings
	Nasutitermes costalis (Isoptera: Termitidae)	Cuba	On saplings
Defoliators	*Ozola minor* (Lepidoptera: Geometridae)	Philippines	
	Dichocrocis megillalis (Lepidoptera: Pyralidae)	Malaysia	
	Pionea aureolalis (Lepidoptera: Pyralidae)	Malaysia	
	Pleuroptya (Sylepta) balteata (Lepidoptera: Pyralidae)	Malaysia	
	Archips sp. (Lepidoptera: Tortricidae)	Malaysia	
	Spilosoma maculata (Lepidoptera: Arctiidae)	Nigeria	
Bark feeders	*Acalolepta cervina* (syn. *Dihammus cervinus*) (Coleoptera: Cerambycidae)	Malaysia	
Stem borers	*Alcidodes ludificator* (syn. *Alcides gmelinae*) (Coleoptera: Curculionidae)	Indonesia	On green shoots of saplings
	Prionoxystus sp. (Lpidoptera: Cossidae)	Indonesia	Mostly on saplings
	Xyleutes ceramicus (Lepidoptera: Cossidae)	Malaysia	

Myanmar and Thailand. The cerambycid *Acalolepta* (=*Dihammus*) *rusticator* bores longitudinal galleries along the stems of saplings, leading sometimes to breakage of the stem (Mathew, 1986; Chey, 1996). The bark of the tree is damaged by the canker grub *Acalolepta cervina*, the hepialid borer *Sahyadrassus malabaricus* (see pest profile under teak) and the indarbelid *Indarbela quadrinotata*, in India.

In a study in a three-year-old plantation in Kerala, India, Nair and Mathew (1988) found that major damage was caused by only two species of insects, *Craspedonta leayana* and *Tingis beesoni*, although 34 species of insects were found to be associated with the tree. The caterpillars of *Epiplema fulvilinea* were abundant at times, characteristically webbing together the tender leaves and the growing shoot. Also, the scolytine beetle, *Euwallacea* (= *Xyleborus*) *fornicatus*, commonly known as the shothole borer of tea, was found to attack the saplings during the summer months, which appears to be linked with moisture stress. In a study of 3 to 8-year-old plantations of *G. arborea* in western Maharashtra, India, Meshram *et al.* (2001) found that varying degrees of damage was caused by 12 species of insects, of which the defoliator *C. leayana*, the bark feeders *Acalolepta cervina, Indarbela quadrinotata* and *Sahyadrassus malabaricus*, and the sap sucker *Tingis beesoni* were rated as major pests, along with the bark-feeding termite *Odontotermes obesus*.

Pests in exotic plantations

In general, exotic plantations of *G. arborea* do not suffer major pest damage. Some pests of native plantations are also present in exotic plantations in Asia-Pacific. These include the bark feeder *Acalolepta cervina*; the stem borer *Xyleutes ceramicus*; and the shothole borer *Euwallacea fornicatus*, all in Malaysia, and the shoot borer *Alcidodes ludificator* in Indonesia. *X. ceramicus* infested 7–12 % of trees in Malaysia (Gotoh *et al.*, 2003). No major pests have been recorded in Zambia (Selander and Bubala, 1983), other African countries and Brazil. Additional pests noticed in exotic plantations are listed in Table 10.10 and include the following.

In Ghana, seedlings of *G. arborea* are attacked by the shoot-boring scolytine beetle *Hypothenemus pusillus*; the attack appears to be heavy on seedlings weakened by drought (Wagner *et al.*, 1991). Saplings are attacked by root-feeding termites in Malaysia and Cuba but the damage caused is minor.

Major defoliators are conspicuously absent in exotic plantations although the polyphagous leaf-cutting ants are a serious problem in some countries like Brazil, in addition to some other minor defoliators. The minor pests are mostly generalists; an exception is the geometrid caterpillar *Ozola minor* which causes moderate defoliation of out-planted seedlings in the Philippines (Yemane, 1990).

Among stem-boring insects, an unidentified cossid of the genus *Prionoxystus* is common on saplings in Indonesia. In East Kalimantan, 5–70% of the saplings were infested by *Prionoxystus* sp. (Ngatiman and Tangketasik, 1987). In a clonal multiplication nursery at Sebulu in East Kalimantan about 80% of the saplings that were stumped to produce multiple shoots were infested by this borer (Nair, 2000). At the same site, an unidentified small borer was found to damage shoot cuttings maintained in the nursery for rooting (Nair, 2000).

Pest profile

Craspedonta leayana (Coleoptera: Chrysomelidae)

Craspedonta leayana (Latreille) (Coleoptera, Chrysomelidae) (Fig. 10.19a,b), known until recently by the synonym *Calopepla leayana* Latreille, has been recognized since the 1920s as a serious pest of *G. arborea* in northern India, Bangladesh and Myanmar. This chrysomelid beetle of the subfamily Cassidinae, is 12–16 mm long and has a brilliant metallic colouration, with coarsely wrinkled, bluish green to violet blue elytra and pale yellow to reddish brown pronotum and legs. The larva has a characteristic appearance, with lateral spines. As in other cassidines, the excrement, instead of being ejected is extruded in fine, black filaments, longer than the body, and formed into bunches attached to the anal end. The moulted exuviae are also carried attached to the last abdominal segment. When disturbed, the larva flicks the anal filaments up and down and assumes a defensive posture.

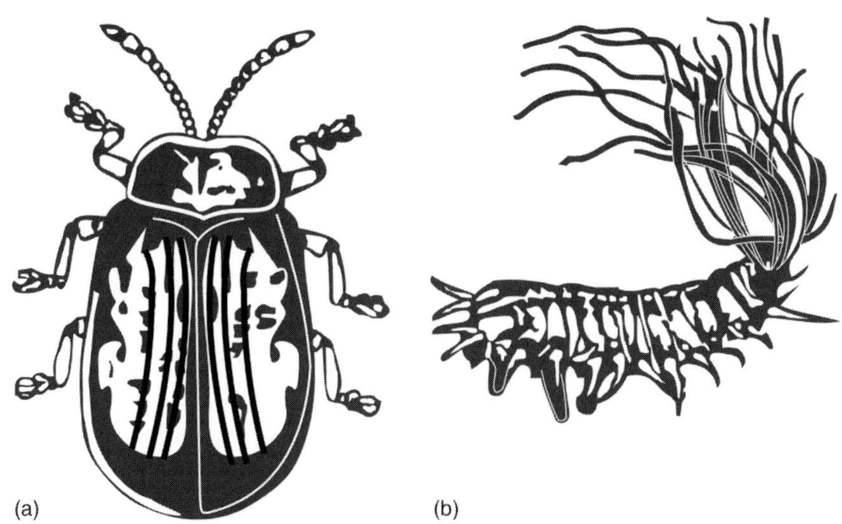

Fig. 10.19 *Craspedonta leayana*. (a) Adult (length 12 mm), (b) larva. After Ahmad and Sen-Sarma (1990).

Life history The biology of *C. leayana* has been studied in detail by Garthwaite (1939) and Ahmad and Sen-Sarma (1990). Under favourable temperatures, the life cycle is completed in 35–50 days, but third generation adults enter hibernation in winter. Eggs are laid in clusters of about 10–100 (average 68), on the under-surface of leaf or on tender stem and are covered by a sticky, frothy secretion which solidifies to form a domed, brownish ootheca. The oviposition period may range up to 45 days, with an average fecundity of 874 (Ahmad and Sen-Sarma, 1990). There are five larval instars. The larvae are gregarious. The early instars feed by scraping the surface of the leaf but later instars and the adult feed by making large, irregular holes on the leaf. Even young shoots are eaten up when the larval density is high. The larval period can be completed in about 18 days under optimal conditions. Pupation occurs on the leaf itself; before pupation the full-grown larva fastens itself to the leaf by the first three abdominal segments.

Seasonal incidence At Dehra Dun, in northern India, the beetles appear in May and pass through three generations, third generation beetles undergoing a quiescent period of about eight months from September–October to May. During this period, the beetles hide in cracks and holes under the dead bark of standing trees, in hollow bamboos, in grass clumps and thatches and in curled dry leaves on the ground (Beeson, 1941). Prior to the resting period, the beetles may disperse up to 2 km away from the plantations in search of suitable shelters. The period of inactivity, generally called hibernation and aestivation, depends on the weather, and its termination coincides with the appearance of new flushes on the host tree after a period of leaflessness in summer, and may vary from region to region.

Host range and geographical distribution There are no records of other hosts for *C. leayana*. The insect has been recorded in India, Bangladesh, Myanmar and Thailand. In India, it is prevalent in the northern region but also occurs in central and southern regions (Meshram *et al.*, 2001; Nair and Mathew, 1988). In a review paper, Suratmo (1996) lists *C. leayana* as a pest of *G. arborea* in Indonesia, where the tree is exotic, but gives no details of the place of occurrence. Since other authors have not listed the insect as occurring in Indonesia, this report needs confirmation (Nair, 2000). There are no reports of its occurrence in Africa or Brazil, where there are extensive exotic plantations of *Gmelina*.

Impact Both the adult and immature stages of *C. leayana* feed on leaves and, when the population density is high, also on shoots. Heavy attack causes total defoliation and drying up of the leader shoots in young trees, leading to severe growth retardation. With two or more consecutive complete defoliations the tree is likely to be killed. It is reported (Beeson, 1941) that in the 1930s over 800 ha of *Gmelina* plantations in the North Shan State of Myanmar were written off due to severe damage caused by this insect. As per reports from the Myanmar Forest Department, re-examination of the abandoned plantations after about 12 years indicated that where trees had survived defoliation they had flourished. However, due to the threat of *C. leayana*, large-scale monoculture plantations of this tree species are not currently favoured in countries where the insect is present.

Natural enemies Natural enemies of *C. leayana* include six species of parasitoids, a pentatomid predator (*Cantheconidia furcella*), a bird and an unidentified nematode parasite of the larva. The two most common parasitoids are *Brachymeria* sp. (Hymenoptera, Chalcidae) and *Tetrastichus* sp. (Hymenoptera, Eulophidae). The chalcid lays eggs on the prepupae and newly formed pupae and appears to be host specific. About 30–37% of pupae were found parasitized at Myanmar and 8–50%, at Dehra Dun in India, but hyper-parasitism has been noted. The eulophid is an egg parasitoid and has been recorded in India and Myanmar. It lays eggs by piercing the ootheca.

Control Although trapping of adults in artificial hibernation shelters, hand-picking of beetles returning to the plantation after over-wintering and mixed cropping (instead of monoculture) have been suggested in the past (Garthwaite, 1939), their effectiveness is limited. No control measures are generally practiced in countries where the tree is native and the pest is not present in exotic plantations (as noted earlier, its reported presence in Indonesia needs confirmation).

Several chemical insecticides, a commercial preparation of *Bacillus thuringiensis* subsp. *kurstaki* and the fungus *Beauveria bassiana* have been shown to be effective against the larvae (Sankaran *et al.*, 1989; Gupta *et al.*, 1989; Sharma *et al.*, 2001).

Knowledge gaps The biology of *C. leayana* has been studied only in regions where there is a well-defined winter season. It has been shown that the third generation adults enter a quiescent stage at the beginning of the winter, returning to the plantation only towards the end of the hot summer when the trees put forth a new flush of leaves. The period of rest, which may last about eight months, has been called hibernation and aestivation. More field observations are necessary on the resting habits, the dispersal of the beetles to and from the resting sites, the physiological state during the resting period (hibernation, aestivation or diapause) etc. The behaviour of the beetles in places where there is no clear-cut winter season also needs to be studied.

Pest profile

Tingis beesoni Drake (Hemiptera: Tingidae)

Tingis beesoni Drake (Hemiptera, Tingidae) (Fig. 10.20) is an occasionally serious pest of young *Gmelina arborea* saplings. The small, dark, lace bugs, 4.5 mm by 1.7 mm, aggregate in large numbers on the stems and branches of saplings and feed gregariously at the base of the leaf blade, sucking the sap from the larger veins.

Life history and seasonal incidence The biology of *T. beesoni* has been studied in detail by Mathur (1979). Eggs are inserted in a vertical row into the tender shoot tissue. The nymphs congregate at the base of the leaf lamina and the axils. There are five nymphal instars and the life cycle is completed in 11–40 days between April and October, depending on the temperature (Mathur, 1979). There is considerable overlap between generations. Up to seven generations may be completed per year and eggs laid in the cold weather overwinter, hatching only in the following March.

Host range and geographical distribution *T. beesoni* has not been recorded on any other host. It occurs in India, Myanmar and Thailand. In India, it has been

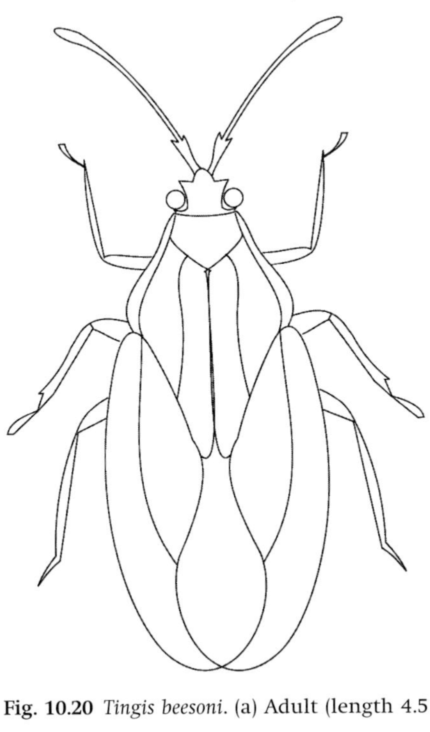

Fig. 10.20 *Tingis beesoni*. (a) Adult (length 4.5 mm), (b) larva. After Mathur (1979).

specifically recorded in Dehra Dun, Madhya Pradesh, Uttar Pradesh, Maharashtra and Kerala.

Nature of damage and impact As a result of feeding by the adults and nymphs, the leaves become spotted and discoloured and wither. Eventually the shoots die back. In an outbreak in a year-old, 10 ha plantation in Kerala, southern India in 1978, 67% of the saplings were infested, of which 21% suffered total defoliation while the remainder suffered varying degrees of leaf fall. Heavy infestation was concentrated over a patch of about two hectares where there was total defoliation (Nair and Mathew, 1988). In this patch, most saplings later showed dieback of shoots and epicormic branching. In Madhya Pradesh, India, Harsh *et al.* (1992) found that the insect attack was followed by infection of the plant by the fungus *Hendersonula toruloidea* (*Nattrassia mangiferae*) which was characterized by black necrotic lesions at leaf bases, followed by defoliation and drying of young shoots. They found that spraying a mixture of insecticide and fungicide (0.02% monocrotophos and 0.1% carbendazim) controlled the damage.

Natural enemies No natural enemies are on record.

Control Nair and Mathew (1988) found the systemic insecticide, dimethoate, ineffective against *T. beesoni* in field trial, but lindane was effective. As noted above, Harsh *et al.* (1992) recommended a combination of insecticide and fungicide because of the additional infection by a fungus. Meshram and Tiwari (2003) recommended application of the synthetic pyrethroid, deltamethrin (0.005%) and the fungicide, carbendazim (0.1%), at 15-days interval.

Knowledge gaps There is little information on the natural control agents of *T. beesoni* and on the seasonal population trend of the pest in areas where there is no well-defined winter season. The circumstances under which outbreaks of *T. beesoni* population occur are not understood.

10.10 *Leucaena leucocephala* (Fabaceae: Mimosoideae) (Common name: leucaena)

Tree profile

Leucaena leucocephala (Lam.) de Wit is a multipurpose legume native to Mexico and some parts of Central America, within 16°N to 30°N latitude. Because it is cultivated throughout region, its true natural distribution is obscure (CABI, 2005). Two major varieties are recognized – the Hawaiian shrubby variety that grows up to 8 m tall and the giant or Salvador variety that grows up to 16 m tall. Due to its several uses – for fodder, green manure, fuel, shade for estate crops, erosion control, nitrogen fixing etc., as well as fast growth and ease of propagation – the species has been widely planted outside its native distribution range. It was introduced to many countries in Latin America prior to 1500, to the Philippines in the early 1600s and to most other tropical countries in the late 1900s (López-Bellido and Fuentes, 1997; CABI, 2005). Most early introductions were of the shrubby variety; the giant variety has been introduced outside Central America only since 1960.

Plantations of *L. leucocephala* have been raised in almost all tropical countries, for various purposes, particularly in agroforestry planting programmes since the 1960s. CABI (2005) lists over 130 countries where it is planted. Large areas have been planted in many countries; for example, Indonesia had 1.2 million ha of leucaena plantations by 1990 (Oka, 1990) and the Philippines over 300 000 ha by 1986 (CABI, 2005).

Overview of pests

Insect pests do not pose a major threat to *L. leucocephala* in its native habitat. A psyllid bug, *Heteropsylla cubana*, which has become a serious pest in exotic plantations of this species (see below) occurs in some places in Mexico where the tree is indigenous. However, it is not a major pest in Mexico, where its population fluctuates between low and high densities at small spatial scales, with the damage always confined to the younger leaves and no loss of older foliage (McClay, 1990; Waage, 1990). Other minor pests in Mexico include an unidentified arctiid caterpillar, a coreid bug that feeds gregariously on shoot tips, a membracid bug and a thrips on young leaves, and bruchid seed beetles (Waage, 1990). Another minor pest in the neotropics is *Semiothisa abydata* (Lepidoptera: Geometridae), polyphagous on *Leucaena* spp. and other tree legumes. It has also recently spread across the Pacific and into Southeast Asia (Waage, 1990). The full range of insects associated with *L. leucocephala* in its natural habitat has not been well documented, but there is no major pest problem.

No serious pest problems have been recorded, either, in small-scale plantations raised in countries where the tree is native. Over the past 500 years, *L. leucocephala* has become naturalized in the broader region of tropical America, including the West Indies and Florida. In this region, *H. cubana* is not considered a major pest although significant damage to plantations has occurred in Florida, Cuba and Colombia. A microlepidopteran, *Ithome lassula* (Cosmopterygidae) has also been reported from Florida; its larva bores into the flower bud. It also occurs in Australia (Beattie, 1981) and India (Pillai and Thakur, 1990).

There are over 40 species of indigenous insects that feed on exotic plantations of *L. leucocephala* in various countries. These have been listed by Nair (2001a). They include leaf-feeding curculionid and chrysomelid beetles and grasshoppers;

sap-sucking coreid, pentatomid, aleurodid, coccid, psuedococcid, eurybrachyid and membracid bugs; root-feeding whitegrubs and termites; and stem or branch-boring cerambycid beetles and a cossid caterpillar. In addition, many seed beetles feed on the seeds of leucaena in Africa, India and the Philippines. Outbreak of the psuedococcid, *Ferrisia virgata* was reported in a three-year-old, 25 ha plantation at Salem in Tamil Nadu in India (Pillai and Gopi, 1990c). However, none of these insects causes consistently serious damage. On the other hand, the psyllid bug *Heteropsylla cubana*, which has found its way from its natural habitat in tropical America to many exotic locations, after nearly 25 years since leucaena planting began, has become a serious pest. A pest profile of this species is given below.

Pest profile

Heteropsylla cubana (Hemiptera: Psyllidae)

Heteropsylla cubana D.L. Crawford (Hemiptera, Psyllidae), first described in 1914 in Cuba, has emerged as a devastating pest of exotic leucaena plantations since the mid 1980s. It is a small bug measuring 1.5–2 mm in length, usually yellowish green, some with shades of brown. The nymphs are also usually yellowish green, but other shades of colour may also be seen. The nymphs and adults feed gregariously on the terminal shoot (Fig. 10.21), sucking the sap of developing leaves. Now commonly known as the leucaena psyllid, this insect has become a typical example of the risk of pest outbreaks in forest plantations of exotics, with a series of devastating outbreaks in exotic *L. leucocephala* plantations across the tropics.

Life history The adult female *H. cubana* lays an average of about 240 eggs (Rauf *et al.*, 1990) which are attached to the tender, unopened pinnules of the new flush of leaves. Up to 21 eggs per pinnule have been recorded (Joseph and Venkitesan, 1996). The eggs hatch in about 3 days and the nymphs suck the sap of developing leaves. They pass through 5 instars in about 8 days. Adults live for about 10–15 days. Rauf *et al.* (1990) estimate a mean generation time of 14.92 days, net reproductive rate (R_0) of 51.35 and an intrinsic rate of increase (r_m) of 0.264. The population doubling time was estimated as 2.52 days by Napompeth and Maneeratana (1990). The insect passes through many overlapping generations per year and all life stages can usually be found together on terminal shoots.

Host range and geographical distribution In addition to *L. leucocephala*, *H. cubana* can survive on *L. diversifolia*, *L. pulverulenta*, *L. trichodes* and *L. salvadorensis*.

Fig. 10.21 Terminal shoot of *Leucaena leucocephala* infested by the psyllid *Heteropsylla cubana*.

It can feed on all 13 species of *Leucaena* to some degree and also on *Samanea saman* (Geiger *et al.*, 1995; CABI, 2005).

From its native habitat in Latin America, the spread of *H. cubana* across the tropics has been dramatic; Napompeth (1994) gives full details of the chronology of the spread. In natural stands of leucaena in Mexico, the insect occurs at varying densities, from sparse to dense populations at small spatial scales. The first noticeable outbreak occurred in Florida in late 1983. It then appeared in Hawaii in April 1984. Since then there has been a progressive westward movement across the globe (Fig. 10.22). By 1985, it spread throughout several small islands in the Pacific and reached the Philippines and Taiwan. In 1986 it was noticed in Indonesia, Malaysia, Thailand, southern Myanmar, southern China and neighbouring countries. In 1987 it appeared in the Andaman Islands in India and in Sri Lanka, and the next year in southern peninsular India. The westward movement continued, and in 1992 infestations were noticed in the African continent, in Tanzania, Kenya, Uganda, and Burundi and by 1994 in Sudan and Zambia (Geiger *et al.*, 1995; Ogol and Spence, 1997). Thus in less than 10 years, this pest has spread from its native range in tropical America, across the Pacific to Asia and Africa – an unusual spread for an insect, in recent history.

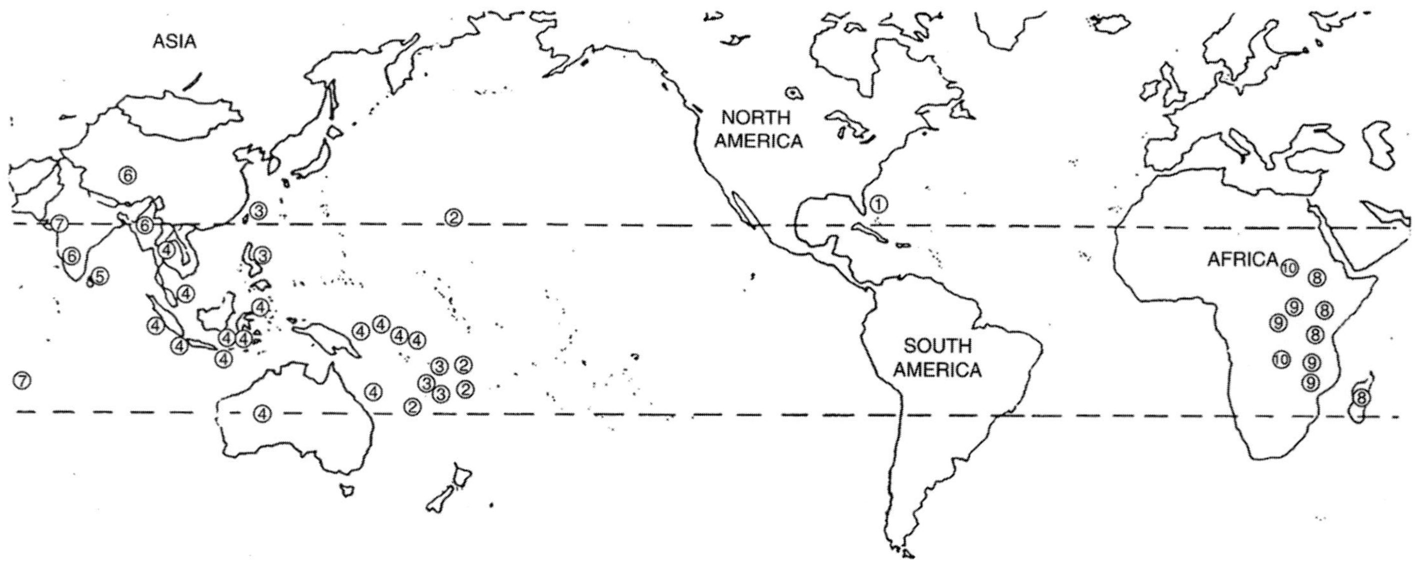

Fig. 10.22 Map showing the westward spread of the leucaena psyllid *Heteropsylla cubana* across the globe. The serial number within the circle shows the year in which the pest was first noticed in the location: 1 – 1983, 2 – 1984, 3 – 1985, 4 – 1986, 5 – 1987, 6 – 1988, 7 – 1991, 8 – 1992, 9 – 1993, 10 – 1994.

It is likely to continue to expand its distribution to other suitable areas of Asia and Africa, and even Europe, where its host plants exist in sufficient numbers (CABI, 2005).

Nature of damage and impact *H. cubana* infestation of terminal shoots is usually heavy; up to 3000 nymphs and adults have been recorded per 15 cm of shoot terminal. As a result of gregarious feeding by the bug, the developing terminal leaves become chlorotic and deformed, or the leaflets become yellow and drop. The petioles become black and the shoots dry up. Pollarding the trees for green manure or fodder results in the development of profuse new young shoots, creating an ideal food supply for the bug. The damage occurs in about a week of infestation. Heavy infestation usually results in complete defoliation and the growth of the tree is stunted. Repeated defoliation sometimes leads to death of the trees although generally they recover. Tree mortality is suspected to be due to secondary infection by pathogenic microorganisms (Napompeth, 1990a).

The devastation caused by *H. cubana* to exotic leucaena plantations during the initial outbreaks has been heavy. It has had significant economic, political and scientific repercussions. The Philippines and Indonesia, which had vast areas under leucaena, were the worst hit and many farmers became reluctant to continue its cultivation. The impacts are well documented in several country reports presented in a 1989 Workshop at Bogor, Indonesia (Napompeth and MacDicken, 1990) and summarised by Napompeth (1994) and Geiger *et al.* (1995). CABI (2005) also gives a concise summary of the impact in some countries, summarised from various sources. It indicates the following. The damage depended on the purpose for which leucaena was cultivated and had a sequential effect on many products and values. Indonesia, which had 1.2 million ha under leucaena when the outbreak started, suffered badly and the Government declared the outbreak a national disaster and established a task force for its control. In Indonesia, leucaena fodder had made it possible to raise tethered cattle which released land from pasture for cultivation. The outbreak affected the export of cattle and their products. In large estates of cocoa, coffee, vanilla and oil palm, where leucaena was used for providing shade to the crops, lack of shade resulted in fall of productivity; for example, cocoa yield fell by 40%. Export income of the country from these crops fell drastically. The projected loss to estate crops, animal production and the forestry sector from 1991–1996 was US $1.5 billion. In the Philippines, where more than 300 000 ha were planted to leucaena by small-scale farmers for production of fuel, fodder and leaf meal, and the local and export demand for leaf meal was 57 and 194 000 metric tons per year respectively (valued at US $109 per ton), damage reached 80% of total leucaena leaf meal production. It was estimated that the monthly income of farmers from leucaena planting fell from 1046 pesos in 1984 to 489 pesos in 1987. In other countries like Thailand, Malaysia, Indonesia and Australia, where the leucaena planting was less extensive, the impacts were many, but less serious.

The psyllid problem also triggered several international meetings and research initiatives. Research was initiated on breeding for pest resistance and biological control using introduced natural enemies.

Population dynamics *H. cubana* is a species which exhibits population outbreaks in exotic plantations where it is newly introduced but not in natural stands in counties where it is indigenous. Large-scale plantations do not exist in countries where the host plant is indigenous, to indicate whether outbreaks are a consequence of raising plantations. The key factors that control the population dynamics of *H. cubana* are not fully understood but it appears that natural enemies play an important role. As noted earlier, in natural stands of leucaena in Mexico the insect occurs at varying densities, from sparse to dense populations at small spatial scales. Here the insect population displays a strong seasonality, apparently related to the synchronous flushing of its host plant and the depression of flushing during the flowering and fruiting season (Waage, 1990). Coccinellid predators and other natural enemies are believed to keep the psyllid populations below economic injury level in agroecosystems in Cuba, although there are seasonal fluctuations (Valenciaga *et al.*, 1999).

Napompeth (1994) observed that seasonality is also exhibited in exotic locations as in Thailand, Laos and Vietnam. In Thailand, the population begins to increase in the cooler months at the end of the wet season, and during warm periods, the insect can be found only in pockets with cooler microclimate and the cooler highlands. In Hawaii, with a generally cool climate, the insect is prevalent throughout the year. In countries near the equator (e.g. Malaysia, Indonesia), the insect can be detected throughout the year, but at low population densities at times. These observations led Napompeth (1994) to suggest that the ups and downs of the psyllid populations are related to an optimum cooler temperature range and the availability of tender shoots. Geiger and Gutierrez (2000) showed that the psyllid infestation was greater at a cool highland than at a warm valley site in north Thailand, that the lower thermal threshold for psyllid development was 9.6° C and that there was a dramatic decrease in its abundance when maximum temperatures exceeded 36° C. Several workers have studied the seasonal fluctuations of *H. cubana* populations in exotic plantations, but no consistent trend has emerged, partly because of the complication introduced by pollarding, which brings about a sudden decline of the population followed by its increase as new flushes come up.

Well-defined population peaks were recorded in maize-leucaena agroforestry plantations in Kenya, where the abundance was lowest during months with little or no rain and months with heavy rainfall, and higher during the intervening period of moderate rainfall (Fig. 10.23) (Ogol and Spence, 1997). They found no clear correlation with temperature, within the natural range obtained. In Yogyakarta, Indonesia, *H. cubana* populations were higher during the dry season than during the rainy season, which was attributed to incidence of fungal disease during the wet season (Mangoendihardjo *et al.*, 1990). In pollarded leucaena plantations in the Philippines, the psyllid population was present throughout the year, with wide fluctuations (Fig. 10.24), extreme wet and dry periods reducing the numbers except in the cool and moist mountainous areas (Villacarlos *et al.*, 1990). It is evident that weather, pollarding and natural enemies are the important factors influencing the leucaena psyllid populations. Weather, mainly rain and temperature, in addition to their direct effect on the dispersal and growth of the insect, exert indirect influence through effects on the growth of the plant and the fungal pathogens of the insect. The increase in the psyllid population following the monsoon rainfall is probably also due to dispersal and arrival of the psyllids through the monsoon wind system.

On a larger temporal and spatial scale, there has been a gradual decline in the abundance of the leucaena psyllid since the outbreaks began. This has been well documented in Thailand (Van Den Beldt and Napompeth, 1992). Following the

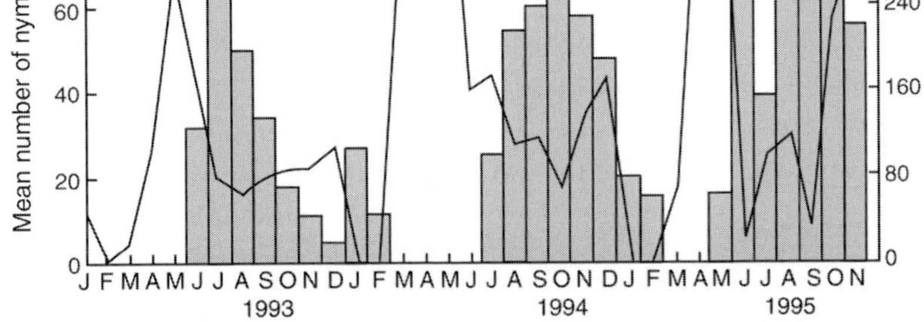

Fig. 10.23 Seasonal abundance of the leucaena psyllid *Heteropsylla cubana*, in relation to rainfall at Mtwapa, Kenya. From *Insect Science and its Application* (Ogol and Spence, 1997).

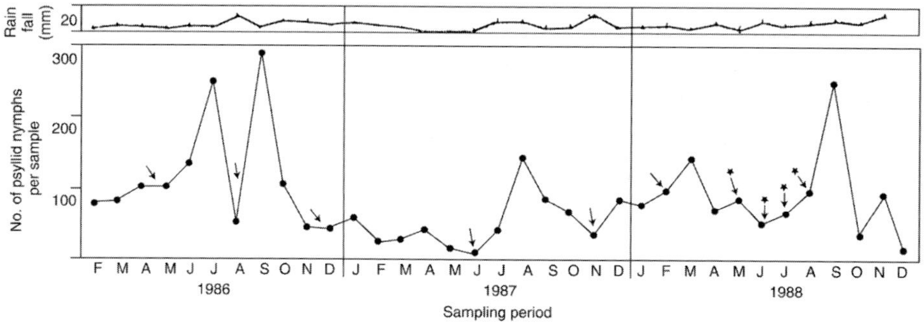

Fig. 10.24 Seasonal abundance of the leucaena psyllid, *Heteropsylla cubana*, in a pollarded leucaena plantation in a hilly humid region in the Philippines. Arrows indicate pollarding and starred arrows, staggered pollarding. (Adapted from Villacarlos *et al.* (1990)).

first invasion of *H. cubana* into Thailand in September 1986, the relative level of damage during the peak infestation period in December–January fell to about 20% of the original over a period of seven years. In an update of the situation in Asia-Pacific, Geiger *et al.* (1995) concluded that the damage is generally heavy in about the first two years of invasion and then gradually weakens in duration and severity. Such has been the case in Indonesia and also the Philippines. The exact reason for this decline is not known, but it is believed that over time the indigenous and introduced natural enemies have played the major role. The vacant niche created by the expanding psyllid populations has now been filled by their natural enemies. The most important component of this might be the inoculum load of indigenous fungal pathogens.

Natural enemies In Mexico, where *L. leucocephala* and *H. cubana* are native, the insect is attacked by several groups of general predators – spiders, syrphids, chrysopids, reduviids, anthocoreids and coccinellids (Mc Clay, 1990). Investigations over a wider area in the Neotropical region showed that the most common natural enemies were the parasitoids *Tamarixia leucaenae* (Eulophidae) and *Phyllaephagus* sp. nr. *rotundiformis* (Encyrtidae) and the predator *Curinus coeruleus* (Coccinellidae). The natural enemies are believed to keep the pest population in check in the native habitat.

In exotic plantations, endemic, general predators such as spiders, dragonflies, ants, coccinellids and birds have been reported from countries like Thailand, Indonesia, India and the Philippines (Mangoendihardjo *et al.*, 1990; Napompeth, 1990b; Villacarlos *et al.*, 1990; Joseph and Venkitesan, 1996; Misra *et al.*, 2001). Many entomopathogenic fungi were also found on *H. cubana* in Taiwan, the Philippines and Thailand; the most dominant in Thailand were *Entomophthora* sp.

and *Conidiobolus coronatus* (Napompeth, 1990b), and epizootics mainly due to the former have been frequently observed in the Philippines (Villacarlos *et al.*, 1990). Other fungal pathogens of the psyllid found in the Philippines were *Fusarium* sp., *Paecilomyces farinosus* and *Hirsutella citriformis*. Villacarlos and Wilding (1994) recorded four new species of Entomophthorales attacking *H. cubana* in the Philippines, and epizootics of one of them, *Neozygites heteropsyllae* sp. nov. occurred commonly in moist areas where the psyllid populations were dense.

Control Although the leucaena psyllid has caused substantial damage during the initial years of its arrival in exotic locations, a combination of several factors has put a brake on the escalating pest population and the crisis is now under control. The control effort has been massive, but it is not clear to what the success has been due.

Many chemical insecticides have been tested and found only partially effective in field applications, with the systemic ones being more effective. They have not generally been used as the risk and cost are prohibitive for a forage crop like leucaena.

Some psyllid resistant varieties or genotypes were identified by screening a large number of accessions of leucaena at the University of Hawaii and other places, and by hybridisation between *L. leucocephala* and other species of the genus at the Taiwan Forestry Research Institute. However, the resistance, attributed to some secondary metabolites of leucaena, has not been found to be stable and the results are largely inconclusive at present (CABI, 2005).

Based on research at the Hawaii Department of Agriculture, four promising natural enemies from tropical America were introduced to some countries in Asia during the 1980s. These were the coccinellid predators *Curinus coeruleus* and *Olla v-nigrum*, an encyrtid parasitoid *Psyllaephagus yaseeni* and an eulophid ecto-parasitoid *Tamarixia leucaenae*. *C. coeruleus* and *P. yaseeni* have become successfully established in several countries and have exerted pressure on *H. cubana* populations. *C. coeruleus* has been introduced into Guam, India, Indonesia, Myanmar, Papua New Guinea, the Philippines, Thailand and Vietnam. It has established itself in most countries although its role in the suppression of the psyllid populations is not confirmed in all the countries. *C. coeruleus*, like most other coccinellid predators, has little prey specificity; it feeds also on mealy bugs, scale insects, aphids and whiteflies. For this reason, as well as its longer generation time, lower rate of fecundity and poorer dispersing ability compared to *H. cubana*, Speight and Wylie (2001) argue that it cannot be an effective biological control agent for *H. cubana*. Over the years, several native predators, parasitoids and pathogens, originally present on other insect hosts, have also attacked the leucaena psyllid, checking its population build-up. As noted earlier, epizootics due to native entomopathogenic fungi have also occurred and several reports indicate that they may have played the major role in the natural collapse of *H. cubana* outbreaks in exotic locations.

Knowledge gaps Unfortunately, research on *H. cubana* has practically come to a halt worldwide, with the decline of its outbreaks in exotic leucaena plantations. The reasons which led to this welcome decline are poorly understood, as discussed above. The present situation offers an opportunity for well-planned ecological studies to elucidate the factors controlling the population dynamics of this typical outbreak species. The status and relative roles of the introduced predator and parasitoid, and the adaptive response of the native predators, parasitoids and entomopathogenic fungi need to be investigated.

10.11 ***Manglietia conifera*** **(Magnoliaceae)**

Tree profile

Manglietia conifera Dandy is an evergreen tree endemic to Vietnam and the southern parts of China (Guangdong, Guangxi and Yunnan). This tree, which is widely planted in north Vietnam, has often been incorrectly identified as *Manglietia glauca* which is found in Indonesia (CABI, 2005). *M. conifera* yields high quality furniture timber which is also used for veneer and pulp. The tree grows up to 25 m in height and 50 cm in diameter. In Vietnam, about 85 000 ha of plantations have been raised, the largest for any single tree species in the country. Only natural stands occur in China.

Overview of pests

Outbreak of a sawfly species (see pest profile below) is common in north Vietnam. There is scanty information on other pests although CABI (2005) lists another sawfly, *Sterictiphora* (Hymenoptera: Argidae) and *Zeuzera* (Lepidoptera: Cossidae) as also feeding on *M. conifera*.

Pest profile

Shizocera sp. (Hymenoptera: Argidae)

Outbreaks of a sawfly, *Shizocera* sp. (Hymenoptera: Argidae), commonly called 'Mo' by the local people, has been noticed in pure stands of *Manglietia conifera* since 1966. The insect is also found in natural forest where the tree is scattered. It feeds on leaves. Because the tree and its pest have a limited distribution, the literature on the pest is also limited. Tin (1990) reported on the results obtained in a study carried out by the Forest Research Institute of

Vietnam from 1971–76, in an experimental forest station in Vinh Phu Province. The following information is based on this report.

Life history The sawfly deposits eggs under the epidermis of the leaf, in two rows on either side of the midrib. The incubation period depends on the temperature – it may range from 3 days at 28 °C to 26 days at 16 °C. As the eggs develop, they increase in size, as do the eggs of other sawflies, by absorbing water from the leaf. The larvae pass through five instars in the male and six in the female. The larval period ranges from about 17 days at 24 °C to 32 days at 19 °C. The mature larva falls down and creeps into the ground to make a cocoon. As in other sawflies, three developmental stages are passed within the cocoon – eonymph, pronymph and pupa. The duration of the eonymph stage is short, pronymph stage is 13–15 days and pupa, 8 days at 27 °C to 15 days at 18 °C. The life cycle is completed in 59–65 days at 22–24 °C and 85–86% RH.

The Mo sawfly appears to prefer a temperate climate: a temperature of 21–24 °C, monthly precipitation of 50–150 mm and RH of 85–88%. Under some combinations of climatic factors, the Mo sawfly enters diapause and/or aestivation (see below).

Host range and geographical distribution *M. conifera* is the only known host of this sawfly and its distribution is coincident with that of the host tree (i.e. Vietnam and southern China).

Impact Although outbreak of the sawfly is reported to cause severe damage to leaves, its quantitative impact has not been studied.

Natural enemies No information is available on natural enemies.

Population dynamics In northern Vietnam, emergence of the adult Mo sawfly occurs twice a year – from spring to early summer (March to May) and from autumn to early winter (late August to early January). Emergence during autumn–winter takes place in waves. The favourable period for the growth of the sawfly is about five months during the spring–summer and about three months during the autumn. The insect undergoes aestivation and/or diapause under certain combinations of climatic factors which are not clearly understood. Diapause occurs in the eonymph stage. Prolonged diapause in cocoons may sometimes continue through the winter, spring and summer, with the adults emerging in autumn–winter of the following year, or the diapause may continue over the summer, autumn and winter, with the adults emerging in the spring of the following year. In the active season, the Mo sawfly may have two generations – one which takes about two months to complete and another which takes about 9–11 months. Depending on the temperature, in some places and some years there may be only one generation per year, due to diapause.

Control No effective control measures are known.

Knowledge gaps Most sawflies are found in the temperate regions and much remains to be learnt about the ecology, impact and control of this oriental sawfly whose taxonomic identity also needs to be established.

10.12 *Milicia* species (Moraceae)

Tree profile

Milicia excelsa (Welw.) C.C. Berg and *Milicia regia* (A. Chev.) C.C. Berg (Moraceae), formerly included under the genus *Chlorophora*, are highly valued African timber species. Together they are known in trade as iroko. They occur in dry, moist and wet forest types, at low elevation. *M. excelsa* is distributed in a wide belt from Senegal in the west to Tanzania in the east while *M. regia* is restricted to West Africa (CABI, 2005). Trees can attain heights up to 30–50 m. The timber is strong and resistant to insect attack and decay, and equivalent in value to teak. *Milicia* is dioecious, with male trees having a narrower crown, with lighter coloured foliage. In natural forests *Milicia* trees occur at very low density; in Ghana, the density ranges from 0.2 trees per ha in rain forest areas to 2.4 trees per ha in dry semi-deciduous forest (CABI, 2005). Plantations of *Milicia* have not been successful, largely because of attack by a psyllid pest (see below).

Overview of pests

The gall-forming psyllid, *Phytolyma* sp. (Hemiptera: Psyllidae) is a major pest of *Milicia* spp. in plantations as well as natural forests. A pest profile is given below. There are no other major pests.

Pest profile

Phytolyma species (Hemiptera: Psyllidae)

Two species of *Phytolyma* are important pests of *Milicia*. Although the specific name *P. lata* had been applied earlier to the psyllid infesting both *M. excelsa* and *M. regia*, recent taxonomic studies indicate that the one infesting *M. excelsa* is *P. fusca* and the one infesting *M. regia* is *P. lata* Walker (Scott). Both are small insects, measuring 3–4 mm in length. The adults are active and move rapidly in a jumping flight. The nymphs make galls on leaves and live inside.

Life history and seasonal incidence The adult psyllid lays eggs in rows, or rarely scattered singly, on the buds, leaves or shoots of *Milicia*. After about eight days of incubation, the first instar nymphs, known as crawlers, emerge and crawl on the plant surface. The nymph then burrows into the leaf tissues. A gall is formed within two days, completely enclosing the nymph. The galls are globular, more than 3 mm in diameter, and occur most commonly on the midrib; some galls may also form on tender stems. The nymph feeds within the gall tissue. Several such galls on young leaves and shoots may coalesce and become one bunched mass of gall tissue. The nymph passes through five nymphal instars within two to three weeks (Wagner *et al.*, 1991). When development is complete, the gall becomes turgid and bursts open, releasing the adult. After this, saprophytic fungi usually colonize the injured leaf tissue, causing decay and eventual dieback of the terminal shoot. Ten or more generations of the insect may occur per year (CABI, 2005).

Impact Plantation programmes of *Milicia* spp., the most valuable timber species of tropical Africa, have been seriously hampered by the attack of *Phytolyma* spp. The damage is more serious in nurseries and young plantations. Heavily infested shoots become a putrefying mass and the stems die back. Repeated attacks damage the auxiliary shoots also. In nurseries, 100% failures have often been reported in Ghana (Wagner *et al.*, 1991). Attack occurs throughout the year but is more severe during the rainy season from April to October. Trees in natural forests are also attacked, but crowded seedlings in nurseries are the worst hit.

Host range and geographical distribution Although *P. lata* was earlier thought to attack several species of *Milicia*, recent literature suggests that different species of *Phytolyma* attack the two main species of *Milicia*, as mentioned above. *Phytolyma* is distributed widely in tropical Africa, from west to east, coinciding with the distribution of *Milicia* species.

Natural enemies Natural enemies of *Phytolyma* are limited to relatively few species (CABI, 2005). Encyrtid and eulophid (Hymenoptera) parasitoids have been recorded on nymphs. At least 10 generalist predators including mantids and reduvids have also been recorded.

Control Although several chemical insecticides were tested against *Phytolyma* in Ghana and Nigeria, and some systemic ones gave promising results, control using pesticides has been largely ineffective and uneconomic.

Some resistance has been noted in provenances of *Milicia* species against *Phytolyma*, although there is no absolute resistance (Cobbinah and Wagner, 2001). The resistant lines produce small and hard galls which do not open to release the adult psyllid, which becomes trapped and dies as a result (CABI, 2005).

Recent studies indicate that planting *Milicia* with other tree species reduces the psyllid damage. Integrated pest management involving vegetatively propagated psyllid resistant clones on which parasitism was found to be higher than on susceptible clones, and planting in mixture with other tree species is currently showing promise for raising successful plantations (Cobbinah and Wagner, 2001).

10.13 *Neolamarckia cadamba* (= *Anthocephalus cadamba*) (Rubiaceae)

Tree profile

Neolamarckia cadamba (Roxb.) Bosser, known until recently as *Anthocephalus cadamba* (syn. *A. chinensis*,) is a fast-growing, medium to large deciduous tree. It is commonly known as Kadam in India, Laran in Malaysia and Jabon in Indonesia. It has a light-coloured wood used for plywood, light construction and pulping. It is widely distributed from India through Southeast Asia to New Guinea and is common in logged-over lowland dipterocarp forests and thrives well in freshwater swamps. Plantations have been raised in India, Sri Lanka, Myanmar, Indonesia, Malaysia and the Philippines. In Indonesia, it is planted in Java to replace poor teak plantations after harvest, and in North Sumatra, Riau and Central Kalimantan as industrial plantations for pulpwood (Nair and Sumardi, 2000). It has also been introduced to and planted in other tropical and subtropical countries including South Africa, Puerto Rico, Surinam, and Taiwan (CABI, 2005).

Overview of pests

More than half a dozen species of defoliators have been recorded on *N. cadamba* (Chey, 2001). The caterpillar *Arthroschista hilaralis*, for which a pest profile is given below, is the most serious pest. Another defoliator is the hornworm, *Daphnis hypothous* (syn. *Deilephila hypothous* (Lepidoptera: Sphingidae) which is common in Malaysia and the Philippines. Although it does not build-up in large numbers, the damage caused is substantial because of its voracious feeding. *Eupterote fabia* (Lepidoptera: Eupterotidae) also causes heavy defoliation of young and old trees in the Philippines (Quinones and Zamora, 1987). Leaf damage is also caused by a few other, less common lepidopteran caterpillars and curculionid and scarabaeid beetles. Some sap-sucking cicadellids have also been recorded. Whitegrubs damage one to two-year-old seedlings in Indonesia (Intari and Natawiria, 1973) and the hepialid caterpillar *Sahyadrassus malabaricus* (see pest profile under teak) bores into the stem of saplings in India (Nair, 1987b). In general, pests other than *A. hilaralis* have not posed a serious threat to *A. cadamba* in plantations.

Pest profile

Arthroschista hilaralis (Walker) (Lepidoptera: Pyralidae)

Arthroschista hilaralis (Walker) (syn. *Margaronia hilaralis, Daphnia hilaralis*) (Lepidoptera: Pyralidae) (Fig. 10.25a,b) is an important defoliator of young plantations of *Neolamarckia cadamba*. The bluish green moth has a wingspan of about 34 mm. The mature larva is pale green, with a dark brown head capsule, and about 25 mm long, with inconspicuous hairs.

Life history and seasonal incidence The life history has been studied in Sabah, Malaysia by Thapa (1970) and in West Bengal, India by Thapa and Bhandari (1976). The female moth lays 60–70 eggs, singly or in groups of two or three, on leaves. There are five larval instars. The first and second instar larvae feed on soft leaf tissue under cover of a silken web. The later instars eat out the entire

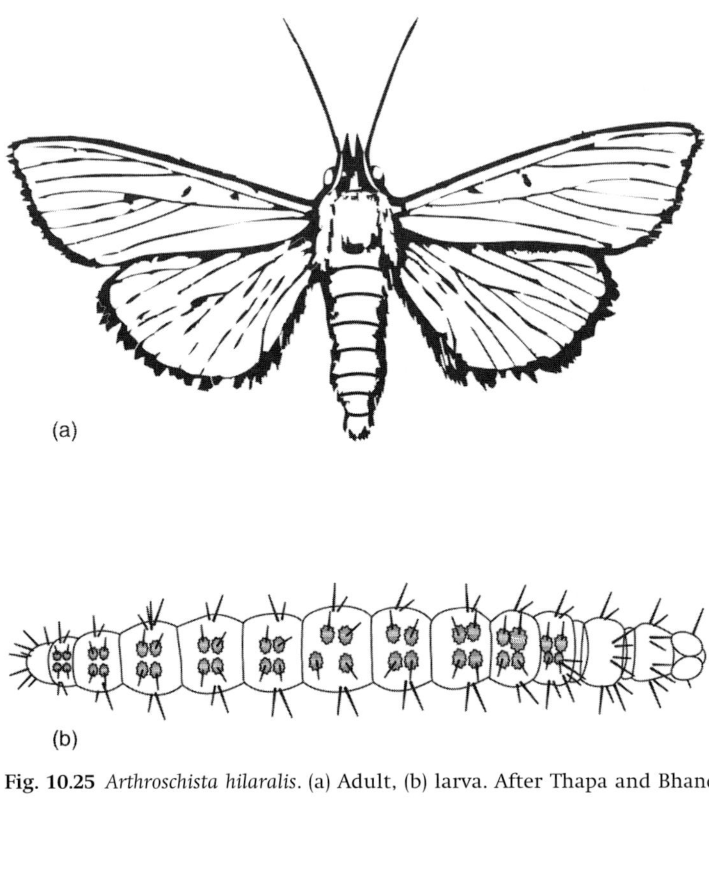

(a)

(b)

Fig. 10.25 *Arthroschista hilaralis*. (a) Adult, (b) larva. After Thapa and Bhandari (1976).

leaf blade between the veins, under cover of a partial leaf fold. The larval development is completed in about 15 days and pupation takes place inside the silken web. The total life cycle is completed in about 21–26 days. In India, the insect can complete 11–12 generations a year in West Bengal and 8–9 at Dehra Dun, where the larval period is prolonged in the winter.

Observations made in young plantations at Chilapata in West Bengal, India (Thapa and Bhandari, 1976) showed that peak infestation occurs during the post-monsoon period in August–September, during which moderate to heavy defoliation occurs in all plantations. A low population persists during the rest of the year. At Sabah in Malaysia, population peaks have been recorded twice a year, in April–June and November–January (Thapa, 1970).

Impact Feeding of the early instars on the leaf surface causes browning of leaves, while consumption of the leaf blade by older larvae leads to shedding of leaves. In defoliated trees, the larvae feed on the soft terminal shoot, causing dieback and formation of epicormic branches. Thus the growth of saplings is adversely affected by *A. hilaralis*, although the plants seldom die.

Host range and geographical distribution The only confirmed host of *A. hilaralis* is *N. cadamba*. Although Beeson (1941) also listed *Duabanga grandiflora* (Sonneratiaceae) as a host, Thapa and Bhandari (1976) reported that the larvae failed to feed on its leaves. *A. hilaralis* has been recorded in India, Malaysia and the Philippines. An undetermined species of *Arthroschista*, probably, *A. hilaralis*, has also been recorded on *N. cadamba* in Indonesia (Suratmo, 1987).

Natural enemies Natural enemies include six hymenopteran larval parasitoids, three hymenopteran pupal parsitoids and a few reduvid, carabid and ant predators. *Apanteles balteata* (Braconidae) was reported to parasitize up to 60% of larvae during peak incidence of the pest in West Bengal, India (Thapa and Bhandari, 1976) and *A. stantoni* up to 50% of larvae in Sabah, Malaysia (Thapa, 1970). *Litomastrix* sp. (Encyrtidae) also causes substantial parasitism in Malaysia. Other larval parasitoids include *Cedria paradoxa* and *Macrocentrus philippinensis* (Braconidae) and *Sympiesis* sp. (Eulophidae).

Control No effective control methods have been developed against *A. hilaralis*.

Knowledge gaps *A hilaralis* is probably a serious pest of *N. cadamba* in Indonesia too (Suratmo, 1987), where this tree species has been raised in industrial plantations in North Sumatra, Riau and Central Kalimantan

(Nair, 2000). Suratmo (1996) observed that serious damage by an undetermined defoliator has prevented expansion of *N. cadamba* plantations in Indonesia.

10.14 *Pinus* species (Pinaceae)

Tree profile

The genus *Pinus* contains over 90 species and constitutes an important group of conifers. Most of the pine species are distributed in the temperate and alpine regions but there are a few tropical pines distributed mostly in the cooler high altitudes of the tropics. In addition to providing good quality timber, pine wood is an established source of long-fibred raw material for pulp and paper. Therefore industrial plantations of pines have been attempted in most tropical countries. However, in spite of the early enthusiasm, exotic pine plantations have not performed well in most tropical countries, primarily because of their dependence on the presence in the soil of suitable mycorrhizal fungi. Alternative pulpwood species such as eucalypts and acacias have also played a part in the decline of interest in tropical pines in recent years. In spite of this, extensive pine plantations already exist in the tropics. The most widely planted species in the tropics are *Pinus caribaea*, *P. kesiya* and *P. merkusii*.

Pinus caribaea Morelet, commonly called Caribbean pine, is indigenous to the Latin American region, between latitudes 12°N and 27°N, and the variety *hondurensis* from the eastern half of Central America (Belize, Guatemala, Honduras, Nicaragua) has been widely planted in the American, Asian and African tropics and subtropics, covering over 65 countries (CABI, 2005). There were 300 000 ha of *P. caribaea* plantations in tropical America in 1990, and 40 000 ha in Fiji.

Pinus kesiya Royle ex Gordon, commonly called Khasi pine, is naturally distributed in Southeast Asia between 10°N and 30°N, in India, Myanmar, southern China, Laos, Vietnam, Thailand and the northern Philippines (CABI, 2005). The tree grows best at elevations between 700 and 1200 m above sea level. In India it is confined to the hilly regions in the east. Extensive patches of natural stands occur in the Khasi hills of Meghalaya. It also occurs throughout Assam and in Arunachal Pradesh, Nagaland and Manipur at elevations ranging from 1500 to 3000 m. *P. kesiya* is grown in plantations in India, Malaysia, the Philippines, Sri Lanka and Thailand. Plantations have also been raised in Africa, South and Central America, Australia and some Oceanic islands. Plantations are very successful in Zambia and Madagascar (CABI, 2005).

Pinus merkusii Jungh. and de Vriese, commonly known as Tenasserim pine or Sumatran pine, is the most tropical of all the pines and is distributed disjunctly between latitudes 21°N and 3°S, in continental Southeast Asia, Indonesia, and the Philippines (CABI, 2005). On the Asian mainland, it is found primarily in the southern Shan States of eastern Myanmar and Chiang Mai Province of north-western Thailand, but is also found scattered in other parts of Thailand and the greater part of Laos, Cambodia and Vietnam. The tree is encountered at elevations from sea level to over 1200 m, growing on various types of soil. In Indonesia, it grows naturally on mountain ridges in Sumatra, at high elevations of 800–2000 m above sea level. It has been planted extensively in Indonesia for afforestation, protection of watersheds and for tapping resin. Indonesia has about 700 000 ha of *P. merkusii* plantations (Nambiar *et al.*, 1998), distributed in the Provinces of Aceh, North Sumatra and West, Central and East Java. About 584 000 ha of pine plantations in Java are tapped for resin (Perum Perhutani, 1995). There have been only limited introductions of *P. merkusii* to areas outside its natural habitat. This includes Papua New Guinea, Sri Lanka and some southern African countries (CABI, 2005).

Overview of pests

Pest problems of the three species of pines are more or less similar and therefore they are dealt with together. There are four major groups of pine pests – shoot moths, bark beetles, aphids and a lepidopteran caterpillar. Shoot moths are pests of all the three pine species in Asia, Africa and Latin America although the species of moths may differ. Bark beetles are serious pests of pines in Latin America, but not in Africa and Asia (except in the Philippines). Exotic aphids are important pests in Africa. Pest profiles of these three groups are given separately below. The fourth, the lepidopteran caterpillar *Dendrolimus punctatus* (Lasiocampidae), is mainly a pest of masson pine *Pinus massioniana* in China and Vietnam. It also attacks *P. merkusii* and a few other species. The female moths lay their eggs in groups on needles and small branches, and the caterpillars feed gregariously on the needles. Two to five generations may occur per year, depending on the climate. When the population is high, complete defoliation may occur and repeated defoliation may cause the death of trees. Frequent outbreaks have been reported in young plantations of *P. massioniana* and *P. merkusii* in Vietnam (Billings, 1991) and annual outbreaks covering about a million ha of *P. massioniana* are common in southern China (CABI, 2005). A large number of techniques including large-scale release of the egg parasitoid *Trichogramma dendrolimi*, and use of fungal, bacterial and viral pathogens are in practice in China for control (CABI, 2005) and an IPM approach involving a combination of mechanical, biological, silvicultural and chemical methods has been advocated for Vietnam (Billings, 1991).

Apart from the above groups of major pests there are other minor pests. These include leaf-feeding sawflies, beetles and other lepidopteran caterpillars, sap-sucking bugs, leaf-cutting ants, and termites. Altogether 26 species of insects have been recorded on *P. caribaea* in Central America (CATIE, 1992a). In Indonesia, the pine looper, *Miliona basalis* (Lepidoptera: Geometridae) feeds on the needles of young *P. merkusii* trees. Frequent but short-lived outbreaks occurred in the 1950s in plantations in North Sumatra (Supriana and Natawiria, 1987). Sporadic outbreaks continued in the 1970s and 1980s (Nair, 2000). A sawfly, *Nesodiprion* nr. *biremis* (Hymenoptera: Diprionidae) also causes sporadic light defoliation of the native *P. merkusii* in North Sumatra. Groups of 5–25 larvae feed on the distal three-quarters of the needles and six months to 10-year-old plants may be affected. However, generally the damage level is not serious (Supriana and Natawiria, 1987). This sawfly has also been reported to attack seedlings and saplings of *P. kesiya* in Thailand.

Other sawflies reported from Thailand include *Diprion hutacharerne, Gilpinia leksawadii* and *G. marshalli* on *Pinus kesiya* and *P. merkusii* (Hutacharern and Tubtim, 1995). Sawflies have also been reported from other countries – *Neodiprion insularis* on *P. caribaea* in Cuba (Hochmut, 1972a), *N. merkeli* on *P. caribaea* in the Bahamas (Greenbaum, 1975) and *Diprion* spp. on *P. caribaea* and *P. merkusii* in Vietnam (Speechly, 1978). In exotic plantations of *P. caribaea* in Malaysia, the subterranean termite *Coptotermes curvignathus* attacks pines over five years old, making tunnels inside the trunk and often causing death of the trees (Abe, 1983). Termites also cause serious damage to pine trees in Australia. In India, young plants in nurseries are attacked and often killed by other root-feeding termites, whitegrubs or cutworms. Some native, wingless grasshoppers have become serious pests of exotic pines in Africa, particularly *P. patula*; aerial spraying of insecticides has been carried out in Malawi to control grasshopper outbreaks (Schabel *et al.*, 1999). In Central America, the giant grasshopper *Tropidacaris dux* (Orthoptera: Acrididae) has been observed completely defoliating native pines in Honduras and Nicaragua during local outbreaks (Billings, personal communication, 2006). It also has been reported as feeding on banana, citrus and mango trees. Adults are up to 12.5 cm long, the world's largest grasshopper.

Pest profile

Pine shoot moths (Lepidoptera: Pyralidae & Tortricidae)

Moths whose larvae tunnel into the shoots of pines are generally known as 'pine shoot moths' although most of them also attack the cones. They are also called 'tip moths' or 'shoot borers'. They belong to two families of Lepidoptera – Pyralidae and Tortricidae. Several species of two genera, *Dioryctria* (Pyralidae) and

Rhyacionia (= *Petrova*) (Tortricidae) are involved. Speight and Speechly (1982 a,b) have reviewed the biology, impact and control of pine shoot moths in Southeast Asia. *Dioryctria* species tend to be predominantly cone borers while *Rhyacionia* species are predominantly bud or shoot borers. Mixed infestation of the two groups of moths may sometimes occur on the same tree. For example, in a plantation of *Pinus caribaea* in the Philippines, out of about 400 infested shoots examined, half were infested by both *D. rubella* and *R. cristata* and a quarter each by *D. rubella* alone and *R. cristata* alone (Lucero, 1987).

Dioryctria species

Several species of *Dioryctria* occur on pines in different geographical regions. *Dioryctria abietella* is the dominant species, present throughout the Palaearctic region. Although it has been reported from North America and Europe from a wide range of hosts including pines, firs, cedars, larches and spruces, according to CABI (2005), it has formerly been misidentified and confused with two very closely related species, *D. abietivorella* which occurs in North and Central America and *D. mutatella* which occurs in northern Europe. *Dioryctria* species recorded on pines in the tropics are listed in Table 10.11.

The biology and habits of *Diocryctria* species vary slightly; the details given below are primarily applicable to *D. abietella*. The moth has a wingspan of 25–35 mm; its forewing is grey, mottled with black and contrasts markedly with the lighter hindwing. The larva varies in colour from reddish to greenish, with a black head, and is about 25 mm long when mature.

Table 10.11. Dioryctria *species recorded on tropical pines*

Species	Country	Pine hosts	Refs
Dioryctria abietella	India	*P. kesiya*, *P. roxburghii*	1
	Thailand	*P. kesiya*, *P. merkusii*	2
D. rubella	Philippines	*P. kesiya*, *P. merkusii*, *P. caribaea*	3
	Indonesia	*P. merkusii*	4
D. sylvestrella	Vietnam	*P. caribaea*	5
	Thailand	*P. kesiya*, *P. merkusii*	2
D. assamensis	India	*P. kesiya*	1
D. castanea	India	*P. kesiya*	1
D. raoi	India	*P. kesiya*	1
D. horneana	Cuba	*P. caribaea*	6
D. clarioralis	Cuba	*P. caribaea*	6

1, Singh *et al.* (1982); 2, Hutacharern and Tubtim (1995); 3, Lapis (1987) and Lucero (1987); 4, Natawiria (1990); 5, Speight and Speechly (1982a); 6, Hochmut (1972b).

Life history The female moth lays eggs singly at the base of needles on young shoots or on the scales of young cones. Each female may lay 30–50 eggs. The newly hatched larva feeds externally for a few days, up to a week, and later bores into the shoot or cone. The larva spins a small silken tent which becomes covered with resin and frass. The larva may come out of its hole occasionally. Boring on the shoot can occur in both directions, towards the shoot tip or downwards. The larval tunnel of *D. rubella* may extend up to 30 cm into the stem and therefore Matsumoto (1994) considered it a stem borer rather than a shoot borer. Pupation occurs in a papery silk cocoon within the cone or shoot or in soil when the infested cone falls to the ground. In Himachal Pradesh in India, *D. abietella* can complete the life cycle in one and a half to two months and the insect may pass through two complete and a partial third generations per year; the mature larvae hibernate in winter (Verma and Gaur, 1994). The number of generations will be reduced in cooler regions.

Impact Infestation by *Dioryctria* causes yellowing or browning of needles or shoot tips initially, followed by dieback of infested leading and lateral shoots. When the infestation is severe, the saplings become stunted and bushy. The insect causes economic damage to seed production by feeding on the cones and even seeds in seed orchards. In India, Bhandari (1988) observed that *D. abietella* caused complete loss of seeds in nearly 30% of cones of *P. wallichiana* in one year at Chakrata in Uttar Pradesh. Singh *et al.* (1988) recorded that during an outbreak of *D. castanea* on *Pinus kesiya* in Arunachal Pradesh, India, all trees of all age groups in a 900 ha plantation were infested. Small patches of *P. patula* and *P. wallichiana* escaped the attack. *Dioryctria horniana* in Cuba attacks older shoots, the inner bark of stems, branches and cones of pines and injury to the stem is often sufficient to cause breakage in the upper part of the crown at the point of attack (Hochmut, 1972b). At Luzon in the northern Philippines, *D. rubella* causes substantial damage to pines both in the natural forest and plantations. Almost all young pine plantations are usually infested. Infestation generally starts in the second year when the shoots are robust and healthy. In 1980, about 80% of two to three-year-old plantations over 1000 ha in Abra were infested (Lapis, 1987). *D. rubella* also attacks young cones, reducing seed production. The devastation caused by shoot moths led to the slowing down or even suspension of planting pines in Luzon. An outbreak of *D. rubella* severely damaging 1000 ha of young *P. merkusii* plantations in North Sumatra, Indonesia was reported by Supriana and Natawiria (1987). About 85% of the trees were infested, with infestation occurring in the trunk, leader shoot or lateral shoots. Involvement of more than one borer species was suspected. It is noteworthy that shoot moth infestation has not been reported on the extensive *P. merkusii* plantations in Java, unlike Sumatra (Suratmo, 1987).

Natural enemies An ichneumonid parasitoid, *Syzeuctus* sp. has been reported to cause about 36% parasitism of *D. abietella* infesting *P. girardiana* (chilgoza pine) in Himachal Pradesh, India (Thakur, 2000). A bacterium *Bacillus licheniformis* has also been isolated from diseased larvae (Thakur, 2000). From *D. rubella* in the Philippines, two ichneumonid larval parasitoids and two chalcid pupal parasitoids have been recorded; the ichneumonid *Eriborus* sp. was found to infest about 54% of the larvae in some seasons (Lapis, 1987).

Control There is no effective control against *Dioryctia* spp. Insecticidal sprays have been suggested for controlling infestation of shoots (Lapis, 1987; Singh *et al.*, 1988) and cones in seed orchards (Thakur, 2000).

The primary pheromone component of *D. abietella* infesting spruce cones has been reported as (9Z,11E)-9,11-tetradecadienyl acetate, but in trapping experiments the synthetic pheromone component was only weakly attractive (CABI, 2005).

Studies in Cuba have shown some genetic differences in susceptibility of pines to *D. horneana*. *Pinus caribaea* var. *caribaea, P. cubensis* and *P. maestrensis* were the most damaged, while *P. caribaea* var. *bahamensis, P. kesiya* and *P. tropicalis* showed no significant damage (Echevarria, 1985). Halos *et al.* (1985) also showed that *Pinus caribaea* var. *bahamensis* was the most resistant of 11 pines screened against shoot moths (including *Dioryctria rubella* and *Rhyacionia cristata*) in the Philippines; similar results were reported by Lapis (1987). However, *Pinus caribaea* var. *bahamensis* is comparatively slow growing.

Rhyacionia species (Tortricidae)

Rhyacionia (= *Petrova*) spp. are similar to *Dioryctria* spp. in habits, but smaller. A typical example is *Rhyacionia cristata*, present in Southeast Asia. The adult moth has a wingspan of 12 mm; the forewings are light orange to light brown, with whitish bands along the length. Young larvae are yellowish and turn brown as they mature (Lapis, 1987). The following species have been recorded on tropical pines – *R. cristata* in the Philippines (Lapis, 1987), *R. salweenensis* (probably synonymous with *R. cristata*, according to Speight and Speechly, 1982a) and *R. khasiensis* in Thailand (Hutacharern and Tubtim, 1995), *R. subtropica* in Guatemala (CATIE, 1992a) and Cuba (Hochmut, 1972b) and *R. frustrana*, common to all Central American countries (CATIE, 1992a). *R. frustrana* is also common in the eastern United States where it is known as the 'Nantucket pine tip moth' (it was first discovered and studied on Nantucket Island, Massachusetts) (Berisford, 1988).

Life history Eggs are laid on the needles. Newly hatched larvae feed externally at the base of needles for a few days and then bore into the upper part of the shoot. The larval stage lasts about 20–25 days and the total developmental period is about one to one and a half months (Lucero, 1987; Lapis, 1987; Berisford, 1988). *R. frustrana* can undergo more than eight generations per year in Central America (CATIE, 1992a) and two to five in the United States where it overwinters in the infested shoots (Berisford, 1988).

Impact *Rhyacionia* damage is usually most severe on saplings under five years of age and the infestation intensity declines as the tree age advances. Infested trees are usually less than 3 m in height. In some plantations in Costa Rica up to 91% of the trees were attacked by *R. frustrana* (Salazar, 1984). In young plantations, severe infestation can cause loss of increment and growth form. In Costa Rica, Ford (1986) observed that 38% of trees in a plantation are likely to be forked as a result of damage by *R. frustrana*. *Rhyacionia* also attacks cones.

Natural enemies Most studies on natural enemies were carried out for *R. frustrana* in the United States, on which about 64 species of parasitoids were recorded (CABI, 2005). A world survey of parasitoids and predators of the genus *Rhyacionia* is also available (Harman and Kulman, 1973). Egg parasitoids (*Trichogramma* spp.) are considered to be important in regulating the populations (Berisford, 1988).

Control In the United States insecticides are most commonly used to protect high-value stands such as Christmas tree plantations, seed orchards, progeny tests, and/or short-rotation sawtimber and pulpwood stands against *R. frustrana* (CABI, 2005). Systemic insecticides are more useful as the larvae are usually concealed within their tunnels in shoots or bark except when they are newly hatched or are very young. Many systemic insecticides such as furadan, dimethoate, azinphos-methyl and carbofuran have been shown to be effective (Speight and Speechly, 1982b). Application of dimethoate caused 80% larval mortality in *R. frustrana* in Cuba eight days after treatment (Salazar, 1984). However, the use of insecticide will be uneconomical in most tropical plantations.

Some control has also been achieved by release of a trichogrammatid egg parasitoid (Berisford, 1988).

Sex pheromones of *Rhyacionia* species have been isolated and shown to be straight-chain 12-carbon acetates or alcohols, and that of *R. frustrana* was identified as a mixture of (E)-9-dodecenyl acetate and (E)-9,11-dodecenyl acetate in the ratio of 96:4, but only weak attraction was found (Berisford, 1988).

Application of an aqueous spray containing conidia of the fungus *Metarhizium anisopliae* at monthly intervals was reported to control attack by *R. frustrana* in

Cuba (Duarte *et al.*, 1992). Commercial preparations of Bt have also been shown to be effective for control of *R. frustana* in Cuba (Menendez *et al.*, 1986).

P. caribaea var. *bahamensis* is reported to be virtually completely resistant to *Rhyacionia* attacks (Baylis and Barnes, 1989).

Knowledge gaps There are indications that certain species and provenances of tropical pine are more resistant to shoot moth attack than others. More critical studies are needed on genetic resistance to shoot moths in tropical pines.

Peak infestations of shoot moths occur three to five years after the establishment of pine plantations and the incidence declines as their age advances and the canopy closes. Explanations vary from age-related host resistance to stabilization of the natural enemy complex. The actual reasons remain unknown.

Some authors (see Speight, 1996) have suspected a link between poor site conditions and high incidence of shoot moth attacks in the northern Philippines where pines are often planted in suboptimal sites, but no conclusive proof exists.

Research for management of pine pests is not considered a priority in the tropics now because of the decline of interest in planting of exotic pines, due to various reasons including the pest problems.

Pest profile

Pine bark beetles (Coleoptera: Curculionidae: Scolytinae)

Bark beetles are very destructive pests of pines in temperate forests. In Europe and North America, periodic outbreaks of several species of the genera *Dendroctonus*, *Ips* and *Scolytus* are known to kill millions of hectares of pines and other conifers. Although not as destructive as in the temperate forests, some bark beetles attack pines in the tropics. The most damaging attacks in the tropics have been recorded in Central America, including Mexico.

Bark beetles are small beetles, 3–6 mm long and semi-cylindrical in shape. The adult beetles bore through the bark of trees and feed and oviposit in the phloem. The larvae develop in the phloem and the emerging beetles may reinfest the tree. The beetles carry a fungus which grows on the tunnels and hastens the death of the trees. In the tropics, *Dendroctonus frontalis* (Fig. 10.26) attacks pines in Honduras and other Central American countries, and *Ips calligraphus* does so in the Philippines, Mexico, the Caribbean Islands and Central America. (In Central America, *I. calligraphus* has been recently redescribed as *I. apache*.) Other *Dendroctonus* species that infest pines in Central America include *D. adjunctus*, *D. mexicanus*, *D. valens*, *D. approximatus*, *D. vitei* and *D. parallelocollis*, of which the first has been the most destructive (Cibrian-Tovar *et al.*, 1995).

Fig. 10.26 Larvae of the southern pine beetle *Dendroctonus frontalis*, in galleries under bark of *Pinus caribaea*. Courtesy: R. F. Billings, Texas Forest Service, USA.

Life history and habits In *Dendroctonus frontalis*, female beetles are the first to attack a tree. The infestation process and the role of pheromones are described in detail by Flamm *et al.* (1988). If the host tree is suitable, the pioneer females release attractive pheromones which set in motion an aggregation phase. The primary aggregation pheromone, frontalin, along with host tree odours (mainly α-pinene), attracts large numbers of beetles, especially males. The arriving males release a pheromone that is attractive to females. A complex of pheromones is produced by the male and female beetles, some of which, at higher concentrations, induce dispersal of beetles to new trees. Mating takes place in a nuptial chamber formed by the female in the inner bark. The mated female makes an S-shaped gallery and deposits eggs at irregular intervals on opposite sides of the gallery. Larvae make their galleries in the phloem (Fig. 10.26), perpendicular to the egg gallery, and when nearly mature, bore into the outer bark. In Honduras, *D. frontalis* completes its life cycle in less than a month (Billings and Espino, 2005), but in winter months in northern regions the life cycle may last over two months (Flamm *et al.*, 1988). After completion of egg-laying, adult beetles either die in the gallery or re-emerge and attack other trees. The bark beetles have symbiotic relationships with fungi that are thought to be important in larval nutrition and overcoming host resistance. The beetles have specialized body structures called mycangia in which the fungi are carried. The identities and role of the associated fungi are not fully known. A blue-stain fungus *Ceratocystis minor* (Ascomycetes), carried externally by *D. frontalis* but not found in the mycangium, is considered a major tree-killing agent (Flamm *et al.*, 1988).

The pine bark beetle which infests *P. caribaea* and *P. oocarpa* in Central America is recognized as the subspecies *D. frontalis arizonicus*, as distinct from *D. frontalis frontalis* found in the southeastern United States (Billings and Espino, 2005). *Ips calligraphus* infests *Pinus kesiya, P. merkusii, P. caribaea* and *P. oocarpa* in the Philippines. It completes its life cycle in 17–30 days (Quinones and Zamora, 1987). Infested trees are easily recognized by holes in the bark with resin exudation and frass. A smaller, 3 mm long unidentified scolytine with similar habits has also been found associated with pines in the Philippines (Quinones and Zamora, 1987).

Impact Outbreaks of *D. frontalis* causing death of pine trees have occurred frequently in the pine forests of Honduras and other Central American countries, and vast areas of natural pine forests have been devastated (Billings *et al.*, 2004). An outbreak during the years 2000–02 affected mainly *P. caribaea* and *P. oocarpa* in Honduras, Belize, Guatemala, El Salvador and Nicaragua, killing millions of trees. *D. adjunctus* has caused extensive timber loss on *P. hartwegii* (= *P. rudis*) in Guatemala and Mexico at elevations above 2800 m. From 1975–80, it killed an estimated 100 000 ha of the same species in Guatemala's Altiplano Region (Billings *et al.*, 2004). *Ips calligraphus*, which can kill 3-year-old saplings to mature trees, is considered to be a dangerous pest in the Philippines (Quinones and Zamora, 1987). Widespread infestation is common in the dry season although it is less severe in the rainy season.

Normally, the bark beetles attack trees that have been weakened by various causes like drought, fire, lightning strike, overstocking etc. But healthy trees also succumb when the beetle population is large enough to overwhelm the tree's defence system.

Host range and geographical distribution As noted earlier, there are several species of scolytine bark beetles that attack a wide variety of conifers, but most tree-killing bark beetles are distributed in temperate forests.

D. frontalis will attack a wide range of *Pinus* species but *P. palustris* (longleaf pine) is reported to be relatively resistant (CABI, 2005). It will also attack some *Picea* (spruce) and *Tsuga* (hemlock) species. *D. frontalis* has a wide distribution covering both the temperate regions in the southern United States (hence known as 'the southern pine beetle') and tropical Central America. Other *Dendroctonus* species occur elsewhere in North America and Europe.

Ips calligraphus (= *I. apache* in Central America) will also attack a wide range of *Pinus* species. Its distribution similarly covers temperate and tropical regions in North and Central America and several of the Caribbean Islands (e.g. the Dominican Republic, Haiti, Cuba). An outbreak of *I. calligraphus* in native stands of *Pinus occidentalis* in the Dominican Republic in 1988 resulted from several years of drought (Haack *et al.*, 1989). This species is also present in the Philippines, but is thought to be introduced (CABI, 2005).

Several other species of scolytine bark beetle occur in the tropics, both at temperate high-elevation sites and in the plains, but most of them attack dead or dying trees or felled timber and have not become serious pests of living trees. Examples are *Ips longifolia* and *Polygraphus longifolia* which occur on pines in the temperate Himalayan forests of India (Beeson, 1941) and *Xyleborus* and *Xylosandrus* species attacking hardwoods in the tropical plains. An exception is *Euwallacea* (= *Xyleborus*) *fornicatus*, the 'shothole borer' of tea, which breeds on the living tree. Bark beetles which attack felled timber are discussed in Chapter 6.

Control Several methods are employed to control bark beetle attack (Flamm *et al.*, 1988; Billings and Espino, 2005; CABI, 2005), but mostly in the developed countries of the temperate regions. It is generally accepted that low vigour encourages bark beetle outbreaks and therefore silvicultural operations, especially timely thinning and removal of fire or cyclone damaged trees, are carried out to enhance stand health. Sanitation cutting is the most commonly practiced method in Central America and the Philippines. It consists of rapid removal of all infested trees, along with a buffer strip of uninfested trees adjacent to the most recently attacked trees, to prevent the build-up and spread of the beetles. Generally, cut trees are either removed from the site or burnt. When the outbreak is extensive and the terrain is mountainous and less accessible, the cut trees are left at site (the cut-and-leave method). Cut-and-leave is only recommended for control of *D. frontalis* infestations and not for those caused by *Ips* spp. (Billings and Espino, 2005). Actual practice depends on the management constraints dictated by forest ownership, the value of the infested stock, market conditions and availability of labour and equipment.

Chemical control of beetles, use of aggregation pheromones for trapping beetles or use of inhibitory compounds to halt the spread of bark beetle infestations have also been tried in the developed countries in the past, with varying degrees of success.

Knowledge gaps Why bark beetles have not become serious pests of living trees, including pines, in most of the tropics, in spite of their great biodiversity and importance as pests of felled timber in the tropics, is not known.

Pest profile

Pine aphids (Hemiptera: Adelgidae and Aphididae)

Two species of exotic aphids (order Hemiptera, superfamily Aphidoidea), *Pineus pini* (family Adelgidae) and *Eulachnus rileyi* (family Aphididae), are important pests of pines in eastern and southern Africa. *P. pini*, known as the 'pine woolly aphid' or 'pine adelgid', is a native of Europe and is believed to have been accidentally introduced to Africa, via Australia, in the 1960s. It spread rapidly into several countries in eastern and southern Africa, affecting many species of exotic as well as native pines. There is some confusion on its taxonomy and it has been referred to in the literature sometimes as *P. laevis* and confused with *P. boerneri*, which is probably of East Asian origin and difficult to differentiate morphologically from *P. pini* (CABI, 2005). A pine adelgid identified as *P. laevis* has also been recorded in pine plantations at high elevation sites in Kerala and Tamil Nadu in southern India where it is believed to be an inadvertent introduction (Singh *et al.*, 1982). *E. rileyi*, known as pine needle aphid, is also native to Europe from where it has spread to North America and Africa.

Life history, nature of damage and impact Aphids are sucking insects which feed on plant sap. Both *P. pini* and *E. rileyi* attack a wide range of pines including *P. caribaea*, *P. kesiya*, *P. merkusii* and *P. patula*. Infestation causes the needles to turn yellow and drop prematurely.

Aphids have complicated life histories in northern temperate zones, with both winged and apterous adults and an alternation of asexual and sexual generations. In Africa, where they multiply throughout the year, the life cycle is simpler and reproduction is parthenogenetic, although both winged and apterous forms are produced. The young *P. pini*, called 'crawlers', insert their tubular mouthparts into the tissues and suck the sap from the base of the needles and young bark. They go through several moults and complete the life cycle within a few weeks. There are many generations per year and there is considerable overlap between generations. The apterous form of *P. pini* produces waxy thread-like secretions which form a woolly covering over its body, giving it the name 'woolly aphid'. The apterous *P. pini* adult is about 1 mm in length (Murphy *et al.*, 1991). In Kenya population density of *P. pini* is influenced by weather, the density being lowest during the rainy period (Mailu *et al.*, 1980). The adult *E. rileyi* is about 2.5 mm in length; all stages feed on pine needles, both young and old (Murphy *et al.*, 1991). *E. rileyi* is relatively uncommon within its native geographic range and is not considered to be a pest, but it has acquired pest status in Africa where it has been introduced (CABI, 2005).

The two invasive pine aphids, together with a third invasive cypress aphid *Cinara cupressi*, have caused substantial damage to pines and cypress in

Africa where these conifers had been free of major pests until these aphids arrived. Their arrival resulted in severe growth retardation and sometimes tree mortality. Aphids were estimated to cause 50% loss of growth increment and up to 20% tree mortality (CABI, 2005). According to Murphy (1996) the two pine aphids were causing an annual loss of £1.5 million by way of increment loss in plantations across Kenya, Malawi and Uganda. Outbreak of the conifer aphids was characterized as a crisis in African forestry (FAO, 1991) and in 1991 the Kenya Forestry Research Institute organized a regional workshop in technical collaboration with FAO and the International Institute of Biological Control to address the problem and develop a regional programme for conifer aphid management.

Control Control attempts have mostly relied on classical biological control although chemical control has also been tried with varying degrees of success (Day *et al.*, 2003). These aphids are not serious pests within their natural geographic range and it is assumed that this is because of the pressure exerted by indigenous natural enemies. They become pests in exotic locations when released from the grip of natural enemies. *P. boerneri* has been successfully controlled by introduced natural enemies in Hawaii, New Zealand and Chile. The most effective natural enemy in Hawaii is the dipteran predator, *Leucopsis obscura*. However, biological control with introduced alien predators has been mostly unsuccessful in eastern Africa (CABI, 2005). In Kenya, indigenous predatory coccinellid beetles have given some degree of control (Mailu *et al.*, 1980). Over the years, the severity of the problem has been reduced, apparently due to a combination of factors including stabilization of the aphid population due to the action of indigenous and introduced natural enemies, and the slowing down of pine plantation establishment.

Knowledge gaps There is little published information on the current status of pine aphids in Africa. This is partly because the problem has become less severe after the initial escalation following the introduction of the exotic aphids into Africa. There has been some confusion initially in the taxonomy of pine aphids which affected the progress of biological control efforts (Day *et al.*, 2003). Obviously more research is needed on the taxonomy of conifer aphids.

10.15 *Shorea* species (Dipterocarpaceae)

Tree profile

Shorea is an important genus of commercial timber species of the family Dipterocarpaceae, a dominant family in the lowland rain forests of Indonesia,

Malaysia and the Philippines (see Chapter 1). The genus comprises about 350 species (CABI, 2005), but they have not received much attention as plantation species. Most planting in the past has been experimental, mainly as enrichment planting in logged-over forests, using wildlings. However, small-scale, conventional plantations of a few species have been raised, since the 1950s, in Indonesia, Malaysia and India. The species planted include the relatively fast-growing *Shorea javanica*, *S. leprosula*, *S. parviflora*, *S. selanica*, and *S. smithiana* in Indonesia (Cossalter and Nair, 2000), about a dozen species including *S. leprosula* and *S. parviflora* in Malaysia (Appanah and Weinland, 1993) and *S. robusta* in India.

Shorea robusta C.F. Gaertn. (commonly called 'sal' in India) which has a major pest problem, is described in some detail here. The tree is distributed in over 10 million ha of forests in central and northern India, between latitudes 18°N and 32°N, extending into the subtropical zone (Fig. 10.29a). It also occurs in the sub-Himalayan tract of Nepal and Pakistan and in Bangladesh. The tree is gregarious in habit. Under favourable conditions, the tree attains a height of about 30 m. It grows at altitudes as low as 10 m to over 1500 m, and 1000–3000 mm rainfall. It can tolerate temperatures as high as 45°C and as low as 0°C. It produces a hard and durable timber, used for various construction works, railway sleepers and mining operations. When injured, the tree exudes a resin called sal dammer, which is used as incense. Generally, sal has been managed under a shelterwood system with natural regeneration. The tree coppices well and coppice rotations of 40, 60 or 80 years are practised with periodic thinning. Sal has been planted within its native distribution range in India as well as in Hainan Island in southern China and Zimbabwe in Africa (CABI, 2005).

Overview of pests

Insect pests of *Shorea* species include defoliators, sap-sucking bugs and stem borers. Most information is available for *Shorea robusta* in India on which about 145 species of insects have been recorded. However, except for the periodic outbreaks of a cerambycid trunk borer *Hoplocerambyx spinicornis* on *Shorea robusta* in India (see pest profile below) there are no major problems for *Shorea* species. The most important pests are the following.

In the nursery, seedlings of *S. javanica* are killed by a sap-sucking bug, *Mucanum* sp. (Hemiptera: Pentatomidae) in Sumatra, Indonesia (Intari, 1996) and seedlings of *S. robusta* by the 'seed and seedling borer' *Pammene theristis* (Lepidoptera: Eucosmidae) in India (Beeson, 1941). The latter hollows out the tap root and part of the stem above ground; it also attacks young growing shoots, causing dieback.

The more important defoliators are the following. Unidentified caterpillars, including bagworms, and scarabaeid beetles feed on the leaves of *S. leprosula, S. selanica* and other *Shorea* spp. in West Java and East Kalimantan in Indonesia, with small-scale outbreaks on some occasions (Nair, 2000; Rahayu *et al.*, 1998). *Calliteara cerigoides* (Lepidoptera: Lymantriidae), a polyphagous caterpillar, defoliates *S. leprosula, S. pinanga, S. selanica* and *S. stenoptera* in Indonesia (Messer *et al.*, 1992; Matsumoto, 1994). The caterpillar of *Lymantria mathura* (Lepidoptera: Lymantriidae) occasionally builds up in large numbers on *S. robusta* in Assam and Madhya Pradesh in India, causing defoliation (Beeson, 1941; Dey and Tiwari, 1997). Small-scale outbreaks of *Ascotis selenaria imparata* (Lepidoptera: Geometridae) have also occurred periodically on *S. robusta* in India. In an outbreak in 1975 at Dehra Dun, in a nearly pure natural stand, trees on about one hectare were totally defoliated (Singh and Thapa, 1988). A generation of the insect is completed in 40–65 days. Similar defoliations over an area of lesser extent were caused by two subsequent generations, before the insect population abruptly collapsed by the end of July due to a nucleopolyhedrosis virus infection of the larvae. *A. selenaria* is polyphagous and understorey trees of *Mallotus philippinensis* and *Murraya koenigi* also suffered total defoliation during the outbreaks. Other defoliators of *Shorea* include the pyralids *Omiodes* sp. on *S. argentifolia* and *Lista* sp. on *S. parviflora* in Malaysia (Chey, 1996).

A mealy bug, *Drosicha stebbingii* (Hemiptera: Coccidae) attacks *S. robusta* in India (Beeson, 1941). Early instar nymphs of this mealy bug cluster on the leaves near the veins and suck the sap. They excrete a sticky liquid which dries up rapidly and coats the surface of the leaves. When about two months old, the nymphs move from the foliage to young shoots where they continue to feed. The insect breaks out periodically in epidemics, causing drying up of twigs and branches. A thrips, *Araeothrips longisetis* (Thysanoptera: Tubulifera) is a minor pest of *S. robusta* in India; it causes curling of the margin of leaves and their subsequent withering (Srivastava *et al.*, 1984). The sap sucking cicada *Lawana candida* is an occasional pest of seven to nine-year-old *S. leprosula* trees in East Kalimantan, Indonesia (Rahayu *et al.*, 1998).

Apart from the sal borer which is a serious pest of *S. robusta* in India, a cerambycid borer *Cyriopalus wallacei* is known to attack living trees of *Shorea leprosula, S. leptoclados* and a few other dipterocarps in Malaysia (Chey, 1996). It tunnels between the sapwood and heartwood from the top downwards and has a two-year life cycle, but seldom kills trees.

Large-scale plantations have not been raised for most species of *Shorea* and the pest problems are likely to be aggravated in future as the plantation area increases.

Pest profile

Hoplocerambyx spinicornis (Coleoptera: Cerambycidae)

Hoplocerambyx spinicornis Newman (Coleoptera: Cerambycidae) (Fig. 10.27a,b), known as sal borer, is a pest of *Shorea robusta* (sal) in India. It bores into the stem of sal trees and is the most notorious forest pest of India because of its periodic outbreaks, during which millions of sal trees are killed. The adult beetle is dark brown and variable in size, measuring 20–65 mm in length. In the male, the antennae are much longer than the body. The full-grown larva is large, measuring up to 9 cm in length.

Life history The life history, ecology and control of *H. spinicornis* on sal in India have been the subject of several studies since the 1900s and the literature is extensive (Stebbing, 1906; Beeson and Chatterjee, 1924; Beeson, 1941; Roonwal, 1978; Bhandari and Rawat, 2001). The beetles appear every year soon after the monsoon rainfall in June or July, with fresh batches of beetles emerging with each bout of rain, until within about two months almost all beetles have emerged from the tree trunks. The beetles pair soon after emergence and lay eggs about a week later, on cuts or holes in the bark of sal trees. Normally, the trees chosen for egg laying are freshly dead or highly weakened by various causes, but during outbreaks even healthy trees are attacked. Each female will lay 100–300 eggs over a lifespan of about a month. High humidity favours

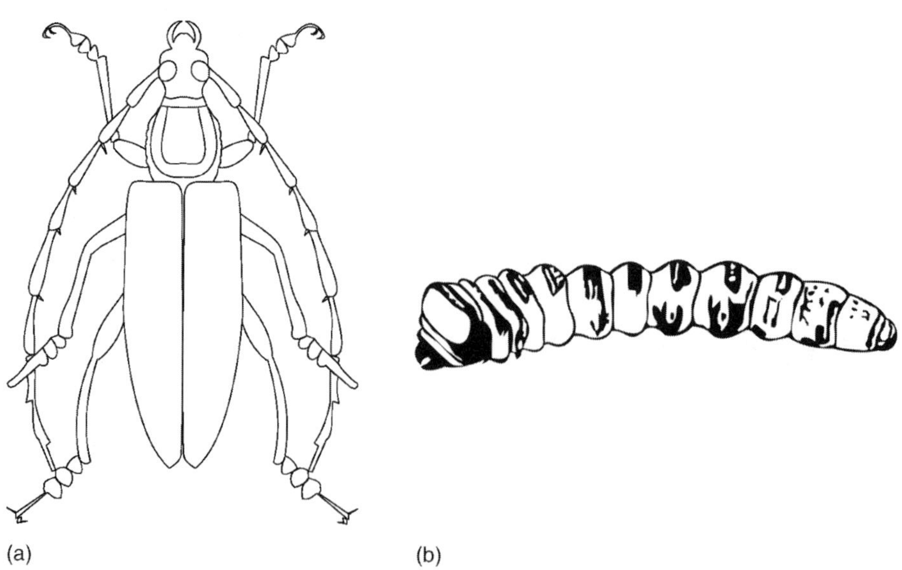

Fig. 10.27 *Hoplocerambyx spinicornis*. (a) Adult female (length 40 mm). After Thakur (2000); (b) larva. After Stebbing (1914).

oviposition; the number of eggs laid may reach upto 465 per female at 91% RH (Beeson, 1941).

The newly hatched larvae feed under the bark initially, then in the sapwood and finally bore into the heartwood (Fig. 10.28a,b). Many young larvae are trapped in the exuding resinous sap and die, the proportion of surviving larvae

(a)

(b)

Fig. 10.28 (a) A log of *Shorea robusta* infested by *Hoplocerambyx spinicornis*, split to show the larvae and the damage caused by them. (b) Cross section of a log of *Shorea robusta* infested by *Hoplocerambyx spinicornis*, showing the large larval tunnels.

depending on the health of the tree and the density of larvae, which influence the tree's ability to defend the attack. A big sal tree may often support the development of about 300 beetles, although more than a thousand eggs may be laid on the tree. As the attack progresses, coarse dust is thrown out of holes in the bark, which accumulates at the base of the tree in large heaps. The larval development is usually completed by November when the larva constructs a chamber in the heartwood with an adult exit hole, and turns into a pre-pupa. Then it moults into a pupa and later by May–June into an adult beetle, and remains quiescent until it emerges with the onset of rainfall. The life cycle is thus annual.

Extensive galleries in the sapwood made by several larvae cause partial or complete girdling of the tree, leading to its death. Although the tree offers resistance by the outflow of resin, mass attack during epidemics kills even vigorous trees. Both the main trunk and crown branches are attacked. In a typical dead tree, 60–70% of the borer population occurs in the main trunk and 40–30%, in the crown branches (Beeson, 1941).

Host range and geographical distribution *H. spinicornis* also attacks some other dipterocarps such as *Dipterocarpus tuberculatus*, *Shorea assamica*, *S. obtusa*, *Parashorea* sp. and *Pentacme* sp. as well as some trees of other families, i.e. *Duabanga grandiflora* (Sonneratiaceae) and *Hevea brasiliensis* (Euphorbiaceae). However, population outbreaks have occurred only on *S. robusta*.

H. spinicornis is distributed in Central, South and Southeast Asia – in eastern Afghanistan, Pakistan, India, Nepal, Bangladesh, Myanmar, Thailand, Indonesia and the Philippines (Beeson 1941, Roonwal 1978, Hutacharern and Tubtim 1995). In India, its distribution is confined to the northern and north-eastern sal belt.

History of outbreaks, impact and population dynamics Since the year 1897, when *H. spinicornis* was first recorded as a pest of sal at Chota Nagpur in Bihar, India (Stebbing, 1914), a series of outbreaks has occurred in different parts of its distribution range in India. These outbreaks have ranged from mild ones limited to small areas over a year, to heavy and devastating ones such as the 1923 outbreak in Madhya Pradesh which persisted over a five-year period, killing about seven million sal trees. While the insect is normally endemic and attacks only a small number of unhealthy or overmature trees, during outbreaks large numbers of healthy standing trees are attacked and killed. The chronology and basic details of some of these recorded outbreaks are given in Table 10.12.

It may be seen that a large number of outbreaks has occurred since 1900, but there has been no regularity in their occurrence. Outbreaks have been reported from several states where sal occurs naturally, Assam, Bihar, Himachal Pradesh,

Table 10.12. *Chronology of sal borer* (Hoplocerambyx spinicornis) *outbreaks in India*

Year(s) of outbreak	State and place of initiation	Area infested (ha)	No. of trees killed	Remarks
1897	Bihar: Chota Nagpur, Singbhum	-	-	Mild
1905	MP: Balaghat	-	-	Mild
1906	Assam: Kachagaon, Goalpara	-	-	Mild
1914-5	MP: Banjar	-	400	Mild
1916-24	UP: Dehra Dun	1 800	80 000	
1923-8	MP: Mandla and adjacent areas	-	7 000 000	V. heavy
1924-5	UP: Kalagarh	-	-	Mild
1927-8	MP: Supkhar and Baihar	12 200	45 000	
1931-4	WB: Sevoke, Kurseong	700	3 000	
1934-7	UP: Kalagarh	-	-	Mild
1948-52	HP: Nahan	8 500	7 000	
1948-52	MP: Supkhar and Mukki	7 000	3 000	Moderate
1950-5	MP: Mandla	-	57 000	
1958-60	UP: Timli	-	-	
1959-62	MP: Mandla (south)	32 400	50 000	
1961	Assam: Nowgong and Goalpara	-	-	Mild
1961	Bihar: Palamau	49 100	-	
1961	UP: Lachhiwala	-	-	Mild
1965	UP: Thano	500	2 000	Mild
1974	WB: Bhatkhawa	1 400	23 000	
1976-81	MP: Pachmarhi	-	-	
1979-82	MP: Pachmarhi	5 200	8 000	
1994-2000	MP: Mandla and adjacent areas	500 000	>3 000 000	V. heavy
1995	UP: Thano	500	8 000	

HP = Himachal Pradesh, MP = Madhya Pradesh, UP = Uttar Pradesh, WB = West Bengal
Data from Bhandari and Singh (1988), Thakur (2000), Bhandari and Rawat (2001) and Dey (2001)

Madhya Pradesh, Uttar Pradesh and West Bengal, but not from Haryana, Meghalaya, Orissa, Sikkim and Tripura. The gap between the outbreaks at one place varies markedly. For example, in Madla District in the State of Madhya Pradesh, one of the most outbreak prone areas, out of three outbreak episodes in the past 50 years, the first occurred in 1950–55, the second after 4 years in 1959–62 and the third after 32 years in 1994–2000. Although the practice of some routine and emergency control measures (see below) may have influenced the frequency and timing of initiation and termination of the outbreaks, no clear pattern is evident. The gap between outbreaks has lasted between 1 and 32 years and the duration of outbreaks has lasted one to eight years.

To understand the dynamics of outbreaks and their impact, it is instructive to examine one of the outbreaks more closely. The following details of the outbreak that occurred during 1994–2000 in Madhya Pradesh are mainly based on data gathered by Dey (2001) and unpublished reports from the Madhya Pradesh Forest Department. Sal trees cover about 2.78 million ha, spread over 14 districts in the erstwhile undivided State of Madhya Pradesh and account for over one-quarter of the sal forests of India. The sal borer outbreak was first noticed after the 1994 rains, in a few pockets around Chada in the Dindori Forest Division in Madla District. The population increased substantially during 1995. In 1996, district-wide trapping operations using sal logs (see below) carried out by the State Forest Department yielded 2.15 million beetles which increased to 15 million in 1997 and 32.59 million in 1998, before the number declined to 13.4 million in 1999. In Madla District alone, the number of infested trees was 0.884 million in 1997, 1.683 million in 1998 and 0.647 million in 1999. The outbreak progressed in spite of control operations consisting of trapping of beetles and limited cutting and removal of badly infested trees. During the ascending phase of the outbreak in 1997, the infestation spread to the sal forests in five adjoining districts, Balaghat, Bilaspur, Sarguja, Rajnandgaon and Shahdol, covering about half a million hectares (Fig. 10.29).

By early 1998, about three million trees were infested and about 0.8 million badly affected trees felled and removed in an attempt to check the spread of the outbreak. Although removal of heavily infested trees is standard prescription to contain an outbreak, such massive tree felling invited widespread public criticism, particularly against tree felling in wildlife sanctuaries and national parks. This led to public interest litigation and a ruling by the Supreme Court of India suspending felling of sal trees except for certain categories that were considered dead or beyond recovery. The outbreaks declined drastically during the year 2000.

In summary, the sal borer outbreak built up during 1994 and 1995, peaked during 1998, declined thereafter and ended in 2000, infesting over 3 million sal trees and killing a large percentage of them. It is evident that the outbreak had a devastating effect on the sal forests. The timber of the heavily attacked trees is rendered useless, with large criss-cross tunnels, causing enormous economic loss.

The circumstances under which the outbreaks develop are not fully understood. It is believed that the outbreaks begin in dense overmature stands where the conditions are favourable for rapid build-up of the beetle population. The following observations by Dey (2001) during the 1994–2000 outbreak in Madhya Pradesh indicate that the sal borer had a preference for trees of higher girth class. He studied 14 representative, one-hectare plots spread over the

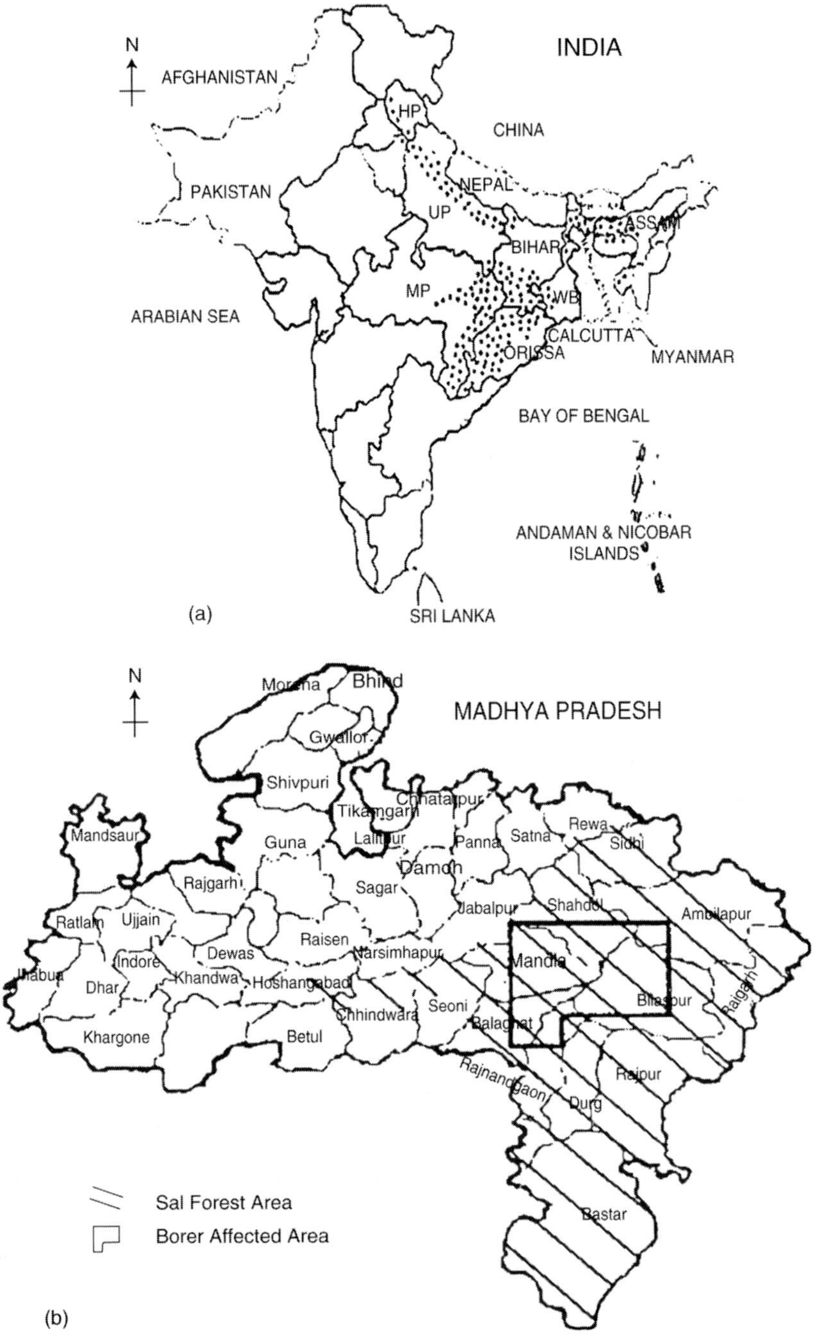

Fig. 10.29 Sal borer outbreak in India. (a) Map of India showing the distribution of the sal tree, *Shorea robusta*; (b) map of the State of Madhya Pradesh (before re-organization) showing the sal area and the borer affected area during the 1994–2000 outbreak. Reproduced with permission from IUFRO (Dey, 2001).

affected districts. In March 1998, out of an average growing stock of 262 sal trees/ha (above 20 cm diameter at breast height), 3–66% were infested. While the infested trees constituted 2.6–9% of trees within the girth class 20–60 cm, it constituted 34.4–55.9% within the girth class 61–120 cm and 69.4–78.3% within the girth class 121–180 cm, showing a definite preference of the insect for trees of higher girth classes. This trend has also been noted by some earlier workers. It appears that any factor which imposes stress on trees, such as drought, crowding, overmaturity etc., which compromises the tree's ability to produce the defensive resin flow, may trigger an outbreak. Singh and Thapa (1988), who reported outbreak of a geometrid caterpillar *Ascotis selenaria imparata* that caused total defoliation of patches of sal stands prior to the beginning of the monsoon, remarked that such patches could prove susceptible to attack by the sal borer beetles which start emerging with the beginning of the rains. Another favourable factor is rain. As noted earlier, sal borer adults emerge with the onset of rains and are active and fly during mild rains. The number of eggs laid and their hatching and survival rate are higher at high humidity. It is obvious that years of high rainfall are very favourable for the build-up of sal borer populations. The circumstances that lead to the collapse of the outbreak are also matters of speculation. Decline in the number of susceptible trees, drought years and build-up of a predatory beetle (see below) are suspected to play a role.

Natural enemies Information is scanty on the natural enemies of the sal borer; apparently, there are not many. An elaterid beetle *Alaus sordidus* is recorded as a predator of *H. spinicornis* and of other cerambycid borers of some trees. The adult *A. sordius* emerges with the onset of monsoon and lays eggs on the bark of trees attacked by the sal borer. The predacious larvae attack the sal borer larvae between the bark and sapwood; older larvae enter the larval tunnels and pupal chambers of the sal borer. One *A. sordidus* larva can destroy up to 10 sal borer larvae/pupae. The predator is not abundant initially but during sal borer epidemics its population builds-up steadily. Up to 10–15% vacant sal borer pupal chambers have been found occupied by *A. sordidus* (Beeson, 1941). The fungus *Beauveria bassiana* was isolated from *H. spinicornis* and in laboratory tests it caused 75–78% mortality of young larvae within six days of exposure (Sharma and Joshi, 2004).

Control The sal borer is a chronic, endemic pest, i.e. the insect is always present in small numbers in sal areas, usually infesting fallen, unhealthy or dying trees. Living, healthy trees are infested only during population outbreaks. Infestation of up to one per cent of the growing stock, i.e. an average of 2.5 trees per hectare, is considered normal. For management purposes, a population density above this level is reckoned as the beginning of an outbreak. Pest management aims at two goals: (1) prevention of outbreaks by keeping the infestation below the above defined tolerable level and (2) remedial actions to limit the damage during outbreaks. Appropriate methods have been developed for each. To facilitate the implementation of these measures, infested trees have traditionally been classified into seven types, as shown in Table 10.13, to represent different intensities of infestation (Beeson, 1941), type one representing an almost dead tree and type seven representing a tree in the very early stage of attack. It was suggested that during control operations types 1, 2, 3 and 6 should always be removed; types 4 and 5 may be omitted in an incomplete clean up; and type 7 should not be felled. These recommendations were incorporated into the working plan of the forest department.

Preventive measures Since the sal borer is believed to preferentially attack trees of higher girth class and of unsound health, preventive measures are aimed at reducing the presence of such trees through silvicultural measures. Preventive measures are also aimed at removing the existing beetle population. To accomplish these ends, the following measures are recommended.

1. During the winter season, the forest staff should carry out regular patrolling to discover fallen, unsound and borer-infested trees. Borer-infestation is indicated by excessive resin flow and ejection of wood dust.
2. Cut and remove borer-infested and unsound trees. Carry out regular thinning so that the stand does not become too dense; fell trees whenever they become commercially exploitable, instead of retaining them to the maximum age.

Table 10.13. *Beeson's classification of borer infested* Shorea robusta *trees*

Type	Characteristics
1	Crown dead, leafless; epicormics leafless, wood dust in large heap.
2	Crown dead, brown; epicormics dead, brown; wood dust in large heap.
3	Crown dead, brown; epicormics or bark dead in upper part, alive in lower part of trunk; wood dust in heap more than 7.5 cm deep or less abundant.
4	Crown entirely alive, green; epicormics green; wood dust in large heap.
5	Crown partly alive, green and partly dead, brown; epicormics green; wood dust scattered, less than 7.5 cm deep.
6	Stump with large heap of wood dust.
7	Crown entirely alive, green; epicormics green; resin abundant or absent; wood dust scattered or scanty.

3. Fellings should be confined to the period, October to March, when egg-laying sal borer adults are not present. The bark should be removed from all stems above 20 cm diameter left in the forest.

Remedial measures Remedial measures have two components: (1) felling and removal of badly infested trees and (2) trapping and destruction of adult beetles, both aimed at reducing the multiplication and spread of the borer population.

1. *Felling and disposal of attacked trees* After the rainy season, mark the infested trees, classifying them into the different types based on the intensity of attack. Then, depending on the severity and extent of the outbreak, as indicated by the enumeration, decide on the proportion of trees that can be felled, converted on-site, transported to storage yards or disposed by burning, taking into account the available manpower and facilities. If all the infested trees cannot be handled properly, concentrate on those trees having the largest numbers of borers per tree. Debarking of the felled trees is sufficient for the destruction of larvae at the early stage of attack when the larvae have not penetrated into the wood. Moderately attacked logs may be stored in depots, sprayed with insecticide and covered with thick polythene sheets, to kill the emerging beetles. Heavily attacked logs must be burned. For this purpose, they should be arranged suitably around stumps, with small wood and good aeration, to ensure good burning.
2. *Trapping and destruction of beetles* Fresh sap from the bark and sapwood of the sal tree is highly attractive to the sal borer adults. The sap is imbibed with avidity until the beetle is engorged and becomes inactive. Taking advantage of this behaviour, an effective 'trap-tree method' has been developed for capturing the beetles. Silviculturally undesirable trees, including lightly infested trees, are felled and cut into billets. The bark at the ends of the billets is beaten and loosened to facilitate oozing of the sap and to provide a hiding place for the beetles. Beetles are attracted in large numbers, from great distances, to feed on the sap. The trap billets are inspected daily and the assembled beetles collected and killed, by pulling off the head. Usually, local labourers are employed to collect the beetles and wages are paid on the basis of beetle head counts. After every three

to four days, the logs are cross cut again and the cut ends beaten, to restore their attractiveness. A freshly cut tree remains attractive for 8–10 days.

Effectiveness of control measures Both the preventive and remedial measures are considered to be effective, if implemented rigorously. The occasional recurrence of outbreaks is attributed to neglect in the implementation of the prescribed preventive measures. The effectiveness of remedial control measures is difficult to assess because the outbreaks do end naturally after a few years, apparently due to the reduction in infestable trees, adverse weather conditions and the build-up of predators. But trapping and killing of tens of millions of beetles should surely exert a negative influence on the progress of outbreaks. For example, during the 1994–2000 outbreak in Madhya Pradesh, about 63 million beetles were caught and destroyed over four years from 1996 (Dey, 2001). One can imagine the havoc that would have been caused if these beetles had not been caught and destroyed. Felling of infested trees, however, has invited criticism both with respect to the necessity of felling all the prescribed categories of trees and the environmental impact of felling large number of trees. During the 1994–2000 outbreak in Madhya Pradesh, some national newspapers commented that the answer to the crisis was as bad as the problem and that at a time when science had made so much progress, it was indeed sad that other solutions have not been suggested by experts. A study by Dey (2001) on the fate of infested trees showed that practically all trees with dead crown (they possessed partially live trunk and green epicormic branches) failed to survive, but 52–70% of infested trees with partially or fully live crown (T5 and T7 trees as per Beeson's classification that had epicormic branches and ejected wood dust) recovered from the injury. Interestingly, he observed that trees which showed resin exudation but no ejection of wood dust (such trees often constituted 40% of the trees enumerated as attacked) were those which had successfully resisted the attack and were not destined to die (unless reinfested). They harboured small dead larvae underneath the bark.

The seven-category classification of infested trees is cumbersome, in practice. It appears that classification into the following four categories would be sufficient: (1) infested trees with fully dead crown, (2) infested trees with partially live crown, (3) infested trees with fully live crown but with ejected wood dust and (4) trees with resin exudation but with no ejected wood dust. The first category represents trees that are destined to die and should be cut and the last, trees which have successfully resisted the attack and should not be cut.

Others are in between and their management may be decided based on the severity of the outbreak situation.

Knowledge gaps The causes of sal borer outbreak largely remain unknown although it is generally believed that dense stands with overmature trees precipitate the outbreaks. It is most likely that events like lightning strikes, storm damage or heavy defoliation by caterpillar outbreaks that weaken a large number of trees, making them susceptible to attack, may provide an epicentre for build-up of the outbreak populations. Most research on the sal borer problem has been conducted during the periods of outbreak. Obviously, systematic population ecological studies covering the non-outbreak periods in the outbreak prone areas, and covering areas where outbreaks are not known to occur, can be expected to throw further light on the factors regulating population build-up and the causes of outbreak.

'Trap-tree operation' is an effective method for attracting and collecting beetles but is cumbersome. Isolation, synthesis and formulation of the attractive components in the sal tree sap should help to develop a more convenient, and perhaps more effective, trapping method. In a recent study, Kaur *et al.* (2003) reported 28 volatile compounds from the bast (cambium and secondary phloem) of sal, of which nine - T-cadinol, alpha-cadinol, globulol, alpha-copaene, gamma-cadinene, viridiflorene, beta-elemene, alpha-terpineol and gamma-muurolene - made up nearly 49%. If a more convenient and effective adult trapping system were developed, it should be possible to use it as a continuous population monitoring tool to warn of impending outbreaks so that suitable preventive measures can be taken in time.

10.16 *Swietenia* species (Meliaceae) (common name: mahogany)

Tree profile

Swietenia species, commonly known as 'mahogany', are native to tropical America, occurring between latitudes 20°N and 18°S (CABI, 2005). Three species are recognized. The most well-known and widely planted is *Swietenia macrophylla* King, commonly called 'big-leaved' or 'broad-leaved' mahogany, to distinguish it from the small-leaved *S. mahagony* (L.) Jacq. The natural distribution of *S. macrophylla* covers south-east Mexico in North America; Belize, Costa Rica, El Salvador, Guatemala, Honduras, Nicaragua and Panama in Central America; and Bolivia, Brazil, Colombia, Ecuador, Peru and Venezuela in South America. *S. mahogany* is common in the Caribbean countries. These evergreen trees grow up to 30–45 m tall. The third species, *S. humilis* Zucc., is a smaller tree, 8–10 m in height, and commonly distributed in Central America. The biological boundaries between the three species are not clear-cut and natural hybrids occur (CABI, 2005). *S. macrophylla* is the most widely distributed and the most widely planted of the trees, both in native and exotic locations. It is moderately fast-growing and is usually grown in a 30–40 year rotation. It produces one of the world's best furniture timbers. *S. mahagony* is also raised in plantations, but to a much lesser extent.

Plantations of *S. macrophylla* have been raised in over 40 countries outside its native range, throughout the lowland humid and sub-humid tropics in South and Southeast Asia, the Pacific Islands, the Caribbean and tropical Africa (CABI, 2005). In 1995, the plantations covered 151 000 ha worldwide (Pandey, 1997). Indonesia has about 55 000 ha (Cossalter and Nair, 2000) and Fiji 26 500 ha (Kamath *et al.*, 1996).

Overview of pests

Pests are common to both *S. macrophylla* and *S. mahagoni*, but most studies have been made of the former. In natural forests in Latin America, mahogany seedlings are attacked by the leaf cutting ants *Atta cephalotes* and *A.* cf. *sexdens* (Larrea, 1999), and saplings by the shoot borer *Hypsipyla grandella* (Lepidoptera: Pyralidae). Both pests also occur in plantations where mahogany is native. Additional pests in native plantations include the mahogany webworm *Macalla thyrsisalis* (Lepidoptera: Pyralidae) which webs the newly flushed leaves and feeds on them and *Phyllocnistis meliacella* (Lepidoptera: Gracillariidae) whose larvae mine in the leaves (Howard and Solis, 1989; Howard, 1995). In exotic plantations of mahogany, the dominant pest is one of the two closely related species of *Hypsipyla* which bore into the shoot of saplings. A pest profile of this shoot borer complex is given below.

Next in importance are some species of termites (Isoptera) of the genus *Neotermes* (Kalotermitidae) and *Kalotermes* (Rhinotermitidae) which attack the wood of living trees. In Fiji, three species of *Neotermes*, *N. samoanus*, *N. papua* and an unidentified species, attack healthy trees of all ages and feed in galleries within the bole, causing swellings on the trunk and hollowing out the tree. Kamath *et al.* (1996) estimated that termites attacked 7.7% of mahogany trees in plantations in Fiji. In Sri Lanka and the Solomon Islands, a species of *Coptotermes* attacks living mahogany trees (Mayhew and Newton, 1998). However, termites have not been recorded as pests of mahogany in other countries. Also in Fiji alone, two species of ambrosia beetles, *Crossotarsus externedentatus* and *Platypus gerstackeri* (Coleoptera: Curculionidae: Platypodinae), infest living trees and tunnel into the wood, making narrow galleries which become visible as pin holes in sawn timber. Heavy infestations of these beetles were reported in the 1970s. Both species are highly polyphagous, attacking over 40 tree species in Fiji, and it is believed that the large build-up of these beetles was facilitated by slow-dying trees in the natural forest which were poison-girdled in preparation for establishment of the mahogany plantations (Roberts, 1978).

Some pests of lesser economic importance have also been recorded in exotic plantations. A few species of scolytine beetles bore into the stem of seedlings in the nursery, excavate galleries and lay eggs in them, leading to collapse of the seedlings as the grubs develop. The galleries become blackish due to growth of an ambrosia fungus. The species include *Xylosandrus compactus* in Indonesia, Thailand and Sri Lanka (Day *et al.*, 1994), *Hypothenemus eruditus* in Malaysia (Mayhew and Newton, 1998) and an unidentified species in Fiji (Anon, 1954). Among these *X. compactus*, known as 'coffee shothole borer' is the most damaging; it also attacks twigs of young saplings in Puerto Rico (Mayhew and Newton, 1998). This species is also a pest of seedlings of the related African species *Khaya grandifoliola* and *Khaya senegalensis* in India (Meshram *et al.*, 1993) and a variety of other forest tree species (Browne, 1968), including *Acacia mangium*. In Malaysia a weevil, *Dysercus longiclaris* ring barks and kills young trees while in Puerto Rico another weevil, *Diaprepes abbreviatus*, feeds on young leaves, with its larvae feeding on the root stalk (Mayhew and Newton, 1998). A coreid bug *Amblypelta cocophaga* causes dieback of the terminal bud of saplings in Solomon Islands (and also attacks *Eucalyptus deglupta*).

Pest profile

Hypsipyla species (Lepidoptera: Pyralidae)

Hypsipyla species (Lepidoptera: Pyralidae) are well-known shoot borers of mahogany and have been the main hindrance to expansion of mahogany plantations throughout the tropics. Two main species are recognized – *H. grandella* (Zeller), present in the Latin American tropics and southern Florida, and *H. robusta* (Moore) (Fig. 10.30a,b) in South and Southeast Asia, Australia and West and East Africa. They attack several genera of Meliaceae within the subfamily Swietenioideae, including *Swietenia, Khaya, Cedrela* and *Toona*. The literature on *Hypsipyla* species is extensive and has been reviewed by Newton *et al.* (1993) and Mayhew and Newton (1998). The life cycle and habits of both species are similar but the details given below apply specifically to *H. robusta*, unless otherwise specified. The moth has a wingspan of 25–50 mm, the female being larger than the male. The moth is brownish, with black zigzag lines and patches on the forewing and a whitish semi-hyaline hindwing. The mature larva measures 20–30 mm and is light blue, with longitudinal rows of black spots, but in the earlier instars the colour may vary.

Fig. 10.30 *Hypsipyla robusta*. (a) Adult (wingspan 25 mm). (b) Larva inside shoot of *Swietenia macrophylla*. Courtesy: Chey Vun Khen, Sabah Forest Department, Malaysia.

Life history and seasonal incidence The life history may vary with the host and the climate. In the tropics, *H. robusta* has continuous generations throughout the year on the shoots of mahogany, with a generation taking four to eight weeks. The moths are most abundant in August to December (Beeson, 1941; Mohanadas, 2000). Most attacks occur in the rainy season, when the trees put forth new growth (Morgan and Suratmo, 1976). The female moth lays eggs singly on tender shoots. A moth may lay 400–600 eggs over a period of 7 to 10 days (Beeson, 1941; Speight and Wylie, 2001). A female *H. grandella* moth lays one to seven eggs at a time, occasionally in clusters of three to four, on one or more plants, and oviposition may extend over a period of six days, with 200–300 eggs laid in all (Newton *et al.*, 1993). The newly hatched larva bores into the shoot at a suitable place after probing at several places. Some larvae may get trapped and killed in the exuding sap. Some feed initially in the veins of leaves or under the bark. The entrance of the tunnel into the shoot is usually marked by a mass of frass bound with silk. There are five to six larval instars and in Kerala in southern India, the larval period lasts 15–20 days and the pupal period about 10 days. Pupation occurs within the tunnel. Many larvae do not complete development in one shoot; they leave the original tunnel and bore into a new shoot, apparently because the first shoot is too short or too lignified (Beeson, 1941).

In the subtropical and temperate regions in northern India, the most common host of *H. robusta* is *Toona ciliata* (syn. *Cedrela toona*) and the insect is known as 'Toon fruit and shoot borer'. Here, the last instar larvae of the fall generation enter hibernation within the shoot before winter leaf shedding begins. The first generation moths lay eggs the following year on the flowering shoots of the host tree. The larvae feed gregariously on all parts of the inflorescence, held together within a loose network of silk threads. The second generation of larvae feed on the young fruits, one larva boring into more than one fruit. Mature first and second generation larvae descend to the ground and pupate in crevices in the trunk of the tree, in cocoons, several of which are often closely packed in layers. The subsequent generations of larvae attack the shoots. In Australia also *H. robusta* is known to feed on the flowers and fruits of *T. ciliata* and this switch from shoots to fruits is thought to be associated with the dry season when unlignified shoots are not available (Speight and Wylie, 2001).

Impact Saplings are the most susceptible to *Hypsipyla* attack. Tunnelling of the leading shoot kills the terminal growth, resulting in the development of lateral shoots which may also be attacked, causing a bushy top and loss of tree form, in addition to growth loss. Attack is usually more severe on trees growing in the open compared with those in shaded areas. Apparently, trees in the open grow more vigorously, producing lush foliage, which may be more attractive to the egg-laying female. Studies on *S. macrophylla* in West Java, Indonesia, showed that the degree of infestation of *H. robusta* decreased with the increasing age and height of the tree. Infestation was about 90% for trees aged 3 years or 2.5 m high, decreasing to less than 5% for trees older than 14 years or taller than 13 m (Morgan and Suratmo, 1976; Suratmo, 1977). Seedlings in nurseries are also often attacked (Beeson, 1941; Ambika-Varma *et al.*, 1996). In exceptional cases, infestation has been found in the crowns of 50-year-old plantation trees, 45 m in height, in north Queensland, Australia (Nair, 2001a). According to Wagner *et al.* (1991), on *Khaya* species in Africa *H. robusta* feeds extensively on the soft, living bark of the terminal stem of saplings, causing heavy sap exudation. But recent morphological and molecular studies indicate that the so-called African *H. robusta* is a separate species and that two different *Hypsipyla* species are present on *Khaya* in Ghana (Marianne Horak, unpublished report, 2000).

In young mahogany plantations, incidence of *Hypsipyla* attack is usually heavy. In plantations in Kerala, in southern India, about 70% of plants in three to four-year-old plantations were attacked, with less damage in younger and older plantations (Mohanadas, 2000). In some plantations in India, 100% infestation has been recorded by the second year (Beeson, 1941).

Retardation of growth in the early years of establishment of a plantation is a serious disadvantage, but more damaging is the formation of forked, crooked or branchy boles. Consequently, many mahogany plantations have been abandoned on account of *Hypsipyla* damage in Asia-Pacific, Latin America and Africa (Beeson, 1941; Newton *et al.*, 1993; Wagner *et al.*, 1991). Nevertheless, since shoot borer incidence is usually confined to the sapling stage, many abandoned plantations have survived and fared well later.

Host range and geographical distribution *Hypsipyla* species are polyphagous on tree species of the subfamily Swietenioideae of Meliaceae. Recorded hosts include *Carapa guianensis*, *C. grandiflora*, *C. procera*, *Cedrela odorata*, *C. lilloi*, *Chukrasia tabularis*, *Entandophragma angolense*, *E. candollei*, *E. cylindricum*, *E. utile*, *Khaya anthotheca*, *K. grandifolia*, *K. ivorensis*, *K. nyasica*, *K. senegalensis*, *Lovoa trichilioides*, *Pseudocedrela kotshyi*, *Soymida febrifuga*, *Swietenia macrophylla*, *S. mahagoni*, *Toona ciliata*, *T. sinensis*, *T. sureni* and *Xylocarpus moluccensis* (Beeson, 1941; Wagner *et al.*, 1991; Speight and Wylie, 2001). Either *H. grandella* or *H. robusta* is present wherever the host trees are grown in the tropical, subtropical and temperate regions, with the exception of Fiji and some smaller islands in the Pacific, which the insect has not reached due to geographic isolation. It was found in the Pacific island of Vanuatu only in the year 2000. *H. grandella* occurs in Latin America and southern Florida and *H. robusta* in South and Southeast Asia, Australia and West and East Africa. A third species,

H. ferrealis (Hampson) is present in tropical America, but it exclusively attacks the fruit of *Carapa guianensis* (Newton *et al.*, 1993). Wide variations have been reported in the biology and behaviour of *H. robusta* in different geographic locations and, as mentioned above, more than one closely related species may be involved.

Natural enemies More than 50 species of parasitoids of *H. robusta* have been recorded in India alone, although the rates of parasitism were low (mostly $>1\%$). They include 17 braconids, 13 ichneumonids, 12 chalcidoids, 2 each of trichogrammatids and tachinids, and 1 each of elasmid, eulophid and eurytomid (Newton *et al.*, 1993). Fewer parasitoids are on record for *H. grandella*. Although the causative agent was not identified, disease levels of 4–16% of sampled larvae were recorded from *H. robusta* in surveys in India (Newton *et al.*, 1993). Misra (1993) recorded the fungal pathogen *Beauveria bassiana* on *H. robusta* in India. In *H. grandella* a fungus, *Cordyceps* sp., was recorded (Newton *et al.*, 1993).

Control It has been generally observed that mahogany saplings growing under partial shade in mixed natural forests are less prone to *Hypsipyla* attack than those growing in the open, although the reasons are not clear. Experimental studies have produced variable results. Mahogany, however, is not immune to attack of *Hypsipyla* in the natural forest. For example, Yamazaki *et al.* (1990) observed that in the Peruvian Amazon, the population of *H. grandella* increased rapidly in the rainy season when food availability increased with the growth of new sprouts. In southeast Mexico, the insect attacked the fastest growing seedlings in the logged-over natural forest (Dickinson and Whigham, 1999). In spite of considerable research, no practical control measure has emerged and the shoot borer continues to be the main factor limiting the cultivation of mahogany. In natural forests, mahogany occurs in very low density. In Mexican forests, its average density is 1–2 mature trees per ha, and the range may vary from 1 tree per ha in Brazil to 20–60 trees per ha in Bolivia (Mayhew and Newton, 1998). It is possible that the comparatively low incidence of *Hypsipyla* attack in natural forests is attributable to both shade effect and the action of many natural enemies. Efforts made to control *Hypsipyla* in plantations are briefly discussed below.

Silvicultural control Several authors have recommended the planting of mahogany under the shade of an overhead canopy of evergreens or with lateral shade given by planting in mixture with a faster growing species (Beeson, 1941). Species suggested for mixing range from *Senna siamea* (syn. *Cassia siamea*), *Cassia timoriensis* and *Leucaena leucocephala* in Indonesia to maize in Honduras. The benefits of such measures have seldom been critically evaluated. It is argued, without proof, that lateral shade offers a mechanical obstacle to moths in search of suitable plants for oviposition, that the slower growth of mahogany under shade makes it less attractive to the gravid moths etc. Newton *et al.* (1993) reviewed several instances of mixed planting of mahogany in Puerto Rico, Surinam, Brazil, Belize and Guatemala, either as enrichment planting in secondary natural forests or when planted in the open in admixture with other species, some with suitable controls. They concluded 'growing mahoganies in mixtures with other species seems to have afforded some degree of protection' (p. 308). They also pointed to several other mixed planting trials in Honduras and Costa Rica where shade or cover did not reduce *Hypsipyla* attack. Matsumoto *et al.* (1997) and Matsumoto and Kotulai (2000) reported that some plantations of *S. macrophylla* in Malaysia and Indonesia surrounded by *Acacia mangium* were not attacked by *H. robusta*. In a study of mahogany seedlings established in clearings within natural forests in Mexico, Snook and Negreros-Castillo (2004) found that in plots cleared of competing vegetation from around the seedlings, 44% of the seedlings were attacked by *H. grandella*, compared with 12% in uncleared plots. From the available literature, we can draw the conclusion that planting mahogany in mixture with other species does not guarantee successful *Hypsipyla* control, although it often does, but underplanting of mahogany in managed natural forests reduces the incidence of attack, for reasons not fully understood.

Timely pruning of affected shoots to destroy the larva is another silvicultural method advocated and found effective (Cornelius, 2001).

Genetic resistance It has been reported that *H. grandella* moths are attracted to *Toona ciliata* and oviposit on it, but the larvae die when they begin to feed, suggesting the presence of some toxic substance (CABI, 2005). This substance is not toxic to *H. robusta* which readily attacks *T. ciliata* in Asia and Africa where both are native. These observations suggest that there is scope for breeding for resistance to *H. grandella* (Newton *et al.*, 1993).

Chemical control *Hypsipyla* larvae concealed within shoots are inaccessible to insecticidal sprays and the infestations usually occur in the rainy period when the sprays get easily washed off the plant. Due to these reasons, conventional insecticidal application has not proved effective. To be effective, spraying needs to be carried out repeatedly to target the exposed young larvae and this is neither economically worthwhile nor ecologically acceptable. However, systemic insecticides like carbofuran applied to soil at the time of planting were found effective in field trials in Costa Rica (Newton *et al.*, 1993). Mohanadas (2000) also reported the effectiveness of phosphamidon and dimethoate against *H. robusta* in field trials in India. However, the duration of effectiveness of systemic insecticides under different kinds of soil and climatic conditions and the cost-effectiveness needs to be established.

Biological control Between the 1960s to 1970s, attempts were made in the Caribbean at classical biological control, by introducing parasitoids from India. Several releases were made of the eulophid *Tetrastichus spirabilis*, the trichogrammatid *Trichogrammatoidea robusta*, the braconid *Phanerotoma* sp. and the chalcid *Anthrocephalis renalis* into some islands in the region, but only the egg parasitoid *T. robusta* became established, and no recognizable control was obtained (Newton *et al.*, 1993). The reasons for the failure are debatable; inadequate effort has been suggested but lack of specificity of *H. robusta* parasitoids against *H. grandella* may have been important.

Pheromones The female sex pheromones of *H. robusta* moths have been identified as (Z, E)-9,12-tetradecadiene-1-ol-acetate, (Z)-9-tetradecen-1-ol-acetate and (Z)-11-hexadecen-1-ol-acetate, but field attempts made in Malaysia to trap the males were not successful (Nakamuta *et al.*, 2002a).

Knowledge gaps Satisfactory control of *Hypsipyla* attack is still elusive. Various approaches such as genetic engineering of the plant by inserting toxin genes from *Bacillus thuringiensis* and hybridization between *Toona ciliata* and *Swietenia* to transfer the toxicity of the former to *H. grandella* to the latter needs to be explored. Taxonomic studies on the *Hypsipyla* species are also needed.

10.17 *Tectona grandis* (Lamiaceae) (common name: teak)

Tree profile

The teak tree, *Tectona grandis* L. F., is well known for its versatile timber. Its heartwood combines several qualities like termite and decay resistance, lightness and strength, drying without warping and splitting, easy workability and attractive appearance, making it one of the world's finest timbers. Teak's position among timbers has been likened to that of gold among metals and diamond among precious stones. The teak tree is native to South and Southeast Asia, more specifically India, Myanmar, Thailand and Laos. Over the past 150 years it has been planted extensively both within its native range and in other tropical and subtropical regions in Asia, Africa and America. It is naturalized in the Indonesian island of Java and some of the smaller islands east of Java, where it is believed to have been introduced some 400–600 years ago. Naturally regenerating teak stands are also present in the western part of the Yunnan Province of China but it is not known whether these stands are indigenous or not. The natural teak area totals about 28 million ha, with Myanmar accounting for 59%, India 32%, Thailand 8.9% and Laos < 0.1% (Teaknet, 1995). In the year 2000, plantations of teak were estimated to occupy an area of 5.7 million ha, with about 92% in Asia, 4.5% in Africa and 3% in Central and South America (Ball *et al.*, 2000; FAO, 2001a). Rapid expansion of commercial teak plantations is now taking place in Central America. While the native plantations are grown on a rotation of 50–80 years, producing a mean annual increment (m.a.i.) of 3–10 m^3 of wood per ha, many exotic plantations are managed at a shorter rotation of 20–30 years, with a m.a.i. of 10–20 m^3 per ha.

Overview of pests

As pointed out in Chapter 2, in India and the neighbouring countries alone at least 174 species of insects have been recorded from the living teak tree – 137 leaf feeders, 16 sap feeders, 14 shoot or stem feeders, 5 root feeders and 3 seed feeders. These were listed in Table 2.3. Additional insects have been recorded from other countries. The majority of species cause only slight or occasional damage. The main types of damage and the important pests are the following.

Defoliators

Leaf feeders constitute the majority of insects associated with the living teak tree, as noted above. Many of them do not cause serious damage. The major defoliators are *Hyblaea puera* (Lepidoptera: Hyblaeidae) and *Eutectona machaeralis* (Lepidoptera: Pyralidae) (or the closely related *Paliga damastesalis* in some countries). Pest profiles of these species are given below. In Kerala in southern India, during the early part of the growth season, an unidentified chrysomelid beetle caused up to 2.5% leaf loss and an unidentified curculionid beetle caused up to 15% leaf loss, both feeding on tender leaves (Nair *et al.*, 1985). In Indonesia, the grasshopper, *Valanga nigricornis* (Orthoptera: Acrididae) causes sporadic defoliation (Nair, 2000) and in the drier parts of Ghana, the cricket *Zonocerus variegatus* (Orthoptera: Acrididae) causes frequent defoliation (Wagner *et al.*, 1991).

Stem borers

Among stem borers, the cossid caterpillar *Xyleutes ceramicus* is the most serious. Another caterpillar, *Sahyadrassus malabaricus* (Lepidoptera: Hepialidae) (or related species), that bores into the stem of the saplings, is also of some importance. Pest profiles of these two species are also given below.

A wood-dwelling termite, *Neotermes tectonae* (Isoptera: Kalotermitidae) is a pest of economic importance in Java, Indonesia. Popularly known as 'inger inger' in

Java, this termite lives within the main trunk or branches of living teak trees and makes galleries in the stem from within. Swellings develop on the stem which become visible about 3–5 years after the infestation. Trees over three years old may be attacked. Crevices in the attacked wood degrade the valuable construction timber to fuel wood, causing economic loss. In some places in Central Java, 10–72% of the trees in plantations were attacked and the production loss was estimated at 9–21% (Subyanto *et al.*, 1992). Cutting and removal of the infested trees is the only practical method of reducing the damage, although introduction of fumigants into the affected portion of the trunk has been tried. This pest has been recorded only in Indonesia and is confined to some endemic patches in Central and East Java.

Also in Indonesia, an ambrosia beetle, *Xyleborus destruens* (Coleoptera: Curculionidae: Scolytinae) attacks the trunk of living teak trees, making branching tunnels that extend into the heartwood. It is prevalent in areas where there is no definite dry season (Kalshoven, 1953). The related *X. morigerus* has been reported as infesting young plantations in Mexico (Vazquez, 1980). Another scolytine beetle, *Hypothenemus pusillus* attacks mainly unhealthy seedlings and twigs of trees in Ghana (Wagner *et al.*, 1991).

Another caterpillar, *Zeuzera coffeae* (Lepidoptera: Cossidae), is an occasional pest noticed predominantly in agroforestry plantations where the plants are vigorous and succulent. It has been recorded in Kerala and Tamil Nadu in India, Central Java in Indonesia and in Thailand. The larva bores into the pith of the stem of saplings and ejects the frass through holes made on the stem. The mature reddish brown larva is about 35 mm long. Commonly known as red coffee borer, it also attacks coffee, tea and cocoa.

The grub of a cerambycid beetle, *Acalolepta cervina* (syn: *Dihammus cervinus*), known as teak canker grub, causes damage to teak saplings in India, Bangladesh, Myanmar, Thailand and Malaysia (Beeson, 1941; Baksha, 1990; Hutacharern and Tubtim, 1995; Chey, 1996). The adult beetle feeds on the bark of saplings and lays eggs on the stem, near ground level. Feeding and tunnelling by the larva, usually below one meter above ground, causes the formation of a bulging canker all around the stem and the stem may break at this point. In some plantations in northern India, more than 50% of the growing stock in two to three-year-old plantations may be attacked, but this percentage falls off as age advances and by 7–8 years there is no attack (Beeson, 1941).

Twig gall

Stem galls on branches of teak, caused by a gall midge, *Asphondylia tectonae* (Diptera: Cecidomyiidae), are common in some plantations in India, particularly of poor class. The insect attacks new shoots and causes formation of globular, multilocular galls that coalesce, harden and surround the stem of twigs. This may retard growth when the infestation is heavy.

Root feeders

Larvae of some families of beetles, known as whitegrubs, feed on the fleshy taproot of teak seedlings and kill the plants, as mentioned under Section 5.1.

Flower and fruit feeders

The caterpillars *Pagyda salvalis* and *Dichocrocis punctiferalis* (Lepidoptera: Pyralidae) and the bug *Leptocentrus* sp. (Hemiptera: Membracidae) feed on the flowering shoots and green fruits of teak in India and Thailand. In Thailand, *Mylabris phalerata* (Coleoptera: Meloidae) and *Machaerota elegans* (Hemiptera: Cercopidae) also cause serious damage (Hutacharern, 1990).

Pest profile

Hyblaea puera (Cramer) (Lepidoptera: Hyblaeidae)

Hyblaea puera (Cramer) (Lepidoptera: Hyblaeidae) (Fig. 10.31a,b), commonly known as the teak defoliator, is the most notable pest of teak in Asia-Pacific and is now becoming increasingly important in Latin America also, where teak is planted as an exotic. The species was first described by Cramer in 1777 as *Phalaena puera* and was originally included under the family Noctuidae. *Noctua saga* Fabricius (1787) is a synonym. The systematics of the genus *Hyblaea* is poorly studied (Kim and Sohn, 2003) and according to CABI (2005) the species *puera* Cramer is a species-complex. *H. puera* was first recognized as a pest of teak plantations in Kerala, India over 150 years ago (Bourdillon, 1898). Since then, vast literature has accumulated on its biology and ecology. The state of knowledge has been summarised and reviewed by Beeson (1941), and more recently by Nair (1988, 1998), Thakur (2000) and CABI (2005).

The moth has a wingspan of 30–40 mm. When at rest, the wings are held slanted and roof-like, giving the moth a triangular shape. The forewings are dull grey or reddish brown and the hindwings are dark brown, with an orange, scarlet-edged transverse band across the middle, which is constricted or sometimes broken into three patches. The abdomen is dark brown with orange segmental bands. The full-grown larva is 35–45 mm long. The larval instars show marked variation in body colouration and pattern. Generally the larvae are greyish-green with white, black and flesh-coloured dorsal longitudinal bands, but uniformly black-coloured larvae predominate on some occasions, particularly during population outbreaks. Both types of larvae often occur in the same population.

(a)

(b)

Fig. 10.31 The teak defoliator *Hyblaea puera*. (a) Moth (wingspan 35 mm). When at rest, the wings are held slanted and roof-like. Courtesy: V.V. Sudheendrakumar, Kerala Forest Research Institute. (b) Larva.

Life history The female moth lays eggs on tender new leaves of teak, attaching them singly near the veins, and usually on the under-surface. Each female lays about 500 eggs, with a recorded maximum of 1000 (Beeson, 1941). The female has an average lifespan of about 13 days in the laboratory and mates only once (Sudheendrakumar, 2003). The males mate with more than one female. The pre-mating period of both sexes is a day and the mean ovisposition period is seven days; most eggs are laid during the first half of the oviposition period, between sunset and midnight.

There are five larval instars. The first and second instars feed mainly on the leaf surface, protecting themselves in a shallow depression on the leaf, under strands of silk. Starting with the third instar, the larva feeds from within leaf folds (Fig. 10.32a,b). The entire leaf, excluding the major veins, is eaten. Early instars cannot feed successfully on old, tough leaves. Under optimal conditions, the larval period lasts 10–12 days, but an average of 21 days has been recorded in the cooler climate of Dehra Dun in northern India.

(a)

(b)

Fig. 10.32 Early instar larvae of *Hyblaea puera* on tender teak leaf. (a) Larval leaf folds on leaf edge. (b) Larval leaf folds on entire leaf.

Mature larvae usually descend on silk thread from the tree crown to the ground, and pupate under a thin layer of leaf litter or soil within a loosely built cocoon made of dry leaf pieces or soil particles held together with silk. During the rainy season, when the ground is wet, or in mangroves, pupation may occur within folded or juxtaposed green leaves of host or non-host plants in the undergrowth. The average pupal period is 6–8 days under optimal conditions, but it may be prolonged to 20–25 days in cooler climates. There is no evidence of hibernation or aestivation of pupa.

The development from egg to adult is completed in a minimum of 18–19 days and a maximum of 36 days and a new batch of eggs can be produced in about 2 days, thus giving a minimum generation time of 20–21 days. In field insectaries in southern India and Myanmar, 14 complete generations and a partial 15th have been obtained (Beeson, 1941). At Dehra Dun, in northern India, where there is a distinct winter season, with chances of occasional frost, the number of generations is reduced to 10, with a partial 11th. Here the moths are believed to hibernate for a period of about three months from December to February (Beeson, 1941), but no details of the hibernation behaviour or the places of hibernation have been reported.

Host range *H. puera* has been recorded on 45 host plants, including some shrubs. Most host plants belong to the families Bignonaceae and Lamiaceae, with some representatives from Verbenaceae, Rhizophoraceae, Oleaceae, Juglandaceae and Araliaceae. Most host records are from Asia. The insect occurs on *Vitex parviflora* and *Tabebuia pentaphylla* in the Caribbean and on the straggling shrub *Vitex trifolia* in Australia. No information could be traced on its host plants in Africa. Outbreaks are common only on teak, although there are rare records of outbreaks on the mangroves *Avicennia marina* on the Bombay coast of India (Chaturvedi, 1995, 2002) and *A. germinans* in Guadeloupe in the Caribbean (Saur *et al.*, 1999). It has also been reported as a pest of the mangroves *Rhizophora*, *Bruguiera* and *Avicennia* in Thailand (Hutacharern, 1990). It is believed that during non-outbreak periods the insect thrives on hosts other than teak but data are not available on the periods of infestation or population levels on most other hosts. Laboratory investigations show that some of the host plants like *Vitex negundo*, *Premna latifolia*, *Spathodea companulata*, *Callicarpa arborea* and *Avicennia officinalis* are as good or even better than teak in supporting the development of *H. puera* (Beeson, 1941; Amin and Upadhyaya, 1976; Baksha and Crawley, 1995).

Geographical distribution *H. puera* is widely distributed across the tropics and subtropics, covering Asia-Pacific, Africa, Central America, the Caribbean and South America (Table 10.14). However, information on its host plants outside

Table 10.14. *Recorded world distribution (country/region) of* Hyblaea puera

Asia-Pacific	Africa	America
Bangladesh	Malawi	**Central America**
Cambodia	South Africa	Costa Rica
China (southern part)	Uganda	Honduras
India		**Caribbean**
Indonesia		Cuba
Japan		Dominican Republic
Laos		Guadeloupe
Malaysia		Jamaica
Myanmar		Puerto Rico
Nepal		Trinidad and Tobago
Philippines		**South America**
Sri Lanka		Brazil
Taiwan		
Thailand		
Vietnam		
Oceania		
American Samoa		
Australia (northern part)		
Papua New Guinea		
Samoa (western)		
Solomon Islands		

For references see Nair (2001b)

Asia is meagre, and as noted above, it has been suspected that what we call *H. puera* may be a species-complex. It has not attacked teak plantations in Africa so far and only very recently has it attacked teak in Latin America, in spite of the long history of teak planting in these regions. Within Latin America, outbreak was noticed first in 1995 in Costa Rica and in 1996 in Brazil.

Seasonal incidence Given the biological attributes described above and the year-round warm temperatures of the tropics, one would expect that populations of *H. puera* would be present continuously in teak plantations. But this is not so. What happens in Kerala, in southern India, is typical. For most of the year there is no visible defoliator activity. The teak trees put forth a new flush of leaves, generally by March–April, following a brief deciduous period. Then about a month later, usually between May and June, widespread infestations covering hundreds of ha suddenly occur, with millions of similar aged caterpillars feeding gregariously on the teak canopy. During these outbreaks, each tender leaf of the infested trees may harbour some 50–100 larvae, and it has been estimated (Nair, 1988) that a 30 ha teak plantation may have over 450 million larvae. When the outbreak is in progress (Fig. 10.33), the faecal pellets falling on dry leaves on the ground can be heard like the sound of a mild drizzle of rain. Multitudes of larvae descend on silk threads from defoliated trees and are wafted to adjacent trees still holding green leaves. Mature larvae pupate on the ground under litter. Within a week or two, extensive areas of plantations are left totally leafless. Although small-scale outbreaks occur about a month prior to these widespread outbreaks, they usually go unnoticed as they cover small areas (usually 0.5–1.5 ha) and are widely separated in space within large plantation areas (Nair and Mohanadas, 1996). They occur soon after the first pre-monsoon rainfall and are characterized by confinement of damage to the tender leaves at the tree top, older leaves at the lower crown level usually escaping attack. This is because the group of egg-laying moths is small. Fig. 10.34 shows the typical temporal sequence of defoliation recorded in four 50-tree observation plots within a large plantation.

A characteristic feature of the outbreak is the concentration of infestation into discrete patches, whether the infested area is small or big. In a large plantation

Fig. 10.33 Appearance of the teak tree during the progression of *Hyblaea puera* outbreak. Except for some larvae still feeding within leaf folds, the leaves are completely eaten up, leaving the major veins. The leaf skeletons will eventually fall off.

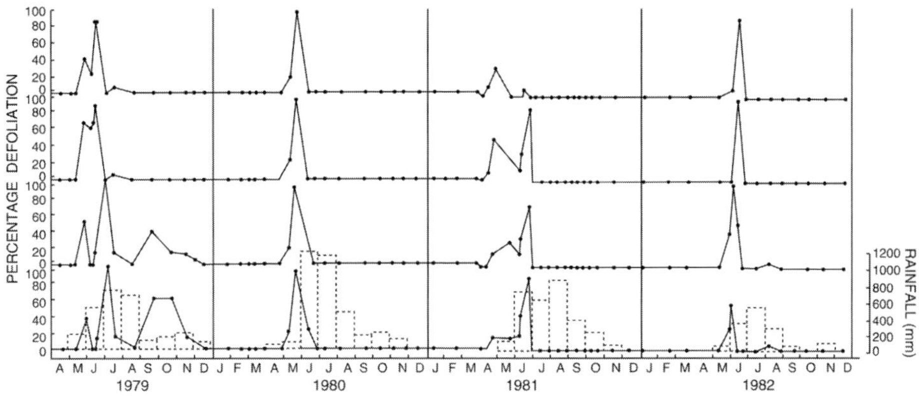

Fig. 10.34 Seasonal incidence of defoliation caused by *Hyblaea puera* infestation in four 50-tree observation plots (400 m^2 in area) within a 2500 ha teak plantation area at Karulai in Nilambur, Kerala, India, over a four-year period. The trees were about five-years old and 8 m tall at the beginning of the observation period. The bars show monthly rainfall. (Data from Nair and Sudheendrakumar, 1986).

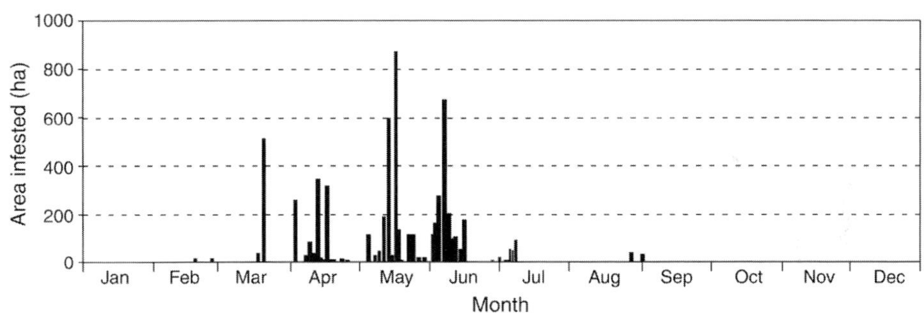

Fig. 10.35 Temporal sequence of *Hyblaea puera* outbreaks within about 10 000 ha of teak plantations at Nilambur in Kerala, India, during the year 1993. Outbreaks occur in a series of infestations over discrete patches. The area infested in each episode of outbreak is shown. On a given date, the infested area is not necessarily contiguous. For example, the first infestation on 19 February occurred in two patches of 12.8 ha and 1.7 ha, separated by a distance of 3 km, but the second infestation on 26 February covered 10 ha in one place. See Fig. 10.36 for spatial sequence. (Data from Nair *et al.*, 1998a).

area, these infestation patches appear in a wave-like succession, at different places, at intervals shorter than the life cycle of the insect. Small permanent plots as in Fig. 10.34 cannot capture all the defoliation episodes that occur in a large plantation area. Fig. 10.35 shows the timing and frequency of such outbreak episodes in about 10 000 ha of teak plantations at Nilambur, in the year 1993. The spatial distribution of the early outbreak episodes is shown in Fig. 10.36.

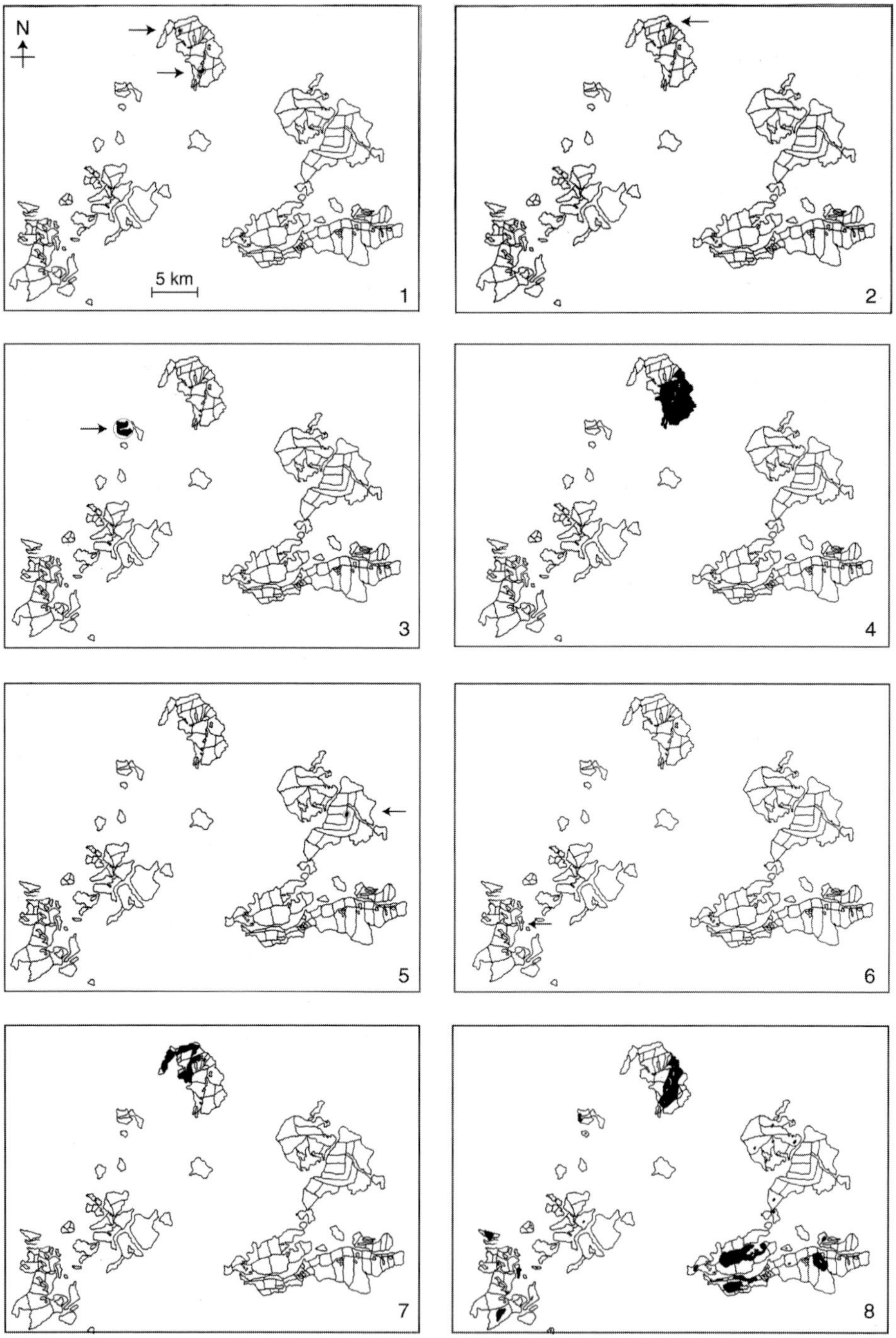

Fig. 10.36 Map of the teak plantations in Nilambur Forest Division in Kerala, India, showing the spatial distribution of the early outbreaks of *Hyblaea puera* in 1993. Black dots or shading indicate the infested areas; where the infested areas are

It may be seen that a series of infestations occurred in discrete, discontinuous patches, at short and irregular intervals, covering small as well as large areas. During the year, re-flushed trees were attacked again during later episodes of the outbreak. Total or near-total defoliation may occur twice or more in the same area, between April and June. In a given area, the outbreaks usually subside after one or two episodes (Fig. 10.34). Incidence of a viral disease, causing large-scale mortality, is usually noticed in the declining phase of the outbreak. In most years, a further peak of low-density infestation occurs in southern India between mid-August and October. Careful observations have revealed the presence of a very sparse population of the insect at other times in teak plantations in Kerala, the lowest density occurring during December to February (Nair, 1988).

In many other places in India and other countries, *H. puera* populations have been noticed only during periods of outbreak. Within India, outbreaks first appear in Kerala in the south, and move slowly towards the north, coinciding with the flushing of teak and the advancement of the monsoon (Bhowmick and Vaishampayan, 1986). While the large-scale outbreaks occur in May–June in Kerala in southern India, they are delayed until about July–August in Madhya Pradesh in central India. Similar large-scale outbreaks are common in Myanmar, Thailand, Bangladesh, Sri Lanka and Indonesia during the main flushing period of teak, although detailed information on the seasonal abundance is not available (Beeson, 1941; Hutcharern, 1990; Tilakaratna, 1991; Nair, 2000).

Population dynamics The population dynamics of *H. puera* are characterized by sudden outbreaks, following a period of near-absence of the insect within teak plantations. The source of the moths that arrive suddenly to cause the early, tree-top outbreaks as well as the subsequent waves of larger outbreaks remains a mystery. There is no evidence of diapause in pupae. Circumstantial evidence suggests that the moths are migratory and arrive through the monsoon wind

Figure 10.36. (cont.) small arrows are used to point to the locations. Outbreaks up to 20th April only are shown. During the entire year, a total of 7260 ha were infested, including re-infested areas.

1. First infestation on 19 February (2 patches, 14.3 ha),
2. Second infestation on 26 February (1 patch, 10 ha),
3. Third infestation on 17 March (1 patch, 38.8 ha),
4. Fourth infestation on 20 March (1 patch, 512 ha),
5. Fifth infestation on 21 March (1 patch, 1.7 ha),
6. Sixth infestation on 26 March (1 patch, 0.12 ha),
7. Seventh infestation on 3 April (3 patches, 254.4 ha),
8. A series of infestations from 7 to 20 April (24 patches, 934.4 ha).

From Nair *et al.* (1998a)

system. A correlation between the monsoon rains and the occurrence of *H. puera* outbreaks in teak is well established by observations from many places in India (Bhowmick and Vaishampayan, 1986; Nair and Sudheendrakumar, 1986; Vaishampayan *et al.*, 1987; Khan *et al.*, 1988b; Loganathan and David, 1999; Loganathan *et al.*, 2001). The earliest tree-top outbreaks in Kerala, during the pre-monsoon period also coincide with the first pre-monsoon rain showers (Nair and Mohanadas, 1996). The behaviour of the moths also indicates migration. Aggregations of moths on shrubs in the understorey of natural forests near teak plantations as well as on ground vegetation in teak plantations, and oriented mass flight of moths have been observed (Nair, 1988; Sajeev, 1999). Apparently, the moths arrive through the monsoon wind system, by a combination of active flight and passive transport at the cloud front and land on hill tops. Monsoon-linked long-range migration has been observed, using radar and aircraft, in other moths like *Choristoneura fumiferana* (spruce budworm) in Canada (Greenbank *et al.*, 1980) and *Spodoptera exempta* (African army worm) in Africa (Riley *et al.*, 1983; Rose *et al.*, 1985). The immigrant *H. puera* moths probably remain aggregated at hill tops and move en masse on successive nights, for egg laying at different sites, until they exhaust themselves.

Nair and Sudheendrakumar (1986) found that the majority of moths that emerge from an outbreak site do not oviposit in the same area even when suitable host plants are available. They congregate and move away (Nair, 1988). Whether they undertake short-range, gypsy-type migration by active flight, in search of suitable egg-laying sites a few kilometres away (Nair and Sudheendrakumar, 1986), embark on long-range migration or both is not known. But migration, whether short or long range, is an essential feature of the life system of *H. puera*. It was suspected that the few tree-top infestation sites during the pre-monsoon period might serve as epicentres where the population builds up and spreads to other areas (Nair and Mohanadas, 1996). If moths originating from these 'epicentres' are responsible for the subsequent large-scale outbreaks in the vicinity, controlling the insects in the epicentres could prevent the subsequent larger outbreaks. Nair *et al.* (1998a) examined this possibility by temporal and spatial mapping of all the infestations that occurred in the entire Nilambur teak plantations (about 10 000 ha) in the year 1993. They found that the locations of the early tree-top infestation patches were not constant over the years and the patches did not represent highly favourable local environments. Thus these sites cannot be considered as conventional epicentres, i.e. specially favourable sites where the pest population multiplies and then spreads to other areas. Moreover, moth populations originating from these locations alone were not sufficient to account for all the local large-scale outbreaks that followed suggesting that immigration of moths from a long distance continues to occur.

Thus the origin of the discrete populations of moths which arrive in waves in a large plantation area remains unknown. Perhaps they include both locally produced short-distance migrants and long-distance migrants. In a given locality the population outbreaks subside after two or three outbreak episodes, apparently due to incidence of a baculovirus disease (see below).

The moths that cause the smaller outbreaks later in the season (mid-August to October) in Kerala, India, are also immigrants because the local population of moths is not sufficient to account for the number. The arrival of these moths appears to coincide with the northeast monsoon rains in Kerala. These smaller outbreaks do not occur in central India, where there is no monsoon rainfall at this time.

As noted earlier, within India outbreaks first appear in Kerala in the south and move slowly towards the north, along with the advancement of the southeast monsoon (Bhowmick and Vaishampayan, 1986). Moths causing outbreaks in the rest of India may therefore originate in Kerala, but the source of moths for the first outbreaks in Kerala is unknown. Nair (1998) suggested two possibilities – monsoon-linked, long-distance displacement of air-borne moth populations from a distant area or wind-aided concentration of dispersed local populations of moths. The first appears more probable; the moths can get into the pre-monsoon cloud front from some remote areas where there is a pre-existing active population. Over large forest areas, where one or other of the 45 species of host plants may provide at least a small supply of tender leaves throughout the year, an active population of *H. puera* could thrive throughout the year. A fairly large population of moths could be built up prior to March, if the host plants flushed earlier due to climatic differences between geographical regions or phenological differences between host tree species. For example, there are reports of *H. puera* outbreaks on the mangrove, *Avicennia marina* on the Bombay coast of India during September–October (Chaturvedi, 1995, 2002), which is not the main flushing season for teak. *H. puera* is reported as a pest of the mangroves, *Avicennia*, *Brugiera* and *Rhizophora* in Thailand (Hutacharern, 1990) although the period of infestation of these species is not known. It must be investigated whether the pre-monsoon source of the moths in southern India is the extensive mangrove forests of Southeast Asia. Several host plants of *H. puera* are mangrove species (e.g. *Avicennia marina*, *A. officinalis*, *A. germinans*, *Brugiera* sp., *Dolichondrone spathacea*, *D. stipulata*, *Rhizophora* sp.) and since mangroves are evergreen they could sustain the insect population throughout the year.

Impact In most teak plantations in Asia, *H. puera* outbreaks occur every year, following the onset of flushing, causing one or more total or near-total defoliations. In some places (e.g Kerala, India) and in some years this is followed by erratic lighter defoliations later in the season. Outbreaks are spectacular events, creating the impression of severe growth loss, although the trees put forth a new flush of leaves within weeks. Several attempts have been made in the past to estimate the loss due to defoliation, based on artificial defoliation experiments, field observations on the frequency of defoliations etc., which put the loss figures variously at 6.6–65% of the normal increment (Mackenzie, 1921; Beeson, 1931a, 1941; Champion, 1934). Nair (1986a) made a critical review of these early attempts and concluded that the available estimates were not reliable as they rested on untenable assumptions.

In a detailed experimental study, Nair *et al.* (1996a) estimated the growth in volume increment in replicated plots in a young teak plantation at Nilambur, in Kerala, India. Over a five-year period sets of plots were either exposed to natural insect defoliation or protected from the insect using insecticide. One set of plots was fully protected by applying insecticide whenever there was threat of damage by either *H. puera* or *E. machaeralis*, while another set of plots was protected only against *H. puera*. Defoliation by the two insects was well separated in time which facilitated such selective protection. A third set of plots with no protection served as an untreated control. Differences in volume increment were estimated by the following method. The experiment was started at the time of routine 4th year mechanical thinning, when as per standard silvicultural practice alternate rows of trees in a plantation are thinned to facilitate growth. Measurements made on the thinned trees were used to calculate the initial volume of the standing experimental trees. There were a total of nine plots, with 100 trees per plot, of which half were felled in the 4th year. The girth under bark of each felled tree was measured at every 50-cm interval, from which the wood volume of the tree was calculated. Then the mathematical relationship between the wood volume, on the one hand, and the girth at breast height and total tree height, on the other, was determined by fitting the most suitable prediction equation. This equation was used to arrive at the initial volume of the standing experimental trees from measurements of their girth at breast height and total tree height. Similar measurements were carried out at the time of the second mechanical thinning, when 50% of the remaining trees were felled, to obtain the volume of the trees at the end of the experimental period.

The study showed that during the experimental period the trees protected against *H. puera* put forth a mean annual volume increment of 6.7 m^3/ha compared with 3.7 m^3/ha of unprotected trees, a gain of 3 m^3 of wood/ha per annum. Thus, in young plantations of teak, loss due to defoliation caused by *H. puera* was estimated at 44% of the potential volume increment. They also projected that protected trees would be ready for harvest at the age of 26 years instead of the usual 60 years, provided other necessary inputs were given.

Thus it is now well established that *H. puera* causes significant economic loss in teak plantations.

In addition to loss in volume increment, defoliation may cause dieback of the leading shoot of saplings and consequent forking. Champion (1934) reported that repeated heavy defoliation of saplings led to forking. Khan and Chatterjee (1944) observed damage to 52% of the saplings in a three-year-old plantation at Tithimatty in Karnataka, India and attributed it to heavy defoliation by *Eutectona machaeralis*. Incidence of dieback in 43% of saplings, with incidence as high as 91–99% in some plots (267–333 trees per plot), was reported in a three-year-old plantation in Kerala, India (Nair *et al.*, 1985). However, heavy incidence of terminal bud damage is a rare event. In young plantations at Nilambur in Kerala, India, although defoliation occurred every year, dieback of leading shoot occurred only in two out of seven years of observation. Detailed observations and artificial defoliation experiments led Nair *et al.* (1985) to conclude that leading shoot damage occurs only under a unique combination of conditions, leading to repeated destruction of buds. Feeding on the terminal bud by the pyralid, *Eutectona machaeralis* is perhaps more important in causing dieback of the leading shoot. They also found that permanent forking occurred in only about 10% of the saplings that suffered leading shoot damage because in many cases one of the shoots took over as the leader.

Natural enemies *H. puera* has a large number of natural enemies – about 45 species of parasitoids (3 of eggs, 15 of larvae and 26 of pupae), 108 predators (mostly of larvae: 27 insects, 31 spiders and 50 birds), 1 nematode and 7 pathogens. Most records are from India and the neighbouring countries (Pakistan, Bangladesh, Myanmar and Sri Lanka). Further details are given under the section on control.

Control Control options for *H. puera* in teak plantations have been reviewed by Nair (1986a, 2001b).

Biological control with parasitoids and predators Early control attempts relied on the many natural enemies of the insect, particularly the insect parasitoids. Based on detailed studies in India and Myanmar on the parasitoids of the two major defoliators, *H. puera* and *Eutectona machaeralis* (see below), their alternative insect hosts and the plant hosts of these caterpillars, as early as 1934 Beeson (1934) developed a package of practices for biological control of the two pests, by adopting silvicultural measures to conserve their natural enemies. The theoretical foundation was that the pests could be kept in check by encouraging the endemic insect parasitoids and predators, through ensuring the presence, in the surroundings, of plants that supported their alternative caterpillar hosts or prey.

For example, the tree *Cassia fistula* can support nine species of caterpillars, which in turn can host 11 species of *H. puera* parasitoids and 12 species of *E. machaeralis* parasitoids (Fig. 10.37). The recommended package of practices included the following steps: (1) subdivide the planting area into blocks of 8–16 ha, leaving

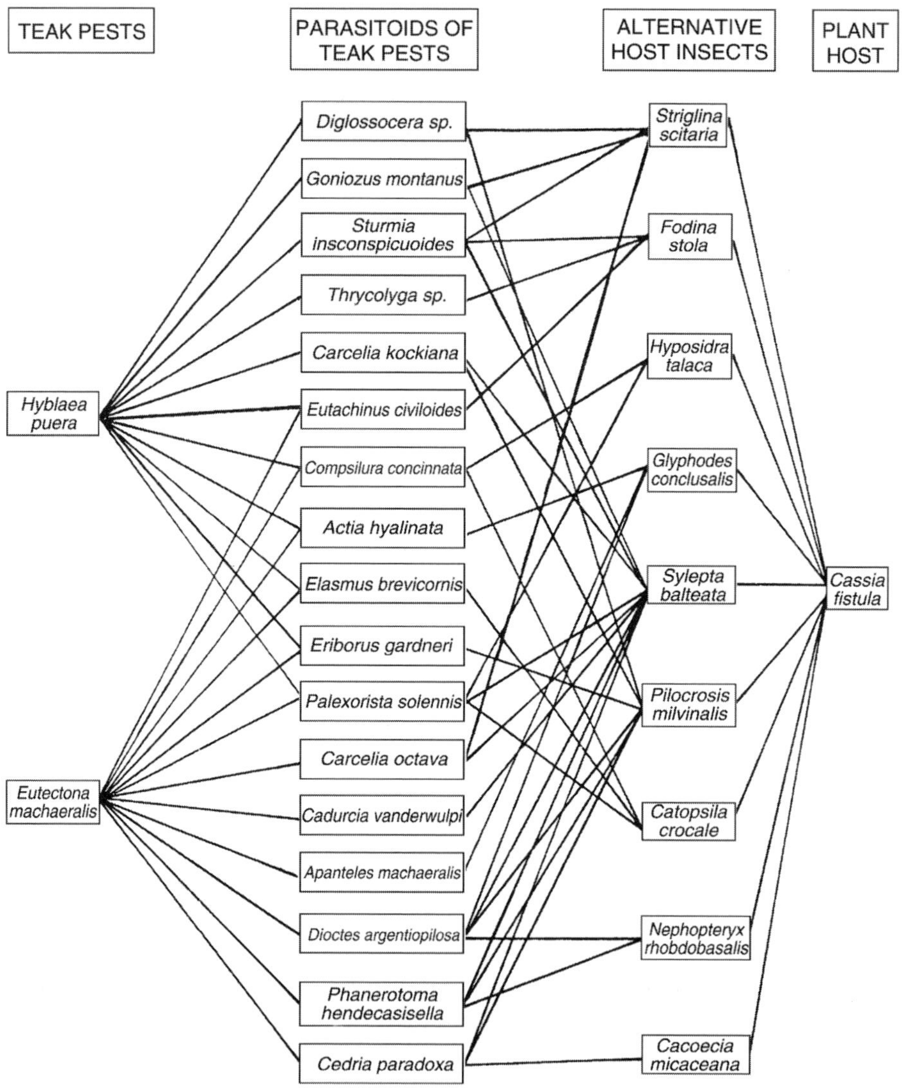

Fig. 10.37 The interrelationships of parasitoids of *Hyblaea puera* and *Eutectona machaeralis* with their alternative host caterpillars supported by the tree species *Cassia fistula*. Retaining trees like *C. fistula* within or in the vicinity of teak plantations was suggested as a means of biological control of the two teak pests as it would help sustain the natural enemies of the pests. From Nair *et al.* (1995)

strips of pre-existing natural forest in between, to serve as reserves for natural enemies; (2) improve the reserves by promoting desirable plant species that support alternative hosts of the parasitoids of *H. puera* and *E. machaeralis* and removing undesirable ones that serve as host plants for the pests themselves; (3) within the teak plantation itself, encourage the natural growth of desirable plant species as an understorey and remove the undesirable ones and (4) introduce selected natural enemies of *H. puera* and *E. machaeralis* where they are deficient.

The initial list of over 100 desirable plant species which support the alternative hosts of parasitoids of both *H. puera* and *E. machaeralis* (Beeson, 1941) was expanded to 213 over the years, through additional research on host–parasitoid relationships (Bhatia, 1948). Although the above biological control measures, based mainly on silvicultural manipulations, appeared ideal and were aggressively recommended and even included in forest working plans (e.g. Vasudevan, 1971), it was not practised for various reasons (Nair, 1991, Nair *et al.*, 1995).

In the meantime research carried out in the 1980s on the population dynamics of *H. puera* (Nair and Sudheendrakumar, 1986; Nair, 1987a, 1988) showed that the proposed biological control cannot succeed because of the unique spatial dynamics of the outbreak populations. Endemic parasitoids would not be able to check the pest outbreaks because of the sudden build-up of high-density larval populations from immigrant moths. Millions of host larvae simply overwhelm the parasitoids and predators and by the time their next generation is built up the host population shifts to another area, thus creating a spatial separation of the natural enemy and host populations (Nair, 1987a). Therefore, the natural enemies will be unable to numerically respond to host populations.

However, the natural enemies must be playing an important role in keeping the non-outbreak populations of *H. puera* in check.

The theoretical feasibility of inundative release of parasitoids was examined by Nair *et al.* (1995) and Sudheendrakumar and Bharathan (2002). In view of the sudden, unpredictable, mass egg laying by immigrant moths and the short incubation period of less than 48 hours timely field release of egg parasitoids like *Trichogramma* spp. is impracticable (Nair *et al.*, 1995). Pupal parasitoids which exert their influence on the next generation are also unsuitable for a migrant moth like *H. puera*. Among the larval parasitoids, the eulophid *Sympiesis hyblaeae* which infests the early instars and displays a high percentage of parasitism during the non-outbreak periods is not suitable for inundative release because it enters diapause during February to June, the period when major outbreaks occur (Sudheendrakumar and Bharathan, 2002). The tachinid, *Palexorista solennis* attacks only late instar larvae and therefore its effectiveness in reducing the damage will be limited. Further study is needed to assess the potential of some bethylid, eulophid, braconid and ichneumonid larval parasitoids (Nair *et al.*, 1995).

Chemical control Several commonly used chemical insecticides have been found effective against *H. puera* in laboratory screening (Gupta and Borse, 1997; Senguttuvan *et al.*, 2000) as well as nursery field trials (Remadevi and Muthukrishnan, 1998). A major plantation field trial was carried out in India, first in 1965, when 76 ha of government-owned teak plantation in Kerala was aerially sprayed with endrin (Basu-Chowdhury, 1971). Again, in 1978, 460 ha of teak plantations in Madhya Pradesh, India were sprayed with carbaryl, using an aircraft (Singh *et al.*, 1978; Singh, 1985). Although the post-spraying evaluations were inadequate, conclusions were drawn that these one-time sprays were effective (Nair, 2001b). However, routine insecticide sprayings have not been carried out in India, although since the 1990s some private teak plantation companies have resorted to occasional insecticidal sprays. In government-owned teak plantations in Thailand, BHC has been applied from ground with a high-power sprayer, between 1966–68, but this practice was later suspended due to harmful effects (Chaiglom, 1990). In the late 1990s, helicopter spraying of chemical insecticides was carried out in a private teak plantation in Costa Rica to control a new outbreak of *H. puera* that spread over 600 ha. In a plantation field trial in Thailand, neem extract containing 0.185% azadirachtin, applied at a concentration of 200–300 ml per 5 litres of spray fluid, using a thermal fogger, gave 79–99% mortality of larvae infesting teak in about six days (Eungwijarnpanya and Yinchareon, 2002).

In general, application of chemical insecticides is one of the effective means of insect control, but even the few reported trials have not brought out critical data on its effectiveness against *H. puera* in teak plantations, under field conditions where the timing and method of application are important. Because of the shifting nature of *H. puera* outbreaks, their sudden and repeated occurrence, and the necessity to resort to aerial spraying, chemical control is not a feasible method under tropical forestry conditions. In addition, there are many well-known long-term disadvantages in the use of chemical insecticides.

Host-plant resistance During *H. puera* outbreaks in teak plantations, it is common to find some trees that have not been attacked in the midst of totally defoliated trees, giving the impression that there are defoliator resistant trees. However, critical investigations (Nair *et al.*, 1997) have shown that the escape of some trees is not due to genetic resistance, but to what may be called 'phenological resistance', caused by the preference of natural populations of *H. puera* moths to lay eggs on trees with tender foliage. Eggs are not laid on a neighbouring tree if it has only mature leaves. Such a tree does not suffer defoliation, unless older larvae migrate to it from adjacent trees after consuming the available foliage. Extensive field search did not give any indication of the existence of *H. puera*-resistant teak trees. Through field observations on marked, escaped trees, it was shown (Nair *et al.*, 1997) that a tree which is not attacked in one year might be attacked in another year. Due to asynchrony between the flushing time of trees and the time of arrival of the immigrant moth populations, different trees may escape defoliation at different times. Based on laboratory screening of excised leaves or observations on susceptibility to leaf damage in clonal orchards (gene banks), some workers have classified various clones as highly resistant, resistant, susceptible etc. (Ahmad, 1987; Jain *et al.*, 1998, 2002), but in the light of the above observations these results should be interpreted with caution. However, there is at least one variety of teak, known as 'teli', in Karnataka, India, which flushes about a month earlier than others and usually escapes defoliation (Kaushik, 1956). Since *H. puera* is unlikely to adapt to an early flushing variety of teak as the moth arrival time is dependent on the arrival of the monsoon, the scope for using this variety in an IPM programme needs further study (Nair, 1998).

Pheromonal control The female moth displays the characteristic calling behaviour prior to mating (Sajeev, 1999) and the male moth possesses the characteristic hair brushes on the hind legs (Sudheendrakumar, 2003), suggesting the presence of a mating pheromone, although this has not been isolated. There is little scope for use of the pheromone for controlling the outbreak populations of *H. puera*, because of the mass influx of the moths. However pheromone may be of use as a population monitoring tool.

Biological control using microbes Commercial formulations of *B. thuringiensis* (Bt) have been found effective in laboratory as well as field trials (Singh and Misra, 1978; Kalia and Lall, 2000; Loganathan and David, 2000; Senguttuvan *et al.*, 2000). Senguttuvan *et al.* (2000) recorded 100% knock-down toxicity of Bt to third and fourth instar larvae within eight hours of their feeding on leaf disc treated with water containing 159 IU/ml of a commercial preparation of Bt. In field trials, 90–99% mortality of larvae were obtained with commercial preparations of Bt at 0.2% of the formulation containing 15 000–55 000 Su/mg spore count and 63–77% mortality at 0.1% of the formulation (Loganathan and David, 2000). At the operational level, commercial preparations of Bt have been used in India in experimental plots as well as some private plantations. In Thailand, Bt has been applied using fogging machines or aircraft, particularly for high value plantations and seed orchards (Chaiglom, 1990; Hutacharern *et al.*, 1993).

The high cost of aerial spraying and the comparatively high cost of the commercial product have prevented its wider use in the developing countries.

The potential of the entomopathogenic fungi *Beauveria bassiana* and *Metarhizium anisopliae* for control of *H. puera* has been evaluated in the laboratory by Sakchoowong (2002). She reported LC_{50} values of 1.4×10^6 and 2.19×10^8 conidia/ml, respectively, for *B. bassiana* and *M. anisopliae*.

The most promising biocontrol agent is the baculovirus, HpNPV. A disease-causing large-scale mortality of larvae, usually during the second or third wave of outbreak at a given site, characterized by liquefaction and rupturing of the body wall was noticed as early as 1903 and confirmed in subsequent observations (Stebbing, 1903; Mathur, 1960). The causative agent was later identified (Sudheendrakumar *et al.*, 1988) as a nuclear polyhedrosis virus. It is a DNA virus, with a genome size of about 99 kbp (Nair *et al.*, 1998b). It belongs to the family Baculoviridae, comprising viruses known to be highly host specific, with no ill effects on non-target organisms. HpNPV was not cross-infective to larvae of *Helicoverpa armigera, Spodoptera litura, Amsacta albistriga* or *Bombyx mori* (Rabindra *et al.*, 1997). Compared to many other baculoviruses, HpNPV is quick acting and causes host mortality in about three days of infection. Preliminary field tests using a crude preparation of HpNPV, containing 1×10^5 PIBs per ml of spray fluid, applied using a high volume sprayer on the foliage of teak trees as soon as infestation became visible (third larval instar stage) gave promising results, reducing the leaf damage up to 76% when there was no rainfall after the application (Nair *et al.*, 1996b). Various parameters for its effective and economic use under field conditions were subsequently standardized (Sudheendrakumar *et al.*, 2001), using the 'control window' concept developed by Evans (1994), in which optimal dosages are determined under laboratory conditions taking into account the important variables. The variables taken into consideration included several factors or conditions related to the insect, the pathogen, the host tree, the physical environment and the spray technology. For example, the larval weight varied from 0.1 mg in the first instar to 110 mg in the fifth instar (more than a 1000-fold difference), with feeding rates of 2 mm^2 leaf area/6 h in the first instar to 300 mm^2 in the fifth instar. Consequently, young larvae were more susceptible to NPV than older larvae; the LD_{50} values for first to fifth instar larvae were about 17, 70, 73, 3932 and 20 125 PIBs per larva, respectively.

The third instar larva was chosen as the best target for NPV spray, because of low LD_{50} and its more open feeding habit, compared with earlier instars. Also, it consumes more foliage per unit time and therefore has a greater likelihood of imbibing the virus dose. Similarly, the distribution of larval stages on the tree, the intensity of UV radiation, wind direction and velocity at the time of spray application, and a host of other conditions are important in arriving at an optimal dosage of NPV for field application. Taking into account a large number of such variables, Sudheendrakumar *et al.* (2001) calculated that theoretically a 90% kill of third instar larvae can be achieved by spraying in the mid to late afternoon using an NPV dose equivalent to 5.49×10^{11} PIBs per ha. Based on field trials carried out using four dosages, 5×10^{10}, 1×10^{11}, 2×10^{11} and 4×10^{11} targeting third instar larvae on teak trees of about 3.5 m height, using an ultra low volume sprayer, it was estimated that over 80% larval mortality could be obtained with a dosage of 2×10^{11} PIB per ha. This dosage represented approximately 1000 larval equivalents per ha and is considered feasible. More refined formulations of HpNPV have since been developed and the scope of preventing large-scale defoliator outbreaks by seeding the early outbreak sites with the virus is being tested (Sudheendrakumar, personal communication, 2004).

Knowledge gaps There are many gaps in our knowledge of the ecology of *H. puera*, some of which were alluded to earlier. The population dynamics of *H. puera* can be understood only at the metapopulation level. A metapopulation is the conceptual assemblage of many spatially distinct populations of a species, some of which may intermingle at times. Further research is needed to elucidate the migratory behaviour of the moths, the role of weather in migration and the source of the moths that initiate the chain of outbreaks in India and elsewhere. One way to study the role of weather is through mathematical modelling of back trajectories for given floating objects using realtime windfield data, and examining the correlation between windfield and moth arrivals determined by insect population sampling at strategic locations in South and Southeast Asia (Nair, 2001b). This requires international and interdisciplinary cooperation.

More investigations are also needed on the ecology of *H. puera* on the mangrove vegetation which may sustain its population when the insect is not active on teak. Research is also necessary to discover the host plants of *H. puera* in Africa and Latin America. As noted earlier, the circumstances under which outbreaks occur on teak in some geographical areas, but not in others in spite of the long history of teak planting in these areas, needs to be elucidated.

Another aspect that needs study is the possible occurrence of moth hibernation in temperate regions. Although no details are available, Beeson (1941) observed that in the cooler climate of Dehra Dun in India the moths hibernate for a period of three months from December to January. If this is true, the possibility needs to be explored whether moths emerging from hibernation in the temperate region could be the source of moths immigrating into the tropical region at the onset of monsoon.

The teak defoliator control problem is similar to confronting a dacoit situation, where we may have the guns ready but cannot pull the trigger until we encounter the enemy (Nair, 1988). We know that the dacoits will strike, but we do not know when and where. Everything is quiet for some time after the flushing of teak plantations. Then suddenly the outbreak appears and spreads in waves into unpredictable patches and we are unable to catch up with control measures. As Beeson (1934) pointed out, mounting a control operation at this time requires the skill and swiftness similar to that of a firefighting organization. It calls for efficient methods for timely detection of outbreaks over extensive areas of plantations and the ability to carry the control agent to the tall teak canopy. At present, timely detection of outbreaks can be accomplished only by ground surveillance deploying large manpower, and the tall tree canopy can be reached only by aerial spraying. Both are highly expensive, by the standards of developing countries. Research into the mechanism of outbreak initiation may suggest alternative approaches to control.

Although it is feasible to control the early episodes of outbreaks by use of baculovirus or inundation with suitable parasitoids, it is difficult to control the extensive subsequent outbreaks. An acceptable management strategy must aim at prevention of widespread outbreak rather than its control, which must await further research into the population dynamics of the insect.

The potential for utilizing phenological resistance needs to be explored further. In theory, an outbreak can be prevented if the trees have mature leaves when the moths arrive for egg laying. Since the moth arrival is dependent on pre-monsoon rainfall, we can break the synchrony between moth arrival and flushing, and therefore the chances of outbreak, by planting early flushing varieties of teak. The early flushing varieties would probably need to be irrigated to retain the leaves until the pre-monsoon showers arrived.

Pest profile

Eutectona machaeralis (Walker) (Lepidoptera: Pyralidae) and related species

Eutectona machaeralis (Walker) (Fig. 10.38a,b) or a closely related species is generally known as the 'teak skeletonizer' or 'teak leaf skeletonizer'. Its larva feeds on the green leaf tissue between the network of veins, leaving the skeleton of veins intact (Fig. 10.39), thus earning the name skeletonizer. Partially damaged leaves are not shed and even the fully skeletonized leaves are retained by the tree for a long time, so that affected trees have a dry, fire-scorched appearance from a distance. The insect causing this damage has long been recognized as *Eutectona machaeralis* (syn. *Hapalia machaeralis, Pyrausta machaeralis*). In a recent paper Intachat (1998), based on studies on wing markings and

(a)

(b)

Fig. 10.38 The teak leaf skeletonizer *Eutectona machaeralis*. (a) Adult (wingspan 22 mm), (b) larva. Courtesy: T.V. Sajeev, Kerala Forest Research Institute.

genitalia, argued that the teak skeletonizer in Malaysia and Indonesia, and possibly also Thailand, is the closely related species *Paliga damastesalis*. Also, according to her, *Eutectona* is a junior synonym of *Paliga* and what has been called *Eutectona machaeralis* should be correctly known as *Paliga machoeralis*. In this context it must be noted that the pattern and colour of the wing markings are

Fig. 10.39 Characteristic skeletonization of teak leaf caused by *Eutectona machaeralis*.

known to be very variable in *E. machaeralis*, and may depend upon the season. Temperature and humidity are believed to influence the colour pattern; light forms have been produced experimentally at high temperatures and dark forms at low temperatures, from the same ancestors (Beeson, 1941). In view of such variability, a more detailed taxonomic study on specimens from India, Indonesia, Malaysia, Myanmar and Thailand is necessary to resolve the species identities.

The moth has a wing-span of 19–26 mm, the males being slightly bigger than the females. Forewings are white to ochreous yellow, with distinct or indistinct pink to crimson zigzag markings; hindwings are paler, with an ochreous or reddish marginal line or band. The full-grown larva is 20–25 mm long. The head is light brown and the body is greenish to brown or purplish, with two pairs of black dots surrounded by a white or yellow margin on each segment. Longitudinal, brown, yellow or green bands appear on the sides in later instars.

Life history Moths rest during the day in shaded places in the undergrowth, especially dry leaves on the ground. Beeson (1941) observed that females often predominate in wild populations and are often twice as numerous as males. This needs verification as data reported by Gopakumar and Prabhu (1981) show that the sex ratio was more or less even in pupal samples collected from teak plantations at Kulathupuzha in Kerala, India in June, September and December. They found an early preponderance of females in the emerging moths, probably an adaptation to prevent inbreeding. However, according to Beeson (1941), a female-dominant strain of the insect occurs in Myanmar. Females are ready to mate on the night of emergence but males do not mate until the third night after emergence. In the laboratory, moths feed on sugar solution or diluted honey. Eggs are laid singly on teak leaves, usually on the underside. Average fecundity ranges between 203 and 374, and maximum between 500 and 550 (Beeson, 1941; Wu *et al.*, 1979; Patil and Thontadarya, 1987a). The oviposition period is 1–2 weeks.

There are five larval instars. The first and second instars feed superficially on the leaf, under protection of strands of silk. Third to fifth instars eat out the entire leaf tissue between the fine network of veins, and thus skeletonize the leaf. Under natural conditions, the larvae feed mainly on older leaves, but given the choice they prefer younger leaves, on which the larval growth is faster, the pupae produced are heavier and the rate of survival is higher (Beeson, 1941; Roychoudhury *et al.*, 1995b, 1997a). The larval shelter on the leaf is characteristic. The larva makes a shelter web and an escape hole on the leaf that permits it to retreat quickly when disturbed to the opposite side of the leaf and drop down on a thread of silk. Pupation occurs on green leaf or on fallen leaf, under cover of a stronger shelter web with small, oval holes round the edges and an emergence hole at one edge. The males live for 9–15 days and females 12–20 days when provided with diluted honey as food.

The duration of the developmental period varies according to the climate. At Nilambur, in Kerala, India, where there is no distinct winter season, the normal developmental period was 2–3 days for eggs, 12–20 days for larva and 5–8 days for pupa (Beeson, 1941). Including a pre-oviposition period of 3 days, the total life cycle lasts from 23–31 days. Thus in field cages, 14 complete generations and a partial 15th were possible per year. At Dehra Dun in north India, where there is a winter season, the larval period is 12–14 days from March to October, but between November and March the larva is reported to enter hibernation which may last for 140–150 days and the pupal period may be prolonged to 27 days (Beeson, 1941). Consequently only 10 generations are completed per year. At Dharwad in Karnataka, India, where there is a mild winter, a variable proportion of larvae enter hibernation in the pre-pupal stage during the winter months. This was shown by Patil and Thontadarya (1986) who collected 30 mature larvae from the field at weekly intervals and maintained them on teak leaves in a field laboratory. Between mid-October and mid-February, 3–67% of the insects entered diapause, with a mean of 38%; the highest was in November–December. Although Patil and Thontadarya (1986) called this phenomenon diapause, since it is promptly terminated at higher temperature it is more similar to hibernation. In laboratory experiments, Patil and Thontadarya (1987b) showed that the majority of mature larvae exposed to 15 °C or 20 °C entered hibernation in the pre-pupal stage. Exposure of eggs or early larval instars to lower temperatures did not induce pre-pupal hibernation. Termination of hibernation depended on the temperature; it occurred in about

3 days at 35 °C and in about 88 days at 15 °C. In Myanmar, 13 generations are completed per year in field cages. Hibernation may be more widespread; it is likely to go unnoticed because only a proportion of the population goes through it.

Host range and geographical distribution *E. machaeralis* has a limited host range. Other than teak, it has been recorded only on *Tectona hamiltoniana* and three species of *Callicarpa*, that is *C. arborea, C. cana* and *C. macrophylla*. No information is available on the extent of damage caused to hosts other than teak. Feeding on *T. hamiltoniana* was negligible in a small number of isolated experimental plantings of this species in the midst of *T. grandis* in Kerala, India (Nair, unpublished observations, made over many years). The rate of development on *Callicarpa* is much slower than on teak (Beeson, 1941).

E. machaeralis or the closely related *Paliga damastesalis* has been recorded in India, Bangladesh, Myanmar, Sri Lanka, Thailand, Malaysia, Laos, Philippines, Indonesia and China. According to Intachat (1998), the teak skeletonizer present in Malaysia, Indonesia and possibly Thailand is *Paliga damastesalis*. The teak skeletonizer in the Andaman Islands (India) has also been identified as *P. damastesalis* (Veenakumari and Mohanraj, 1986). Some books mention its distribution as far as Australia (Beeson, 1941; Browne, 1968; Thakur, 2000), but no primary record could be traced.

Seasonal incidence In spite of the occurrence of 10–14 generations per year in field cages, only a small number of generations can be noticed under natural field conditions. In teak plantations at Jabalpur in Madhya Pradesh, central India, the insect can be seen from April to November, with peak numbers in August and September on in some years, up to October (Khan *et al.*, 1988 a,b; Meshram *et al.*, 1990). Population outbreaks occur regularly during August-September every year (Fig. 10.40), causing moderate to heavy defoliation over wide areas (Khan *et al.*, 1988b). Unfortunately, detailed information is not available on the intensity and sequence of defoliations in the same area, and the interaction with the other major defoliator, *Hyblaea puera*, which causes defoliation during July–August. Data presented by Khan *et al.*, (1988b) indicate a high population of both insects in August. In contrast to the annual population outbreaks in central India as described above, at Nilambur in south India outbreaks are rare. During five years of observations at Nilambur, from 1978–82, measurable defoliation occurred only in two years, and it was confined to the last part of the growth season (November–January) (Fig. 10.41). However, at least a small number of larvae were present at all times, with comparatively larger numbers from May–June and from October–January. Other reports from south

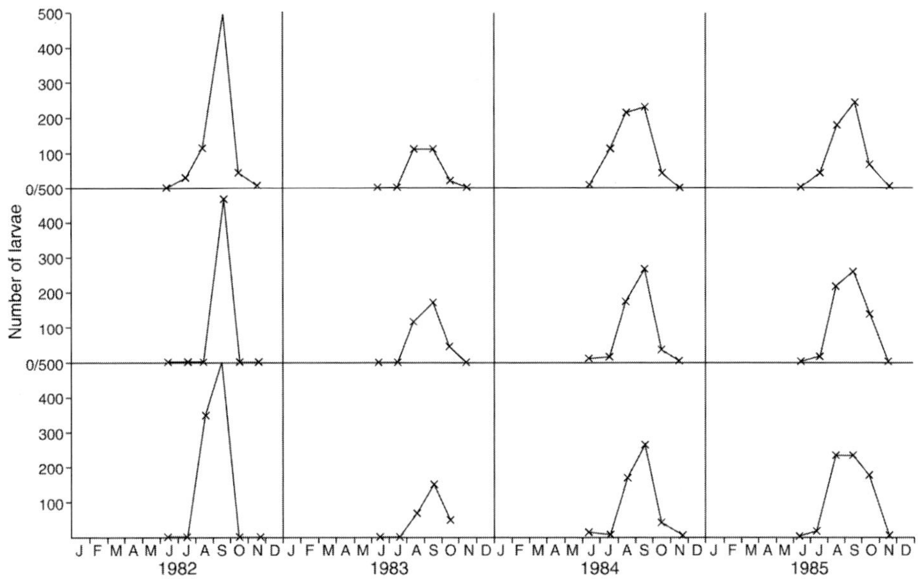

Fig. 10.40 Seasonal incidence of *Eutectona machaeralis* on teak at Jabalpur in central India during the years 1982–5. Data show the number of larvae sampled per 75 terminal leaf pairs. Sampling was carried out only from June to November each year. Note that population outbreak occurred every year in August–September. Data from Khan *et al.* (1988b)

India also indicate low-level infestation in May–June and heavier infestation towards the end of the year (Beeson, 1941; Khan and Chatterjee, 1944; Patil and Thontadarya, 1983a). Continuous presence of the insect has also been noted in teak plantations at Dharwad in Karnataka and Vizhianagaram in Andhra Pradesh (Patil and Thontadarya, 1983a; Loganathan *et al.*, 2001). General observations show that in spite of its continuous presence, widespread population outbreak is a rare event in Kerala. Over a nine-year period from 1976–85, such an outbreak occurred only in 1976, during which year peak defoliation occurred at Nilambur, between the 10th and 15th of November, covering most teak plantations both young and old. At this time, most teak plantations suffered total defoliation and presented the typical spectacle of an outbreak, with larvae wandering everywhere, silken threads hanging from the trees and skeletonized leaves littered all over the ground.

Periodic outbreaks of *E. machaeralis* or *Paliga damastesalis* in teak plantations have also been recorded in other countries where these species are distributed, but details of seasonal variation in abundance are not available, although in general the population increase has been noted towards the end of the growth season.

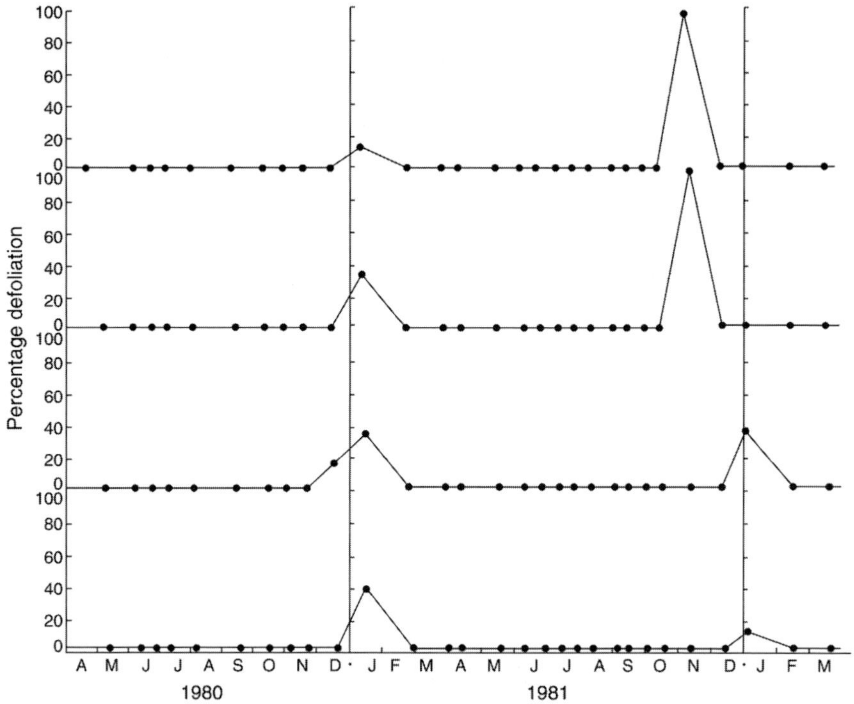

Fig. 10.41 Seasonal incidence of *Eutectona machaeralis* in four 50-tree observation plots within a large teak plantation area at Nilambur in southern India. The data show the percentage of leaf loss. Defoliation occurred only in some years and only between November and January. From Nair *et al.* (1985).

Population dynamics Much remains to be learnt about the population dynamics of *E. machaeralis*. As noted earlier, some confusion exists on the identity of the teak skeletonizer present in the different countries in Asia. Assuming that both *E. machaeralis* and *Paliga damastesalis* have the same habits and population dynamics, the term 'skeletonizer' is used in the present discussion to refer to both species. Although recognized as a major pest of teak and known to cause outbreaks in several countries in Asia, details of its seasonal abundance are not known except for India. Within India, there is a clear difference in seasonal abundance between the south and central parts of the country, which appears to be related to differences in rainfall. For example, Nilambur in the south, where there is no regular outbreak, gets a mean annual rainfall of over 3000 mm, spread over two monsoon seasons, while Jabalpur in central India where there are regular annual outbreaks, gets only about 1270 mm, all of it in one season. Apparently, drier regions are more prone to regular population outbreaks. In south India, although continuous generations of the insect are present, spectacular outbreaks appear to develop suddenly, and not as a result of slow population build-up in the same locality. Apparently, moth migration is involved. Patil and Thontadarya (1983a) recorded a sudden increase in moths caught in the light trap in the last week of September in two consecutive years. In central India, on the other hand, outbreaks occur every year. The circumstances that induce the development of outbreaks are not known. Kalshoven (1953) mentions that although present in Java, Indonesia, *E. machaeralis* does not attack teak there, and feeds only on *Callicarpa cana*, but many recent authors list the insect as a pest of teak in Indonesia (Nair, 2000). Detailed studies are needed on the population dynamics of the teak skeletonizer in central and north India and in other countries in Asia.

Impact It is logical to expect that the outbreaks of *E. machaeralis* which occur at the end of the growth season and cause destruction of the old foliage may have very little impact on the growth of teak. This expectation was shown to be true at Nilambur, in Kerala, India, in the experimental study described earlier under *Hyblaea puera* (Nair *et al.*, 1996a). There was no significant difference in volume increment between trees protected from *E. machaeralis* and those exposed to natural defoliation by this insect. This was attributed to the facts that (1) during the experimental period, *E. machaeralis*, unlike *H. puera*, did not cause defoliation every year; (2) when measurable defoliation did occur in two out of five years, its intensity was low, with less than 40% foliage loss, except on some occasions and (3) the defoliation occurred during the last part of the growth season when the rate of volume increment was very low (Sudheendrakumar *et al.*, 1993). However, we have no information on the impact of *E. machaeralis* defoliation in places like Jabalpur in central India, where *E. machaeralis* causes higher levels of defoliation every year, a little earlier in the growth season. Here the impact may not be negligible as in Kerala although it will be less than that of *H. puera* which destroys the younger foliage. Beeson (1941) suggested that late-season defoliation might affect the growth increment of the following year.

As noted under *Hyblaea puera*, defoliation by *E. machaeralis* may, under some circumstances, contribute to the death of leading shoot of saplings. According to Beeson (1941), *E. machaeralis* feeds on and hollows out the terminal buds of the leader and lateral shoots of teak, under certain conditions.

Dabral and Amin (1975) reported that *E. machaeralis* might also attack flowers, calyces and newly set fruits of teak and thus cause poor fruit formation.

Natural enemies *E. machaeralis* has a large complement of natural enemies, which include 75 species of parasitoids, 31 species of predatory insects, 38 species of predatory spiders and probably many species of predatory birds

(Chatterjee and Misra, 1974; Patil and Thontadarya, 1983b; Sudheendrakumar, 1986). In addition, five species of pathogens have been recorded. The parasitoids include 26 species of tachinids, 19 ichneumonids, 17 braconids, 4 chalcidids, 3 trichogrammatids and one each of bethylid, elasmid, encyrtid, eulophid and scelionid. Important insect predators are praying mantids, reduviids, carabids, coccinellids and ants. A study in Karnataka in south India alone revealed the presence of 43 species of parasitoids and 60 species of predators (Patil and Thontadarya, 1983b), indicating the richness of natural enemies.

The microbial pathogens recorded are the fungi, *Beauveria bassiana*, *B. tenella* and *Fusarium* sp., and the bacteria, *Bacillus cereus* and *Serratia marcescens* (Patil and Thontadarya, 1983b; Agarwal *et al.*, 1985; Singh and Misra, 1987). The LC50 value for *B. bassiana* was 2.9×10^3 conidia for 3rd instar larvae and it increased with increasing larval age (Rajak *et al.*, 1993). This fungus was found to infest 29% of larvae in a teak plantation in Karnataka, India (Patil and Thontadarya, 1981).

Control As discussed under *Hyblaea puera*, the biological control package recommended against the teak defoliators from the 1930s to the 1980s in India was also targeted against *E. machaeralis*, but it was neither put into practice nor tested under field conditions. Similarly, the aerial spraying trials with insecticidal chemicals carried out in India in the past (see under *Hyblaea*) was also intended against *E. machaeralis*. However, no control measures were practised, apparently because there was no proof of effectiveness and the foresters were not convinced of the need for control. On the other hand, erratic infestations in the nursery beds were controlled by remedial insecticidal sprays. For reasons discussed earlier, it is essential to carry out a study on the impact of *E. machaeralis* on the growth of teak plantations in a place where the insect causes regular defoliations, before embarking on its control. The potential of various control agents is examined below.

Parasitoids, predators and microbial pathogens Patil and Thontadarya (1983c) tested 10 species of *Trichogramma* and found that all of them successfully developed in fresh and one-day old eggs of *E. machaeralis*, in the laboratory. They also tested (Patil and Thontadarya, 1984) three exotic species of *Trichogramma*, that is *T. evanescens*, *T. brasiliensis*, and *T. 'pkcal'* (a hybrid), by releasing 5000 parasitoids of each in a moderately infested, three-year-old, 5-ha plantation and obtained high recoveries for 60, 90 and 105 days respectively after release. This suggests that *Trichogramma* spp. could be successfully used for controlling *E. machaeralis*. Other promising parasitoids for biological control are the braconids, *Apanteles machaeralis*, which parasitizes first to third instar larvae, and *Cedria paradoxa* which parasitizes third instar onwards. *C. paradoxa*, which has a limited distribution in northwest India, has been reared successfully in the laboratory in India, released in some places in India and Myanmar during 1937–40 and 1971 and found to establish successfully (Thakur, 2000).

It is obvious that the parasitoids and predators must be playing an important role in keeping the population of *E. machaeralis* in check under natural conditions.

Commercial preparations of *Bacillus thuringiensis* have been shown to be effective against *E. machaeralis* in laboratory tests (Misra and Singh, 1993; Roychoudhury *et al.*, 1994).

Chemical control Several chemical insecticides have been tested against *E. machaeralis* in the laboratory and found to be effective. These include monocrotophos, chlordimeform, quinalphos and formothion and the synthetic pyrethroids, cypermethrin (0.0014%), deltamethrin (0.0018%) and fenvalerate (0.0058%), with the LC50 values shown against them (Singh and Gupta, 1978; Borse and Thakur, 1993, 1994).

Pheromone The sex pheromone of *E. machaeralis* has not been isolated. It is unlikely to be effective for control during outbreaks because of possible migration of the moths and the large numbers of moths present during outbreaks.

Host plant resistance Several papers have examined the differences in susceptibility to *E. machaeralis* among teak clones originating from different Indian states. These are based on damage rating in laboratory feeding trials on excised leaves, or in the field on clones assembled in Germplasm Bank, or both, and the clones have been ranked according to the degree of susceptibility (Ahmad, 1991; Mishra, 1992; Meshram *et al.*, 1994; Roychoudhury *et al.*, 1995 a,b; Roychoudhury and Joshi, 1996; Roychoudhury *et al.*, 1997b). A critical assessment of the methods employed and the results obtained in the above studies show that while some variability exists, there is no practically worthwhile resistance. Leaves of more susceptible clones have been shown to have a higher protein to polyphenol ratio compared to leaves of the less susceptible (Jain *et al.*, 2000). Susceptible leaves also tend to have higher water content than less susceptible leaves (Roychoudhury *et al.*, 1995 a,b). Under natural conditions, outbreaks normally occur late in the growth season when the leaves are mature and tough, but given the choice *E. machaeralis* larvae prefer to feed on younger leaves, on which the larval growth is faster and the rate of survival is higher, as noted earlier. Since the insect will accept leaves of lesser nutritional quality under natural field situations, any resistance based on subtle differences in nutritional quality that has been demonstrated in the above studies will not be of practical value for protection against the insect. However, it is worthwhile to continue the search for resistant trees based on critical field observations.

Knowledge gaps The correct identity of the teak skeletonizer present in different countries in Asia needs to be established by further taxonomic studies. It is possible that both *E. machaeralis* and *Paliga damastesalis* are present in some countries. A taxonomic study is also needed on the seasonal morphs known to occur in the same country.

There is a clear need to ascertain the effect of the skeletonizer-caused late season defoliation on the growth of teak in plantations, particularly in the drier regions of India where regular annual population outbreaks occur. Similarly, there is a need to study the seasonal incidence of the skeletonizer in other countries of Asia. These studies are necessary to determine whether it is worthwhile to control the insect.

More investigations are needed on the hibernation behaviour of *E. machaeralis* as well as on its occasional widespread population outbreaks.

Pest profile

Xyleutes ceramicus (Walker) (Lepidoptera: Cossidae) and related species

Xyleutes ceramicus (Walker) (syn. *Xyleutes ceramica, Duomitus ceramicus*) (Fig. 10.42a,b), commonly called the teak bee hole borer, is a serious pest of teak in some countries, especially Myanmar and Thailand. The larva of this moth bores into the wood of living teak trees. A closely related species, *Alcterogystia cadambae*, causes somewhat similar damage to teak in southern India and is discussed separately below. The *X. ceramicus* moth is fairly large, with a variable wingspan of 8–16 cm (average 10 cm). The body and wings are brownish, with white and black scales making variable, longitudinal streaks or lines. The colouration and pattern somewhat mimic the bark of trees. The mature larva is 6–7 cm long, cylindrical, smooth with sparse hairs and colourful with pink and white transverse bands in each segment.

The ecology and control of *X. ceramicus* have been reviewed by Beeson (1941), Hutacharern (2001) and Gotoh *et al.* (2002).

Life history The life history has been studied in detail by Beeson (1921). The insect has an annual life cycle, but some individuals may take two years to complete development. Moths generally emerge from late February to April and up to August in wet regions (Hutacharern, 2001). The female mates soon after emergence and lays thousands of eggs, attached in strings, in bark crevices. The average fecundity is about 12 500 eggs per female (Gotoh *et al.*, 2002), and

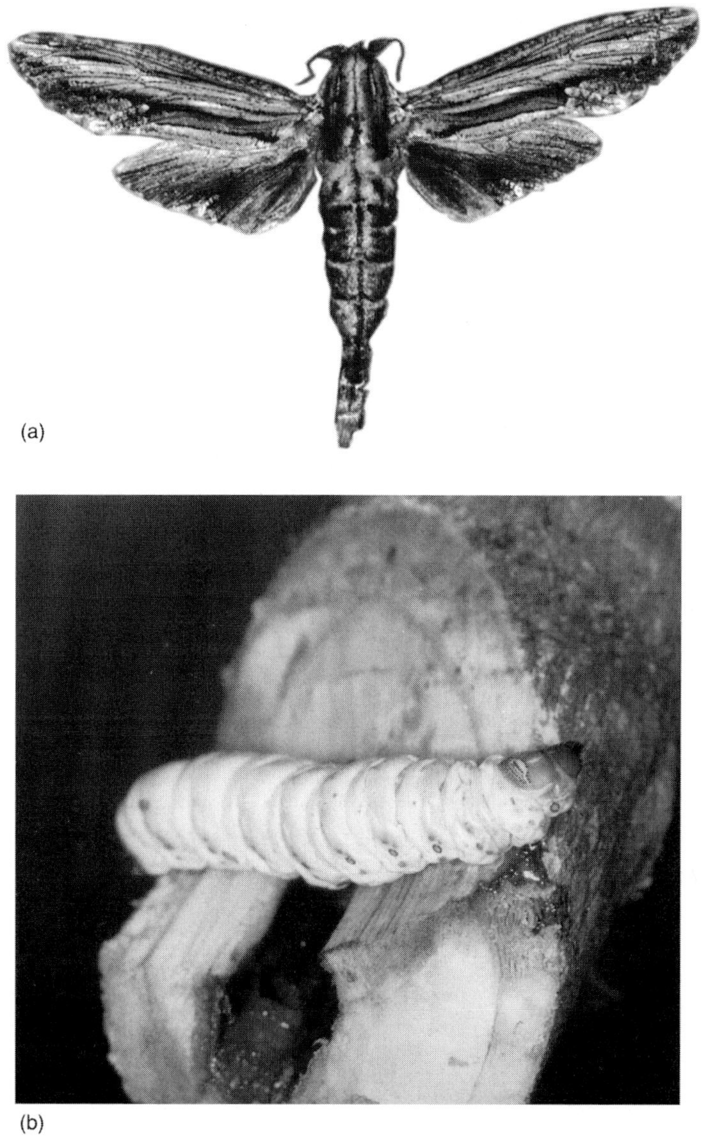

Fig. 10.42 The teak bee hole borer *Xyleutes ceramicus*. (a) Adult (wingspan 100 mm). (b) Larva taken out of its tunnel. Courtesy: Chey Vun Khen, Sabah Forest Department, Malaysia.

according to Beeson (1941) a female may lay up to 50 000 eggs. The female moth has an average lifespan of seven days. Eggs hatch in about 10 days and the newly hatched larvae disperse on silk threads, aided by wind. They can withstand starvation for up to six days (Beeson, 1941). They move into bark crevices, protect themselves under silk web and bore into the tree.

In the sapwood, the larva excavates a shallow patch which is deepened gradually into a tunnel, which in the course of about four months reaches a length of 2.5–5.0 cm and the diameter of a pencil. The tunnel is extended into the heartwood at an upward angle of about 45° for 5–6 cm and then vertically upward for another 15–20 cm. In this manner, a mature larva makes a bee hole which may be more than 25 cm long and 2.5 cm in diameter. The larva feeds on the callus tissue formed from the injured bark, not on wood. Near the mouth of the tunnel, the larva makes a feeding chamber which is a stellate or lobed excavation, the arms of which extend into the living bark and sapwood. On one or more of the arms there are holes, covered with a papery operculum, through which excrement and frass are pushed out. Before pupation, the larva closes the tunnel mouth with a disc of silk and debris and moves to the upper end of the tunnel and shuts itself off with a wad of silk. The mature pupa pushes itself to the tunnel mouth and the empty pupal skin sticks out of the tunnel mouth after moth emergence.

Host range and geographical distribution Other recorded hosts of *X. ceramicus* are *Callicarpa arborea*, *Clerodendron infortunatum*, *Erythrina* sp., *Gmelina arborea*, *Premna* sp., *Sesbania* sp., *Vitex parviflora* and *V. peduncularis* (Fabaceae); *Duabanga grandiflora* and *D. sonneratoides* (Sonneratiaceae); and *Spathodea companulata* (Bignonaceae) (Beeson, 1941; Hutacharern, 2001). In the Philippines, it is recognized as a serious pest in pure stands of the indigenous *Vitex parviflora* (Mesa, 1939).

The countries from where *X. ceramicus* has been reported are Brunei, Indonesia, Malaysia, Myanmar, New Guinea, the Philippines, Sikkim, Singapore, the Solomon Islands and Thailand (Beeson, 1941; Hutacharern, 2001). It does not occur in India.

Ecology and population dynamics The severity of incidence of bee hole borer attack varies from place to place. In Myanmar, infestation is believed to increase with rainfall, within the range of 1750–2750 mm annual rainfall (Beeson, 1941). In Thailand, infestation is common in the northern part of the country where more than 87% of trees were infested in Huay-Tak plantation and 100% in some plantations over 36 years old, while infestation was sparse in the northeastern, central and southern parts (Hutacharern, 2001). In Java, Indonesia, young plantations with dense weed growth suffered greater incidence of the borer, apparently due to favourable moisture conditions (Intari, 1975). Within the same plantation, the infestation has a clumped distribution. Vigorous trees are more prone to attack than suppressed trees. In general, the population density of *X. ceramicus* is low; about 50 moths per ha can be rated as high incidence.

There are indications of high population peaks every 5–6 or 10–12 years in some localities in Myanmar (Beeson, 1941) but hard population data are lacking. Based on various studies carried out in northern Thailand, Gotoh *et al.* (2002) also reported the sudden increase and decline of *X. ceramicus* populations on some occasions. They suggested that this might be caused by the occurrence of fire, which reduces the predator populations, notably of ants. Choldumrongkul and Hutacharern (1990) studied the relationship between infestation and soil properties and found that the infested sites contained a higher concentration of clay, phosphorus, potassium and calcium, a lower concentration of magnesium and manganese, and had a higher, slightly alkaline pH in comparison with uninfested sites. Uninfested trees also had thinner bark, with lower moisture content.

Impact The larval tunnel lies buried in the heartwood because, after moth emergence, the mouth of the tunnel is occluded by callus growth, which extends into the cavity for a short distance, and fresh wood is deposited over the surface. A tree is subject to repeated attacks over the years and the damage accumulates and spreads throughout the bole. Although *X. ceramicus* attack does not cause tree mortality, it causes serious depreciation of wood quality because of the large size and number of bee holes. The bee holes accumulated over the life of the tree will not be discovered until the timber is sawn. Up to 165 holes per tree have been recorded in 40-year-old trees in Myanmar (Beeson, 1941). The timber value decreases in relation to the number of holes. It is considered to be a serious pest of teak in natural forests and plantations in the wetter areas of Thailand and Myanmar, and in plantations in Malaysia.

Natural enemies Among natural enemies, predators are the most prominent. Woodpeckers are believed to account for a large reduction in the larval population. Several species of ants, notably *Anoplolepis longipes*, *Crematogaster* spp., *Monomerium* sp., *Oecophylla smaragdina* and *Tetraponera rufonigra*, are important predators of eggs and young larvae (Hutacharern, 2001; Gotoh *et al.*, 2002). Parasitoids include an ichneumonid *Nemeritus tectonae* and the tachinids *Podomyis adkinsoni* and *Cossidophaga atkinsoni*. Of the fungal pathogens *Cordyceps* sp. has been recorded in Myanmar and *Beauveria bassiana* in Thailand (Hutacharern, 2001).

Control Control options for *X. ceramicus* have been discussed by Hutacharern, (2001).

Silvicultural methods For teak plantations in Myanmar, one of the earliest suggestions for control was to avoid planting in areas with mean annual rainfall within 1750–2750 mm, which was judged to be optimum for infestation, so that the attack is slight to negligible (Beeson, 1941). Intari (1975) suggested weeding plantations to create less favourable moisture conditions for the initial establishment of the larvae.

Physical methods Preventing the emergence of moths from infested trees, by trapping them using a nylon net stapled over the larval hole, was reported to substantially reduce the infestation in Thailand (Hutacharern, 2001).

Scraping off the bark from the infested area of the tree, using a knife, to remove young larvae has been suggested. Frass ejection from a wet bark area indicates an infested site. Hutacharern (2001) estimated that about 30 larvae can be located and removed by a worker in a day. This operation must be carried out during early May to late June in Thailand, when the larvae are still in the outer bark and have not bored into the wood. However, reaching the infested sites higher up in the bole is a difficult task.

Biological control Conservation of natural enemies has been suggested as a means to reduce infestation (Beeson, 1941; Hutacharern, 2001). Gotoh *et al.* (2002) advocated rigorous fire protection to avoid destruction of natural enemies, particularly the ant predators. Chaiglom (1966) tested application of a spore preparation of the fungus *Beauveria bassiana*, by injecting it into the borer hole, and reported 95% mortality of the larvae. Commercial formulation of *Bacillus thuringiensis* has also been tested by the same method and was found effective against early instars (Hutacharern, 2001).

Chemical control The pyrethroid, alpha permethrin, when applied into the borer hole using a pressurized can, gave complete control after 30 days (Hutacharern, 2001).

Pheromonal control In preliminary studies, the female sex pheromone of *X. ceramicus* was isolated and found to belong to the acetate group (Nakamuta *et al.*, 2002b). Since the insect characteristically occurs at a low population density and has a clumped distribution, trapping the moths using pheromone promises to be an effective method for control.

Host plant resistance There is evidence for occurrence of bee hole borer-free teak trees in Thailand. Such trees were found to have thin bark, with low moisture content (Choldumrongkul and Hutacharern, 1990; Hutacharen, 2001). Both the heritability of resistance and the growth performance of such trees need to be investigated.

Knowledge gaps Although removing the larvae by scraping off the infested portion of the bark from trees is an effective and environmentally safe method of control, it is labour intensive. Destruction of moths by trapping them using pheromone appears to be the best option for control because of the low-density, clumped populations of *X. ceramicus*. Research must continue on the identification and synthesis of the sex pheromone and standardization of trapping methods. Critical studies are also needed on the resistance of provenances and their growth performance.

Alcterogystia cadambae (Moore)

As noted earlier, *Alcterogystia cadambae* (syn. *Cossus cadambae*), commonly called teak trunk borer, causes damage in southern India somewhat similar to that of the bee hole borer. The insect is also known as carpenter-worm, a general term used for larvae of moths of the family Cossidae that bore into the wood of living trees. Unlike *X. ceramicus*, *A. cadambae* attacks only older trees and has never been found on saplings and seedlings of teak. Also, unlike *X. ceramicus*, it often causes death of the host trees. This species has been reported only from India.

The life history of *A. cadambae* on teak has been studied by Mathew (1990, 1991). The moth is dull brown and has a wingspan of about 50 mm. The mouthparts are atrophied and evidently the moths do not feed. They live for 5–6 days in the laboratory. The life cycle is annual. The female moth lays eggs in cracks or holes in the bark of trees, either on the main stem or branches, arranged in a row and pasted together with a sticky secretion which later hardens. The newly hatched larvae are very active and move to the axils of side shoots and settle in crevices, injured portions of the bark or on sites of earlier infestation. Under a web of silk, they feed on the bark, and the frass and excreta become attached to the web, concealing the larva. Vigorous feeding of the larva on the bark, callus tissue and outer sapwood causes girdling of the side shoot leading to its death, which is an early symptom of attack (Mathew, 1990). In about three months, the larva attains a length close to 5 cm and by this time, it has made a tunnel, 6–7 cm, in the sapwood. The larva continues to bore into the heartwood. The average larval period is 7–8 months, but larval growth is slow after about three months. The full-grown larva measures about 5 cm.

Although Beeson (1941) mentioned that pupation occurs in the tunnel, Mathew (1990) never encountered pupae on teak trees and he observed hundreds of mature larvae dropping from infested trees to the ground during the pre-monsoon rainfall in May. The larvae crawled over the forest floor and settled at sites with loose soil. Then the larvae burrowed into the soil, prepared horizontal chambers, 3–4 cm below the soil surface, lined them with layers of silk and pupated within. This appears to be the typical pupation behaviour. Most other cossids are known to pupate within a chamber made in the larval gallery itself but in *Cossus cossus*, infesting hardwood trees in Europe and North America, pupation may occur either within the tunnel, near the entrance, in a silken cocoon or at the base of the host tree in an earthen cocoon (Browne, 1968). Bhandari and Upadhyay (1986) who studied the biology of *A. cadambae* infesting the root collar region of young trees of *Diospyros melanoxylon*, mentioned that wood particles are embedded in the pupal cocoon along with faecal pellets, indicating that pupation occurred in the tunnel itself, although this was not explicitly stated. More detailed observations are necessary on the pupation behaviour of this insect infesting different hosts. The average pupal period is about 11 days, and before moth emergence the pupa wriggles to the soil surface and projects out.

A. cadambae has overlapping generations. In light trap collections from a heavily infested teak plantation in Kerala, India, the moths were most abundant during May–June and August–October, but small numbers were present throughout the year. This somewhat reflects the rainfall-linked pupation pattern. Kerala receives two monsoons per year, the first starting in early June and the second in mid-October, with pre-monsoon showers occurring earlier. It is not clear whether the moths collected during the drier period of December–March are indicative of pupation occurring also during the dry period or of dormancy of the pupae that were formed earlier (Mathew, 1990).

Timber quality is degraded by *A. cadambae* infestation. The larval tunnel follows a radial, zigzag course and extends into the heartwood. Because of repeated attack of infested trees, borer holes often occur in a cluster (Fig. 10.43) and these clusters extend throughout the bole. Consequently, planks cut from heavily infested logs will have numerous holes. In addition to such damage, heavily infested trees die in the course of time, possibly aided by associated fungi (Mathew and Rugmini, 1996). The fungus *Phialophora richardsii* has been isolated from borer-infested wood. Studies in Kerala (Mathew *et al.*, 1989) showed that in the initial phase of attack of a plantation, the infestation is clumped. In subsequent years, further deterioration of the already infested trees occurs as a result of reinfestation and there is slow spread of infestation to other trees. A representative survey of the teak plantations of Kerala showed that 4% of the plantations had *A. cadambae* infestation, with 2–40% of the trees within the plantation affected. Trees below 15 years of age were not infested. In affected pockets, the proportion of infested trees increased with age of the trees, obviously due to repeated infestation of the same trees.

Other tree species on which *A. cadambae* has been recorded are *Diospyros melanoxylon* (Bhandari and Upadhyay, 1986), *Grewia tiliaefolia*, *Terminalia bellerica*

Fig. 10.43 Damage caused by the trunk borer *Alcterogystia cadambae* to teak. Courtesy: R.V. Varma, Kerala Forest Research Institute.

(Mathew, 1990), *T. tomentosa* (Kumar *et al.*, 1999) and *Butea monosperma* (Santosh and Kumar, 2003). In *D. melanoxylon*, the larva bores into the stems and roots of young trees. *D. melanoxylon* (locally called 'tendu') is grown in central India for harvesting green leaves which are dried and used for wrapping tobacco to make a kind of cigar ('bidi'). Annual lopping and fire damage are considered as predisposing factors (Bhandari and Upadhyay, 1986). Infestation has been noted in Maharashtra and Madhya Pradesh. Five to nine per cent of plants were infested and some plants died due to repeated and multiple infestations.

Natural enemies include a woodpecker and a barbet which extract the larvae from the tunnels (Mathew, 1990). Some pathogenic micro-organisms were isolated from field-collected, naturally infected larvae. They include the fungi *Aspergillus flavus* and *Paecilomyces fumosoroseus* and the bacterium *Serratia marcescens* (Mathew, 1990).

It is believed that *A. cadambae* attacks only trees in poor health, such as those subjected to lopping, coppicing or burning and have dead wood, fire scars or snags (Beeson, 1941). This is the case with *Diospyros melanoxylon* also. Mathew (1990) reported that attempts to inoculate larvae on healthy teak trees were not successful. He concluded that specially favourable conditions are necessary for the initial establishment and build-up of infestations in plantations.

Callus tissue and coppice shoots formed as a result of mechanical injury provide favourable conditions for establishment.

Trunk injection or implantation of insecticides did not prove effective for control of the borer. Mathew (1990) recommended removal of infested trees during the routine silvicultural thinning operations in the case of low-density infestations and clear-felling of badly affected plantations, in order to prevent further spread of infestation.

As in the case of *X. ceramicus*, *A. cadambae* appears to be an ideal candidate for pheromonal control, because of its generally low population density and clumped distribution. The sex pheromone of this species has not been isolated and identified, although that of the related European species, *Cossus cossus*, has. It is also necessary to unequivocally establish the relationship between tree health and infestation by this borer. An experimental approach using release of newly hatched larvae on healthy and experimentally injured trees, to gauge the success of infestation, may prove useful.

Pest profile

Sahyadrassus malabaricus (Moore) (Lepidoptera: Hepialidae) and related species

Shyadrassus malabaricus (Moore) (syn. *Phassus malabaricus*), commonly called the teak sapling borer, is a pest of teak saplings. It belongs to the family Hepialidae, a family of primitive moths under the lepidopteran group Glossata. The large larvae of this moth tunnels into the central pith of the stem of saplings and is often a conspicuous pest, although the damage caused is seldom serious. The species is most prevalent in southern India; other related species occur elsewhere, as noted below. The moth is greyish brown, with mottled forewings, and is fairly large, with a wingspan of up to 11 cm and body length of 5.5 cm. There is large variation in size, some being about half the above size. When at rest, the moth hangs vertically in a characteristic posture, supported by the first two pairs of legs (Fig. 10.44). The third pair of legs is shorter and non-functional, and the male possesses scent glands which produce a sharp, pungent, mustard-like smell. The mature larva is large and conspicuous, cylindrical, about the thickness of a pencil, and 6–10 cm long. It has a black, hemispherical head and a yellowish white body.

The life history and ecology of the teak sapling borer have been studied by Nair (1987b).

Life history The life cycle is annual. The moths emerge between mid-March and mid-May, during the pre-monsoon season, with small variations between years and regions. In the laboratory, the moths live for three to five days; they have vestigial mouthparts and obviously do not feed. The female moth

Fig. 10.44 The adult teak sapling borer *Sahyadrassus malabaricus*. When at rest, the moth (wingspan up to 110 mm) hangs in a characteristic posture.

lays thousands of eggs, which are believed to be broadcast while in flight. In the laboratory, an unmated female laid 4 166 eggs (Nair, 1987b). Early larval instars have not been observed in the field and it is not known where they live. Although moth emergence is usually completed by mid-May, it is not until some three months later, in mid-August, that the new generation of larvae are found on saplings. During this period, infestation on the saplings builds up suddenly over a few weeks. By then, the larvae are already 15–20 mm long. Obviously the early instars survive elsewhere, probably on litter or humus on the ground or on weedy vegetation, and later migrate to the teak saplings. Larvae of most species of Hepialidae occupy tunnels excavated in soil and feed on roots or ground vegetation, but the early instars of some species pass through a litter phase when they feed on detritus, fungi or fungus-infested wood before moving to living saplings (Grehan, 1987; Tobi *et al.*, 1993). A litter phase has been recorded in the related species *Endoclita sericeus*, while the young larvae of *E. signifier* are known to feed on the stems of grasses. A similar feeding habit can be inferred for the early instars of *S. malabaricus*.

In the sapling, the larva occupies a tunnel in the centre of the stem, along the pith (Fig. 10.45). The tunnel mouth is located at a height of 5–60 cm above

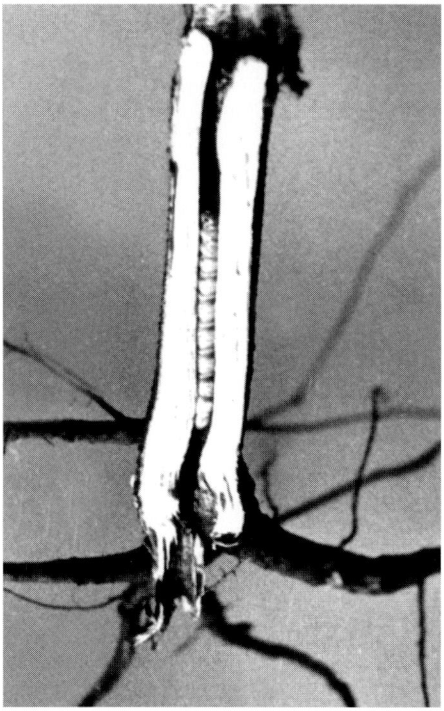

Fig. 10.45 A full-grown larva of *Sahyadrassus malabaricus* inside a longitudinally split stem of the shrub *Clerodendrum viscosum*.

ground, usually at about 30 cm. The mouth is covered by a thick, conspicuous, dome-shaped mat made of coarse, sawdust-like particles of wood and bark, spun together with silk (Fig. 10.46). Faecal pellets are usually attached to this mat. In small saplings, the long, cylindrical tunnel extends into the root. Usually only one larva occurs per sapling. The tunnel is used only as a shelter; the larva feeds on the bark and callus tissue around the tunnel mouth. It browses in such a way that the lower bark layers are left intact at many spots so that sustained regeneration of bark occurs. Feeding takes place at night. Pupation takes place at the bottom of the tunnel. After moth emergence the pupal exuvia sticks out of the tunnel mouth through the mat cover. There is good synchronization in the emergence of the moth population and there is no overlapping of developmental stages.

A favourite host of this polyphagous caterpillar is *Trema orientalis* (Ulmaceae), a soft-timbered pioneering tree species. On this host, multiple infestations are common, unlike on teak saplings. Even the bigger trees of *T. orientalis* are infested and in this case the tunnels do not reach the pith. Observations have shown that bark regeneration is quick and profuse in *T. orientalis*. Larvae collected from teak

Fig. 10.46 *Sahyadrassus malabaricus* attack is characterized by a conspicuous, dome-shaped mass of woody particles held together with silk, covering the tunnel mouth.

saplings were readily rehabilitated on *T. orientalis* by drilling holes and introducing the larvae. The larvae deepened the holes, when necessary, to accommodate their body length and covered the holes with a mat of frass (Nair, 1987b).

Host range and geographic distribution *S. malabaricus* is highly polyphagous; it has been recorded on about 50 plant species belonging to 22 families (Nair, 1987b). Trees most commonly attacked belong to the families Ulmaceae, Fabaceae, Mimosaceae and Myrtaceae. As noted above, *Trema orientalis* (Ulmaceae) is a favourite host, in which both saplings and trees are infested, whereas on other hosts only saplings in the girth range of 4–11 cm at base are infested. Also, multiple infestations are common in *T. orientalis*; two dozen trees examined at one place supported an average of 10 larvae per tree. Another common host is *Clerodendrum viscosum* (syn. *C. infortunatum*) (Fabaceae), a shrubby weed prevalent in open forests. In one instance, out of 27 plants examined (four to seven centimetres girth at base) 21 were attacked, some harbouring two to three larvae.

The geographical distribution of *S. malabaricus* is limited to peninsular India, with other species occurring elsewhere.

Impact In teak saplings, the damage caused by *S. malabaricus* is limited to tunnelling of the pith and feeding on the bark over a small patch or in an incomplete ring around the tunnel mouth. In most cases, this damage is negligible. Rarely, some saplings break off at the point of injury and some become ring-barked, resulting in death of the top portion. Among plantation tree species, *Acacia auriculiformis*, *Neolamarckia cadamba*, *Calliandra callothyrsus*, *Casuarina equisetifolia*, *Eucalyptus* spp., *Gmelina arborea*, *Falcataria moluccana* and *Sterculia companulata* are attacked (Nair, 1987b). A survey in Kerala, India, showed that the incidence of infestation ranged from zero to 61% in teak plantations and zero to 11% in eucalypt plantations. In most plantations, infestations became visible when weed growth was cleared and general observations suggest that plantations with dense weed cover are more prone to attack. *S. malabaricus* attack is not a serious problem except in highly valuable plantations, although the large larva and the conspicuous frass mat covering the tunnel mouth create a scare among growers who may fear further spread of attack, without knowing that the life cycle is annual. *S. malabaricus* accounted for about 22% of all requests for advice on control received by the Kerala Forest Research Institute in India from the State Forest Department, indicating that the perceived impact was much greater than the real impact (Nair *et al.*, 1996c).

Natural enemies Rare instances of predation by woodpeckers, which extract the larvae by making a peck hole at the base of the stem, where the larva rests during the day, were noticed but the larvae are not reachable when in the root portion. Although many species of ants attack the larvae when in the open, the frass mat cover affords protection against them. Rare instances of infestation by the fungus *Metarhizium anisopliae*, which causes mummification of the larvae, have been recorded (Nair, 1987b).

Control In view of the low economic importance of *S. malabaricus* infestation, no control operation is necessary in most large-scale plantations. Control is necessary only in high value plantations.

Generally, it is difficult to control borers because insecticides cannot reach their concealed habitat easily. Methods recommended in the past against this borer included; (1) allowing naturally growing saplings of more attractive host plants to remain in the plantation to act as trap plants, and destroying them later, (2) physical killing by inserting a wire probe through the tunnel mouth, (3) plugging the borer hole with coal tar or (4) injecting an insecticide into the tunnel. The first depends on the occurrence of more favoured hosts within the plantation, which is uncertain. The second will not always succeed because it is difficult to insert a wire probe through the sharply bent initial portion of the tunnel. The third and fourth may prove effective but are cumbersome to practise. Based on observations on the behaviour of the larva, Nair (1987c) tested spot treatment of the tunnel mouth with insecticides, after removing the frass mat cover. When the frass mat cover is removed, the first reaction of the larva is to rebuild it. For this purpose, the larva gnaws out small pieces of bark and wood from the area surrounding the tunnel mouth. During this process, the larva comes into close contact with the treated surface. In addition, if the larva survives this initial contact exposure, further poisoning can take place through the stomach, when it feeds on the treated bark. In a series of experiments in which various insecticides and their formulations were tested by spot application as described above, using a paintbrush, Nair (1987c) found that 0.125% a.i. of the contact-cum-stomach poison, quinalphos, gave effective control. Since the insecticide is brushed over a small area of the stem of infested saplings, environmental contamination is negligible.

Clean cultivation, with timely weeding, can reduce the incidence of attack by creating less favourable conditions for the survival of the early instars.

Knowledge gaps We have no information on the place of occurrence and feeding habits of the early instar larvae.

Related species

Other hepialid species replace *S. malabaricus* in other regions. *Endoclita signifer* attacks teak saplings in eastern India, Myanmar and Thailand; and *E. auratus* and *E. punctimargo* attack other tree species in the east Himalayan region (Beeson, 1941). *E. aroura* and *E. gmelina* occur on teak saplings in Malaysia (Chey, 1996) and *E. hosei* attacks other tree species in Malaysia. *Aenetus* spp. are pests of eucalypts in Australia (Elliott *et al.*, 1998). *Phassus damor* in Indonesia and *Aepytus* sp. in Costa Rica infest saplings of various tree species.

References

Abe, K. (1983). *Plantation Forest Pests in Sabah, FRC Publication No. 8*. Sandakan, Malaysia: Sabah Forest Department.

Agarwal, G. P., Rajak, R. C., Katare, P. and Sandhu, S. S. (1985). Studies on entomogenous fungi parasitizing insect pests of teak. *Journal of Tropical Forestry*, **1**, 91–4.

Ahmad, M. (1987). Relative resistance of different clones of *Tectona grandis* to teak defoliator, *Hyblaea puera* Cram (Lepidoptera: Hyblaeidae) in south India. *Indian Forester*, **113**, 281–6.

Ahmad, M. (1991). Natural resistance in teak clones to leaf skeletoniser, *Eutectona machaeralis* (Lepidoptera: Pyralidae) in south India. *Indian Journal of Forestry*, **14**, 228–31.

Ahmad, S. I. and Sen-Sarma, P. K. (1990). Bionomics of *Calopepla leayana* Latr. (Coleoptera: Chrysomelidae), a serious defoliator of *Gmelina arborea* Roxb. plantations in India. *Indian Forester*, **116**, 71–82.

Ali, M. S. and Chaturvedi, O. P. (1996). Major insect pests of forest trees in North Bihar. In *Impact of Diseases and Insect Pests in Tropical Forests*, ed. K. S. S. Nair, J. K. Sharma and R. V. Varma. Peechi, India: Kerala Forest Research Institute, pp. 464–7.

Ambika-Varma, B., Sudhakara, K. and Beena-Bhaskar. (1996). Insect pests associated with nurseries of selected tree crops in Kerala. In *Impact of Diseases and Insect Pests in Tropical Forests*, ed. K. S. S. Nair, J. K. Sharma and R. V. Varma. Peechi, India: Kerala Forest Research Institute, pp. 468–73.

Amin, P. W. and Upadhyaya, A. K. (1976). Occurrence of teak defoliator, *Hyblaea puera* (Hyblaeidae, Lepidoptera) on the fountain tree *Spathodea companulata* (Bignonaceae). *Indian Forester*, **102**, 306–11.

Ananthakrishnan, T. N. (1996). *Forest Litter Insect Communities: Biology and Chemical Ecology*. New Delhi: Oxford & IBH Publishing Co. Pvt. Ltd.

Anderson, J. M. and Swift, M. J. (1983). Decomposition in tropical forests. In *Tropical Rain Forest: Ecology and Management*, ed. S. L. Sutton, T. S. Whitmore and A. C. Chadwick. Oxford: Blackwell Scientific Publications, pp. 287–309.

Andrews, W. A. (ed.) (1972). *A Guide to the Study of Environmental Pollution*. Ontario: Prentice-Hall of Canada Ltd.

Anonymous. (1954). *Scolytid Beetle Attack on Seedlings*. Fiji: Report of the Forest Department 1953–4, pp. 1–5.

Apolinário, F. E. and Martius, C. (2004). Ecological role of termites (Insecta, Isoptera) in tree trunks in central Amazonian rain forests. *Forest Ecology and Management*, **194**, 23–8.

Appanah, S. and Weinland, G. (1993). *Planting Quality Timber Trees in Peninsular Malaysia: a Review*. Kuala Lumpur: Forest Research Institute, Malaysia.

Atkinson, P. R. (1989). Controlled-release insecticide granules, compared with other soil insecticides, for use against the termite, *Macrotermes natalensis* Haviland, in the establishment of *Eucalyptus* plantations. *Crop Protection*, **8**, 387–96.

Atkinson, P. R. and Laborde, de R. M. (1996). Estimates of the extent of infestation and potential losses caused by common forestry pests in South Africa. In *Impact of Diseases and Insect Pests in Tropical Forests*, ed. K. S. S. Nair, J. K. Sharma and R. V. Varma. Peechi, India: Kerala Forest Research Institute, pp. 274–88.

Atkinson, P. R., Nixon, K. M. and Shaw, M. J. P. (1992). On the susceptibility of *Eucalyptus* species and clones to attack by *Macrotermes natalensis* Haviland (Isoptera: Termitidae). *Forest Ecology and Management*, **48**, 15–30.

Attajerusit, J., Somsook, V. and Nanta, P. (2002). Life cycle, behaviour and control of the bamboo shoot borer, *Cyrtotrachelus dichrous* by entomopathogenic nematode, *Steinernema carpocapsae*. In *Pest Management in Tropical Forest Plantations*, ed. C. Hutacharern, B. Napompeth, G. Allard and F. R. Wylie. Bangkok: FAO Forestry Research Support Programme for Asia and the Pacific, pp. 157–8.

Baksha, M. W. (1990). Some major forest insect pests of Bangladesh and their control. *Bulletin Forest Entomology Series*, 1, 1–19 (Forest Research Institute, Chittagong).

Baksha, M. W. and Crawley, M. J. (1995). Relative preference of different host plants to teak defoliator, *Hyblaea puera* Cram. (Hyblaeidae: Lepidoptera) in Bangladesh. *Bangladesh Journal of Forest Science*, **24**, 21–5.

Balasundaran, M., Mohamed-Ali, M. I., Ghosh, S. K. and Sunderaraman, V. (1988). Presence of sandal spike vector in the insect vector, *Rederatus bimaculatus* and in plants fed with the vector. In *Mycoplasma Diseases of Woody Plants*, ed. S. P. Raychoudhury and N. Rishi. New Delhi: Malhotra Publishing, pp. 23–8.

Ball, J. B., Pandey, D. and Hirai, S. (2000). Global overview of teak plantations. In *Site, Technology and Productivity of Teak Plantations*, ed. T. Enters and C. T. S. Nair. Chiang Mai: FAO Forestry Research Support Programme for Asia and the Pacific, pp. 11–33.

Baltensweiler, W. and Fischlin, A. (1988). The larch budmoth in the alps. In *Dynamics of Forest Insect Populations*, ed. A. A. Berryman. New York: Plenum Press, pp. 332–51.

Baltensweiler, W., Benz, G., Bovey, P. and Delucchi, V. (1977). Dynamics of larch budmoth populations. *Annual Review of Entomology*, **22**, 79–100.

Balut, F. F. and Amante, E. (1971). Note on *Eupseudosoma involuta* (Lepidoptera: Arctiidae), a pest of *Eucalytus* spp. *Biologico*, **37**, 13–16.

Bass, S. (1992). Building from the past: forest plantations in history. In *Plantation Politics*, ed. C. Sargent and S. Bass. London: Earthscan Publications Ltd., pp. 41–75.

Basu-Chowdhury, J. C. (1971). Interim report on the Konny Aerial Spraying Project. In *Working Plan for Nilambur Forest Division, 1967–8 to 1976–7*, ed. K. G. Vasudevan. Trivandrum, India: Government of Kerala, pp. 165–8.

Baylis, W. B. H. and Barnes, R. D. (1989). International provenance trials of *Pinus caribaea* var. *bahamensis*. *Forest Genetic Resources Information*, **17**, 24–5.

Beattie, W. M. (1981). *Ithome lassula* Hodges (Lepidoptera: Cosmopterygidae), a new insect pest of *Leucaena* in Australia. *Leucaena Research Reports*, **2**, 11.

Beeson, C. F. C. (1921). The beehole borer of teak. *Indian Forest Records*, **8**, 1–105.

Beeson, C. F. C. (1924). The economic importance and control of the sal heart-wood borer (*Hoplocerambyx spinicornis*). *Indian Forester*, **50**, 516–24.

Beeson, C. F. C. (1931a). Loss of increment in teak defoliation. *Indian Forester*, **57**, 542–5.

Beeson, C. F. C. (1931b). The life history and control of *Celosterna scabrator* Fab. (Coleoptera: Cerambycidae). *Indian Forest Records (Old Series) (Entomology)*, **16**, 1–16.

Beeson, C. F. C. (1934). The biological control of teak defoliators. *Indian Forester*, **60**, 672–83.

Beeson, C. F. C. (1938). Guide to the insects of *Dalbergia sissoo. Indian Forest Records (Entomology)*, **4**, 8–13.

Beeson, C. F. C. (1941). *The Ecology and Control of the Forest Insects of India and the Neighbouring Countries*, (1961 reprint). New Delhi: Government of India.

Beeson, C. F. C. and Chatterjee, N. C. (1924). The economic importance and control of the sal heartwood borer (*Hoplocerambyx spinicornis* Newm.). *Indian Forest Records (Entomology)*, **11**, 223–69.

Berisford, C. W. (1988). The Nantucket pine tip moth. In *Dynamics of Forest Insect Populations*, ed. A. A. Berryman. New York: Plenum Press, pp. 141–61.

Berisford, Y. C. and Tsao, C. H. (1975). Parasitism, predation and disease in the bagworm, *Thyridopteryx ephemeraeformis* (Haworth) (Lepidoptera: Psychidae). *Environmental Entomology*, **4**, 549–54.

Berryman, A. A. (1986). *Forest Insects: Principles and Practice of Population Management*. New York: Plenum Press.

Berryman, A. A. (1987). The theory and classification of outbreaks. In *Insect Outbreaks*, ed. P. Barbosa and J. C. Schultz. London: Academic Press, pp. 3–30.

Berryman, A. A. (ed.) (1988). *Dynamics of Forest Insect Populations: Patterns, Causes, Implications*. New York: Plenum Press.

Berryman, A. A. (1999). *Principles of Population Dynamics and their Application*. Glasgow: Stanley Thornes (Publishers) Ltd.

Bhandari, R. S. (1988). Insect pests of cone and seed of conifers and their control. In *Tree Protection*, ed. V. K. Gupta and N. K. Sharma. Himachal Pradesh, India: Indian Society of Tree Protection, pp. 342–7.

Bhandari, R. S. and Rawat, J. K. (2001). Sal heartwood borer *Hoplocerambyx spinicornis* Newm. (Coleoptera: Cerambycidae) and its management. *Indian Forester*, **127**, 1387–93.

Bhandari, R.S. and Singh, P. (1988). Epidemic of sal heart-wood borer *Hoplocerambyx spinicornis* Newm. (Coleoptera: Cerambycidae) and its control in Pachmari, Madhya Pradesh. *Indian Forester*, **114**, 152–7.

Bhandari, R.S. and Upadhyay, A.K. (1986). *Cossus cadambae* Moore (Lepidoptera: Cossidae) a new pest of tendu (*Diospyros melanoxylon*). *Indian Forester*, **112**, 169–73.

Bhasin, G.D. and Roonwal, M.L. (1954). A list of insect pests of forest plants in India and the adjacent countries, Parts 1 and 2. *Indian Forest Bulletin (New Series)*, **171**(1), 1–93.

Bhasin, G.D., Roonwal, M.L. and Singh, B. (1958). A list of insect pests of forest plants in India and the adjacent countries, Part 3. *Indian Forest Bulletin (New Series)*, **171**(2), 1–126.

Bhatia, B.M. (1948). On the plant-defoliator-parasite complex, in the biological control of teak defoliators. *Indian Forest Records*, **7**, 193–211.

Bhowmick, A.K. and Vaishampayan, S.M. (1986). Observations on the activity of teak defoliator *Hyblaea puera* Cramer on teak (*Tectona grandis*) influenced by the movement of monsoon. *Journal of Tropical Forestry*, **2**, 27–35.

Bigger, M. (1982). Insect pests associated with forestry plantations in the Solomon Islands. *Commonwealth Forestry Review*, **61**, 249–57.

Billings, R.F. (1991). The pine caterpillar *Dendrolimus punctatus* in Vietnam and recommendations for integrated pest management. *Forest Ecology and Management*, 97–106.

Billings, R.F. and Espino, M.J.V. (1995). Control of a bark beetle outbreak in Honduras. Unpublished paper presented in IUFRO 20th World Congress Working Group S2-07-07 session entitled *Success stories in tropical forest pest control*, Tampere, Finland, 12 pp.

Billings, R.F. and Espino, M.J.V. (2005). The southern pine beetle (*Dendroctonus frontalis*) in Central America: how to recognize, prevent, and control outbreaks. Texas Forest Service Publ. 0605/15000 (available on Internet at http://www.barkbeetles.org/centralamerica/0605e.html)

Billings, R.F., Clarke, S.R., Espino Mendoza, V. *et al.* (2004). Bark beetle outbreaks and fire: a devastating combination for Central America's pine forests. *Unasylva*, **55**, 15–21.

Borse, S.S. and Thakur, M.L. (1993). Relative toxicity of some synthetic pyrethroids against teak skeletonizer, *Eutectona machaeralis* (Walk.) (Lepidoptera: Pyralidae). *Indian Journal of Forestry*, **16**, 193–5.

Borse, S.S. and Thakur, M.L. (1994). Residual toxicity of some commercial synthetic pyrethroids to *Eutectona machaeralis* (Walk.) (Lepidoptera: Pyralidae). *Indian Journal of Forestry*, **17**, 49–52.

Bourdillon, T.F. (1898). Insects attacking teak in southern India. *Indian Forester*, **24**, 126–9 and 424.

Braza, R.D. (1987). Growth and survival of varicose borer (*Agrilus sexsignatus*) larvae in Philippine provenance of bagras (*Eucalyptus deglupta*). *Malaysian Forester*, **50**, 235–8.

Braza, R.D. (1990). Important pests and their control in PICOP forest nurseries. *PCARRD Monitor*, **18**, 8–11.

Broadhead, E. (1983). The assessment of faunal diversity and guild size in tropical forests with particular reference to Psocoptera. In *Tropical Rain Forest: Ecology and Management*, ed. S. L. Sutton, T. S. Whitmore and A. C. Chadwick. Oxford: Blackwell Scientific Publications, pp. 107–19.

Broadhead, E. and Wolda, H. (1985). The diversity of Psocoptera in two tropical forests in Panama. *Journal of Animal Ecology*, **54**, 739–54.

Brown, C. and Ball, J. (2000). World view of plantation grown wood. In *Forests and Society: the Role of Research, Sub-plenary Sessions, vol. 1*, ed. B. Krishnapillay, E. Soepadmo, N. L. Arshad *et al.* Kuala Lumpur: International Union of Forestry Research Organizations, pp. 377–89.

Brown, K. W. (1965). Termite control research in Uganda (with particular reference to the control of attacks in *Eucalyptus* plantations). *East African Agriculture and Forestry Journal*, **31**, 218–23.

Browne, F. G. (1968). *Pests and Diseases of Forest Plantation Trees: An Annotated List of the Principal Species Occurring in the British Commonwalth.* Oxford: Clarendon Press.

Bullock, J. A. and Smith, R. G. (1968). The bionomics and control of *Argyrostagma niobe* (Weym.) and *Dasychira georgiana* Fawc. (Lepidoptera: Lymantriidae) on black wattle in Kenya. *Bulletin of Entomological Research*, **58**, 255–77.

Burckhardt, D., Santana, D. L. Q., Terra, A. L. *et al.* (1999). Psyllid pests (Hemiptera, Psylloidea) in South American eucalypt plantations. *Mitteilungen der Schweizerischen Entomologischen Gesellschaft*, **72**, 1–10.

CABI (Commonwealth Agricultural Bureau International). (2005). *Forestry Compendium*, 2005 edn., (CD version). Wallingford, UK: CAB International.

Cappuccino, N. (1995). Novel approaches to the study of population dynamics. In *Population Dynamics: New Approaches and Synthesis*, ed. N. Cappuccino and P. W. Price. California: Academic Press, pp. 3–16.

Cappuccino, N. and Price, P. W. (eds.) (1995). *Population Dynamics: New Approaches and Synthesis*. New York: Academic Press.

Carson, R. L. (1962). *Silent Spring*. Greenwich, CT: Fawcett Publications Inc.

Carson, W. P. and Root, R. B. (2000). Herbivory and plant species coexistence: community regulation by an outbreaking phytophagous insect. *Ecological Monographs*, **70**, 73–99.

Carson, W. P., Cronin, J. P. and Long, Z. T. (2004). A general rule for predicting when insects have strong top-down effects on plant communities: on the relationship between insect outbreaks and host concentration. In *Insects and Ecosystem Function*, ed. W. W. Weisser and E. Siemann. Berlin: Springer-Verlag, pp. 193–210.

CATIE. (1992a). *Forest Pests in Central America: Field Guide*. Turrialba, Costa Rica: Tropical Agriculture Research and Training Center (CATIE).

CATIE. (1992b). *Forest Pests in Central America: Handbook*. Turrialba, Costa Rica: Tropical Agriculture Research and Training Center (CATIE).

Chacko, K. C. (1998). Termite-aided mesocarp removal of teak (*Tectona grandis* L. F.) fruits for enhanced germination and cost-effective seed handling. *Indian Forester*, **124**, 134–40.

Chaiglom, D. (1966). Teak beehole borer and the control research in Thailand. *Natural History Bulletin of Siam Society*, **21**, 157–60.

Chaiglom, D. (1990). Outbreaks of forest insects and control operations in Thailand. In *Pests and Diseases of Forest Plantations*, ed. C. Hutacharern, K. G. MacDicken, M. H. Ivory and K. S. S. Nair. Bangkok: FAO Regional Office for Asia and the Pacific, pp. 219–23.

Chakrabarti, S. (2001). Gall aphids (Homoptera: Aphididae): their diversity and adaptive features in Northwest and West Himalaya. *Entomon*, **26** (Special Issue), 332–8.

Champion, H. G. (1934). The effect of defoliation on the increment of teak saplings. *Forest Bulletin (Silviculture Series)*, **89**, 1–6. Delhi: Government of India.

Chang, Y. C. (1990). Insect pests in plantation forests of Taiwan. In *Pests and Diseases of Forest Plantations*, ed. C. Hutacharern, K. G. MacDicken, M. H. Ivory and K. S. S. Nair. Bangkok: FAO Regional Office for Asia and the Pacific, pp. 71–4.

Chao, H. W. and Li, H. L. (2004). Investigation on the effect of the mixed plantation of *Pinus massoniana* and *Schima superba*. *Forest Research – Beijing*, **17**, 316–20.

Chase, J. M., Liebhold, M. A., Downing, A. L. and Shurin, J. B. (2000). The effects of productivity, herbivory, and plant species turnover in grassland food webs. *Ecology*, **81**, 2485–97.

Chatterjee, P. N. and Misra, M. P. (1974). Natural enemy and plant host complex of forest insect pests of Indian region. *Indian Forest Bulletin*, **265**, 1–233.

Chatterjee, P. N. and Sebastian, V. O. (1964). Notes on the outbreak of the sap sucker, *Oregma bambusae* in New Forest and measures taken to control them. *Indian Forester*, **90**, 30–1.

Chatterjee, P. N. and Sebastian, V. O. (1966). Additional notes on the outbreak of the sap sucker, *Oregma bambusae* Bucket in New Forest. *Indian Forester*, **92**, 170–2.

Chatterjee, P. N. and Sen-Sarma, P. K. (1968). Important current problems of forest entomology in India. *Indian Forester*, **94**, 112–17.

Chatterjee, P. N. and Singh, P. (1968). *Celosterna scabrator* Fab. (Lamiidae, Coleoptera), a new pest of *Eucalyptus* and its control. *Indian Forester*, **94**, 826–30.

Chaturvedi, N. (1995). A new alternative host plant of teak defoliator *Hyblaea puera* (Hyblaeidae: Lepidoptera). *Journal of the Bombay Natural History Society*, **92**, 431.

Chaturvedi, N. (2002). Some observations on teak defoliator *Hyblaea puera* (Lepidoptera: Hyblaeidae) on secondary host plant *Avicennia marina*. In *Proceedings of the National Seminar on Creeks, Estuaries, and Mangroves: Pollution and Conservation*, ed. G. Quadros. Thane, India: Department of Zoology, B. N. Banddkar College of Science, p. 289.

Cherian, P. T. (2004). Getting the measure of India's insect diversity. In *Perspectives on Biosystematics and Biodiversity*, ed. K. Rajamohana, K. Sudheer, P. Girish Kumar and S. Santhosh. Calicut, India: Department of Zoology, University of Calicut, Systematic Entomology Research Scholars Association, pp. 41–53.

Cherrett, J. M. (1983). Resource conservation by the leaf-cutting ant *Atta cephalotus* in tropical rain forest. In *Tropical Rain Forest: Ecology and Management*, ed. S.L. Sutton, T.S. Whitmore and A.C. Chadwick. Oxford: Blackwell Scientific Publications, pp. 253–63.

Chey, V.K. (1996). *Forest Pest Insects in Sabah*. Sandakan, Malaysia: Sabah Forest Department.

Chey, V.K. (2001). The laran tree and its defoliators. *The Planter*, **77**, 587–92.

Chey, V.K. (2002). Major insect pests and their management in forest plantations in Sabah, Malaysia. In *Pest Management in Tropical Forest Plantations*, ed. C. Hutacharern, B. Napompeth, G. Allard and F.R. Wylie. Bangkok: FAO Forestry Research Support Programme for Asia and the Pacific, pp. 19–23.

Chey, V.K. and Intachat, J. (2000). *Acacia* plantations in Malaysia and their insect pests. *The Planter*, **76**, 171–90.

Chilima, C.Z. (1991). Termite control in young *Eucalyptus* plantations in Malawi using controlled release insecticides. *Commonwealth Forestry Review*, **70**, 237–47.

Choldumrongkul, S. and Hutacharern, C. (1990). Study on soil and properties of teak tree in relation to the number of beeholes. In *Pests and Diseases of Forest Plantations*, ed. C. Hutacharern, K.G. MacDicken, M.H. Ivory and K.S.S. Nair. Bangkok: FAO Regional Office for Asia and the Pacific, pp. 149–54.

Cibrian-Tovar, D., Montiel, J.T.M., Bolanos, R.C., Yates III, H.O. and Lara, J.E.F. (1995). *Forest Insects of Mexico*. Chapingo, Mexico: University Autonoma Chapingo, USDA Forest Service, Natural Resources Canada and North American Forestry Commission.

Ciesla, W.M. (2002). Observations on the life history and habits of a tropical sawfly, *Sericoceros mexicanus* (Kirby) (Hymenoptera: Argidae) on Roatán Island, Honduras. *The Forestry Chronicle*, **78**, 515–21.

Clarke, L.R., Geier, P.W., Hughes, R.D. and Morris, R.F. (1967). *The Ecology of Insect Populations in Theory and Practice*. London: Methuen & Co. Ltd.

Cobbinah, J.R. and Wagner, M.R. (2001). Insect problems in plantations in the equatorial rain forests of Africa. In *Protection of World Forests from Insect Pests: Advances in Research, IUFRO World Series Vol. 11*, ed. R. Alfaro, K. Day, S. Salom. *et al* Malaysia: International Union of Forestry Research Organizations, pp. 37–44.

Cornelius, J.P. (2001). The effectiveness of pruning in mitigating *Hypsipyla grandella* attack on young mahogany (*Swietenia macrophylla* King) trees. *Forest Ecology and Management*, **148**, 287–9.

Cossalter, C. and Nair, K.S.S. (2000). The state of the forest and plantation trends. In *Insect Pests and Diseases in Indonesian Forests: An Assessment of the Major Threats, Research Efforts and Literature*, ed. K.S.S. Nair. Bogor, Indonesia: Center for International Forestry Research, pp. 3–9.

Cowie, R.H., Logan, J.W.M. and Wood, T.G. (1989). Termite (Isoptera) damage and control in tropical forestry with special reference to Africa and Indo-Malaysia: a review. *Bulletin of Entomological Research*, **79**, 173–84.

Curran, L.M. and Leighton, M. (1991). Why mast? The role of generalized insect and vertebrate seed predators on the reproductive biology of Dipterocarpaceae in the Gunung Palung Nature Reserve, West Kalimantan. In *Proceedings of the Fourth Round Table Conference on Dipterocarps, BIOTROP Special Publication No. 41*, ed. I. Soerianegara, S.S. Tjitrosomo, R.C. Umaly and I. Umboh. Bogor, Indonesia: SAEMEO-BIOTROP, pp. 541–2.

Dabral, S.L. and Amin, P.W. (1975). Poor fruit formation in teak in Chanda forests of Maharastra. *Indian Forester*, **101**, 616–20.

Day, R.K., Kairo, M.T.K., Abraham, Y.J. *et al.* (2003). Biological control of homopteran pests of conifers in Africa. *Pineus boerneri*. In *Biological Control in IPM Systems in Africa*, ed. P. Neuenschwander, C. Borgemeister and J. Langewald. Wallingford: CABI Publishing, pp. 104–6.

Day, R.K., Rudgard, S.A. and Nair, K.S.S. (1994). *Asian Tree Pests: An Overview*. Bangkok: FAO Forestry Research Support Programme for Asia and the Pacific.

Dent, D. (1991). *Insect Pest Management*. Wallingford: CAB International.

Dey, R.K. (2001). Management of the sal heartwood borer, *Hoplocerambyx spinicornis* in India. In *Protection of World Forests from Insect Pests: Advances in Research, IUFRO World Series Vol. 11*, ed. R. Alfaro, K. Day, S. Salom *et al.* Malaysia: International Union of Forestry Research Organizations, pp. 241–7.

Dey, R.K. and Tiwari, K.P. (1997). Detection of an imminent defoliator attack on the borer infested sal forests of Madhya Pradesh. *Vaniki Sandesh*, **21**, 21–4.

Diabangouaya, M. and Gillon, Y. (2001). Adaptation of *Helopeltis schoutedeni* Reuter (Heteroptera: Miridae) to planted *Eucalyptus* in the Congo: damage and control. *Bois et Forets des Tropiques*, **267**, 6–20.

Dickinson, M.B. and Whigham, D.F. (1999). Regeneration of mahogany (*Swietenia macrophylla*) in Yucatan. *International Forestry Review*, **1**, 35–9.

Dirzo, R. (1982). Effects of insects on the population dynamics of plants in a Mexican tropical rain forest. In *Proceedings of the Fifth International Symposium on Insect-Plant Relationships*, ed. J.H. Visser and A.K. Minks. Wageningen: Centre for Agricultural Publishing and Documentation, p. 443.

Duarte, A., Menendez, J.M., Fernandez, A. and Martinez, J. (1992). Use of a biopreparation of *Metarhizium anisopliae* strain Nina Bonita in plantations of *Pinus caribaea* for control of *Rhyacionia frustrana*. *Revista Baracoa*, **22**, 17–23.

Echevarria, E. (1985). Observations on the reaction of six pine species to attack by *Dioryctria horneana* (Lepidoptera, Phycitidae). *Revista Forestal Baracoa*, **15**, 35–44.

Eggleton, P., Davies, R.G., Connetable, S., Bignell, D.E. and Rouland, C. (2002). The termites of the Mayombe Forest Reserve, Congo (Brazzaville): transect sampling reveals an extremely high diversity of ground-nesting soil feeders. *Journal of Natural History*, **36**, 1239–46.

Elliott, H.J., Ohmart, C.P. and Wylie, F.R. (1998). *Insect Pests of Australian Forests: Ecology and Management*. Melbourne: Incata Press.

Elouard, C. (1998). Pests and diseases of Dipterocarpaceae. In *A Review of Dipterocarps: Taxonomy, Ecology and Silviculture*, ed. S. Appanah and J.M. Turnbull. Bogor, Indonesia: Center for International Forestry Research, pp. 115–31.

Entwistle, P.F. and Evans, H.F. (1985). Viral control. In *Comprehensive Insect Physiology, Biochemistry and Pharmacology*, ed. G.A. Kerkut and L.I. Gilbert. pp. 348–412.

Erwin, T.L. (1982). Tropical forests: their richness in Coleoptera and other arthropod species. *The Coleopterist's Bulletin*, **36**, 74–5.

Erwin, T.L. (1983a). Tropical forest canopies: the last biotic frontier. *Bulletin of Entomological Society of America*, **29**, 14–19.

Erwin, T.L. (1983b). Beetles and other insects of tropical forest canopies at Manaus, Brazil, sampled by insecticidal fogging. In *Tropical Rain Forest: Ecology and Management*, ed. S.L. Sutton, T.S. Whitmore and A.C. Chadwick. Oxford: Blackwell Scientific Publications, pp. 59–75.

Erwin, T.L. and Scott, J.C. (1980). Seasonal and size patterns, trophic structure, and richness of Coleoptera in the tropical arboreal ecosystem: the fauna of the tree *Luehea seemannii* Triana and Planch in the Canal Zone of Panama. *The Coleopterists Bulletin*, **34**, 305–22.

Eungwijarnpanya, S. and Yinchareon, S. (2002). Control of teak defoliator, *Hyblaea puera* Cramer (Lepidoptera: Hyblaeidae), by thermal fogger application of neem extract. In *Pest Management in Tropical Forest Plantations*, ed. C. Hutacharern, B. Napompeth, G. Allard and F.R. Wylie. Bangkok: FAO Forestry Research Support Programme for Asia and the Pacific, pp. 123–5.

Evans, H.F. (1994). Laboratory and field results with viruses for the control of insects. In *Comparing Glasshouse and Field Pesticide Performance II*, ed. G. Hewitt, J. Caseley, L.G. Copping, B.T. Grayson and D. Tyson. Hampshire: British Crop Protection Council, pp. 285–96.

Evans, J. (1992). *Plantation Forestry in the Tropics*, 2nd edn. Oxford: Oxford University Press.

Fang, H.L., Tong, P.Y., Fang, S.H., Lu, S.C., Mu, H.B. and Dai, J.S. (1997). Study on insect damage in *Casuarina equisetifolia* protective forest in coastal area. *Journal of Zhejiang Forestry Science and Technology*, **17**, 9–14.

FAO. (1991). *Exotic Aphid Pests of Conifers: A Crisis in African Forestry. Proceedings of a Workshop held at Kenya Forestry Research Institute*. Rome: Food and Agriculture Organization of the United Nations.

FAO. (1992). *Mixed and Pure Forest Plantations in the Tropics and Subtropics*, FAO Forestry Paper 103. Rome: Food and Agriculture Organization of the United Nations.

FAO. (2001a). *Global Forest Resources Assessment 2000, FAO Forestry Paper No. 140*. Rome: Food and Agriculture Organization of the United Nations.

FAO. (2001b). *State of the World's Forests 2001*. Rome: Food and Agriculture Organization of the United Nations.

FAO. (2005). *State of the World's Forests 2005*. Rome: Food and Agriculture Organization of the United Nations.

Fauziah, B.A. and Hidaka, T. (1989). Collection of the Albizia bark borer, *Xystrocera festiva* Pasco in Brumas Plantation, Sabah, Malaysia. In *Researches on Behavioural Strategy of Small Animals in the Humid Tropics*, ed. T. Hidaka. Kyoto: Kyoto University, pp. 17–20.

Fernandez, E.E. (1898). Defoliation of teak in the Central Provinces. *Indian Forester*, **24**, 89.

Filho, P.O., Klein, H. and Filho, B.E. (1992). Outbreak of caterpillars on *Eucalyptus* spp. (Myrtaceae) in Mato Grosso, Brazil. *Revista de Agricultura Piracicaba*, **67**, 181–2.

Flamm, R.O., Coulson, R.N. and Payne, T.L. (1988). The southern pine beetle. In *Dynamics of Forest Insect Populations*, ed. A.A. Berryman. New York: Plenum Press, pp. 531–53.

Floyd, R.B., Farrow, R.A., Farrell, G.S. and Court, M.M. (1998). Insect feeding on *Acacia mearnsii* in southeastern mainland Australia. In *Recent Developments in Acacia Planting*, ed. J.W. Turnbull, H.R. Crompton and K. Pinyopusarerk. Canberra: Australian Center for International Agricultural Research, pp. 246–51.

Foahom, B. (2002). *Insect Pest Incidence on Timber Tree Species in Natural Forest in South Cameroon*. Wageningen: The Tropenboz–Cameroon Programme, Tropenboz International.

Fonseca, C.R. (1994). Herbivory and the long-lived leaves of an Amazonian ant tree. *Journal of Ecology*, **82**, 833–42.

Ford, L.B. (1986). The impact of *Rhyacionia frustrana* in Costa Rica. *Turrialba*, **36**, 559–61.

Frankenhuyzen, K. van and Beardmore, T. (2004). Current status and environmental impact of transgenic forest trees. *Canadian Journal of Forestry Research*, **34**, 1163–80.

FSI (Forest Survey of India). (2000). *State of Forest Report 1999*. Dehra Dun, India: Forest Survey of India, Ministry of Environment and Forests.

Garthwaite, P.F. (1939). On the biology of *Calopepla leayana* Latr. (Coleoptera, Chrysomelidae) and the possibilities of control. *Indian Forest Records (New Series) (Entomology)*, **5**, 237–77.

Gaston, K.J. (1991). The magnitude of global insect species richness. *Conservation Biology*, **5**, 283–96.

Geiger, C.A. and Gutierrez, A.P. (2000). Ecology of *Heteropsylla cubana* (Homoptera: Psyllidae): psyllid damage, tree phenology, thermal relations, and parasitism in the field. *Environmental Entomology*, **29**, 76–86.

Geiger, C.A., Napompeth, B. and Van Den Beldt, R.J. (1995). An update on the status of the leucaena psyllid in Southeast Asia. In *Leucaena: Opportunities and Limitations*, ed. H.M. Shelton, C.M. Piggin and J.L. Brewbaker. Canberra: Australian Centre for Agricultural Research, pp. 125–8.

Gentry, A.H. (1988). Changes in plant community diversity and floristic composition on environmental and geographic gradients. *Annals of the Missouri Botanical Garden*, **75**, 1–34.

Gerling, D., Velthuis, H.H.W. and Hefetz, A. (1989). Bionomics of the large carpenter bees of the genus *Xylocopa*. *Annual Review of Entomology*, **34**, 163–90.

Ghosh, R.C. (1977). *Handbook of Afforestation Techniques*. New Delhi: Controller of Publications, Government of India.

Gibson, I.A.S. and Jones, T. (1977). Monoculture as the origin of major forest pests and diseases, especially in the tropics and southern hemisphere. In *Origins of Pest, Parasite, Disease and Weed Problems*, ed. J.M. Cherrett and G.R. Sagar. Oxford: Blackwell Scientific, pp. 139–61.

Gillett, J.D. (1971). *Mosquitos*. London: Weidenfeld and Nicolson.

Goncalves, J. L. M., Poggiani, F., Stape, J. L. *et al.* (1999). Eucalypt plantations in the humid tropics: São Paulo, Brazil. In *Site Management and Productivity in Tropical Plantation Forests*, ed. E. K. S. Nambiar, C. Cossalter and A. Tiarks. Bogor, Indonesia: Center for International Forestry Research, pp. 5–12.

Gopakumar, B. and Prabhu, V. K. K. (1981). Eclosion rhythm and its entrainment by photoperiodism in *Pyrausta (Hapalia) machaeralis* Walker (Lepidoptera: Pyralidae). *Entomon*, **6**, 329–33.

Gotoh, T. (1994). Insect borers of some valuable timber species in Thailand. *Tropical Forestry*, **30**, 30–7.

Gotoh, T., Yincharoen, S., Ungwijarnpanya, S. and Hutacharern, C. (2002). Strategies for the management of the teak beehole borer, *Xyleutes ceramicus* (Walker) (Lepidoptera: Cossidae) in Thailand. In *Pest Management in Tropical Forest Plantations*, ed. C. Hutacharern, B. Napompeth, G. Allard and F. R. Wylie. Bangkok: FAO Forestry Research Support Programme for Asia and the Pacific, pp. 99–103.

Gotoh, T., Kotulai, J. R. and Matsumoto, K. (2003). Stem borers of teak and yemane in Sabah, Malaysia, with analysis of attacks by the teak beehole borer (*Xyleutes ceramica* Wlk.). *Japan Agricultural Research Quarterly*, **37**, 253–61.

Graham, S. A. and Knight, F. B. (1965). *Principles of Forest Entomology*, 4th edn. New York: McGraw Hill Book Company.

Granados, R. R. and Federici, B. A. (ed.) (1986). *The Biology of Bacuoviruses, Vol I and II.* Boca Raton: CRC Press.

Gray, B. (1972). Economic tropical forest entomology. *Annual Review of Entomology*, **17**, 313–54.

Greaves, T. (1959). Termites as forest pests. *Australian Forester*, **23**, 114–20.

Greaves, T. and Florence, R. G. (1966). Incidence of termites in blackbutt regrowth. *Australian Forester*, **30**, 153–61.

Greenbank, D. O., Schaefer, G. W. and Rainey, R. C. (1980). Spruce budworm moth flight and dispersal: new understanding from canopy observations, radar and aircraft. *Memoirs of the Entomological Society of Canada*, **110**, 1–49.

Greenbaum, H. N. (1975). *Neodiprion merkeli* Ross (Hymenoptera: Diprionidae) in the Bahamas. *Florida Entomologist*, **58**, 202.

Grehan, J. R. (1987). Evolution of arboreal tunnelling by larvae of *Aenetus* (Lepidoptera: Hepialidae). *New Zealand Journal of Zoology*, **14**, 441–62.

Grimaldi, D. and Engel, M. S. (2005). *Evolution of the Insects*. Cambridge: Cambridge University Press.

Gupta, B. K. and Borse, S. S. (1997). Relative toxicity of some insecticides as contact poison against third instar larvae of *Hyblaea puera* (Lepidoptera: Hyblaeidae). *Indian Forester*, **123**, 427–9.

Gupta, B. K., Ahmad, S. I. and Veer, V. (1989). Relative toxicity of some conventional insecticides against adult beetles of *Calopepla leayana* Latr. (Chrysomelidae: Coleoptera). *Indian Forester*, **115**, 430–4.

Gupta, T. and Guleria, A. (1982). *Non-wood Forest Products in India*. New Delhi: Oxford and IBH Publishing Co.

Guzman, de E. D. and Nuhamara, S. K. (eds.) (1987). *Forest Pests and Diseases in Southeast Asia*. Bogor, Indonesia: SEAMEO-BIOTROP.

Haack, R. A., Billings, R. F. and Richter A. M. (1989). Life history parameters of bark beetles (Coleoptera: Scolytidae) attacking West Indian pine in the Dominican Republic. *Florida Entomologist*, **72**, 591–603.

Hakkila, P. (1995). *Procurement of Timber for the Finnish Forest Industries*. Vantaa, Finland: Finnish Forest Research Institute.

Halos, S. C., Quiniones, S. S., Guzman, de E. D., Lapis, E. B. and Lucero, R. M. (1985). Towards an integrated control measure against shoot moth and needle blight affecting pines. *PCARRD Monitor*, **13**, 6–8.

Hampson, G. F. (1893–6). *Fauna of British India, Moths, vols 1–4*. London: Taylor & Francis Ltd.

Hanel, H. and Watson, J. A. L. (1983). Preliminary field tests on the use of *Metarhizium anisopliae* for control of *Nasutitermes exitiosus* (Hill) (Isoptera: Termitidae). *Bulletin of Entomological Research*, **94**, 305–13.

Hannelius, S. and Kuusela, K. (1995). *Finland: The Country of Evergreen Forest*. Tampere, Finland: Forssan Kirjapaino Oy.

Haojie, W., Varma, R. V. and Tiansen, X. (1998a). *Insect Pests of Bamboos in Asia: An Illustrated Manual*. New Delhi: International Network for Bamboo and Rattan (INBAR).

Haojie, W., Floyd, R., Farrow, R. *et al.* (1998b). Insect damage on *Acacia mearnsii* in China. In *Recent Developments in Acacia Planting*, ed. J. W. Turnbull, H. R. Crompton and K. Pinyopusarerk. Canberra: Australian Center for International Agricultural Research, 240–5.

Haojie, W., Floyd, R. B., Changfu, F. and Changchun, L. (2002). Evidence for natural control of insect pests of *Acacia mearnsii* in China. In *Pest Management in Tropical Forest Plantations*, ed. C. Hutacharern, B. Napompeth, G. Allard and F. R. Wylie. Bangkok: FAO Forestry Research Support Programme for Asia and the Pacific, 75–82.

Hardi, T. T. W. and Intari, S. E. (1990). Pest control in industrial plantation forest. In *Proceeding Discusi Hutan Tanaman Industri*, ed. Buharman, K. Purba and C. Hediana. Jakarta: Badan Litbang Kehutanan Departemen Kehutanan, 177–86.

Hardi, T. W. H., Husaeni, E. A., Darwiati, W. and Nurtjahjawilasa. (1996). A study of the morphology and morphometrics of adults of *Xystrocera festiva* Pascoe. *Buletin Penelitian Hutan*, **604**, 39–48.

Harman, D. M. and Kulman, H. M. (eds.) (1973). *A World Survey of the Parasites and Predators of the Genus Rhyacionia, parts 1–4*. Maryland: University of Maryland Natural Resources Institute.

Harsh, N. S. K., Jamaluddin, N. S. and Tiwari, C. K. (1992). Top dying and mortality in provenance trial plantations of *Gmelina arborea*. *Journal of Tropical Forestry*, **8**, 55–61.

Hawksworth, D. L. and Kalin-Arroyo, M. T. (1995). Magnitude and distribution of biodiversity. In *Global Biodiversity Assessment*, ed. V. H. Heywood. Cambridge: Cambridge University Press, pp. 107–99.

Hill, D. S. (1997). *The Economic Importance of Insects*. London: Chapman & Hall.

Hochmut, R. (1972a). *Neodiprion insularis*, a defoliator of *Pinus caribaea*. *Baracoa*, **2**(2), 2–18.

Hochmut, R. (1972b). Pests of the genus *Dioryctria* (Lepidoptera, Phycitidae) that attack pines in Cuba. *Baracoa*, **2**(3–4), 2–16.

Howard, F. W. (1995). Reduction of damage to mahogany by mahogany shoot borer and mahogany leaf miner by use of azadirachtin. *Journal of Tropical Forest Science*, **7**, 454–61.

Howard, F. W. and Solis, M. A. (1989). Distribution, life history and host plant relationships of mahogany webworm, *Macalla thyrsisalis* (Lepidoptera: Pyralidae). *Florida Entomologist*, **72**, 469–79.

Hutacharern, C. (1990). Forest insect pests in Thailand. In *Pests and Diseases of Forest Plantations*, ed. C. Hutacharern, K. G. MacDicken, M. H. Ivory and K. S. S. Nair. Bangkok: FAO Regional Office for Asia and the Pacific, pp. 75–9.

Hutacharern, C. (1993). Insect pests. In *Acacia mangium: Growing and Utilisation*, ed. K. Awang and D. Taylor. Bangkok: Winrock International and FAO, pp. 163–202.

Hutacharern, C. (2001). Teak beehole borer, *Xyleutes ceramica*: ecology and management. In *Protection of World Forests from Insect Pests: Advances in Research, IUFRO World Series Vol. 11*, ed. R. Alfaro, K. Day, S. Salom *et al*. Malaysia: International Union of Forestry Research Organizations, pp. 249–53.

Hutacharern, C. and Choldumrongkal, S. (1989). A note on the insect pests of multipurpose tree species in Thailand. *Tropical Forest Science*, **2**, 81–4.

Hutacharern, C. and Panya, S. E. (1996). Biology and control of *Aristobia horridula* (Hope) (Coleoptera, Cerambycidae), a pest of *Pterocarpus macrocarpus*. In *Impact of Diseases and Insect Pests in Tropical Forests*, ed. K. S. S. Nair, J. K. Sharma and R. V. Varma. Peechi, India: Kerala Forest Research Institute, pp. 392–7.

Hutacharern, C. and Tubtim, N. (1995). *Check List of Forest Insects in Thailand*. Bangkok: Office of Environmental Policy and Planning, Ministry of science, Technology and Environment.

Hutacharern, C., Yincharoen, S. and Eungwijarnpanya, S. (1993). Control of the teak defoliation by aerial application of *Bacillus thuringiensis* Berliner var. *kurstaki*. In *Proceedings of the Forest Conference 1993*, ed. B. Boonthawee. Bangkok: Royal Forest Department, pp. 5–19.

Huxley, C. R. (1986). Evolution of benevolent ant–plant relationships. In *Insects and the Plant Surface*, ed. B. Juniper and R. Southwood. London: Edward Arnold, pp. 257–82.

IFGTB (Institute of Forest Genetics and Tree Breeding). (1995). *Pest Problems of Babul (*Acacia nilotica ssp. indica*) and Their Management*. Coimbatore, India: Institute of Forest Genetics and Tree Breeding, Indian Council of Forestry Research and Education.

Intachat, J. (1998). The identity of a Malaysian teak skeletonizer, *Paliga damastesalis* Walker (Lepidoptera: Pyralidae). *Journal of Tropical Forest Science*, **10**, 561–3.

Intari, S. E. (1975). Observations on beehole borer (*Duomitus ceramicus*) in teak plantations in Kendal and Ciamis forest districts, Java. [Hasil pengamatan hamaoleng oleng (*Duomitus ceramicus* Wlk.) pada tegakan jati di KPH Kendal dan Ciamis]. *Laporan Lembaga Penelitian Hutan*, **204**, 1–14.

Intari, S. E. (1996). The control of *Mucanum* sp. causing damage to seedlings of *Shorea javanica* in Krui, West Lampung. *Buletin Penelitian Hutan*, **604**, 31–7.

Intari, S. E. and Natawiria, D. (1973). Whitegrubs in forest tree nurseries and young plantations. *Laporan Lembaga Penelitian Hutan*, **167**, 1–22.

Intari, S. E. and Santoso, E. (1990). Planting pattern of industrial plantation forest as an effort to control pests and diseases. In *Proceedings Discusi Hutan Tanaman Industri*, ed. Buharman, K. Purba and C. Hediana. Jakarta: Badan Litbang Kehutanen Departemen Kehutanan, pp. 85–94.

Irianto, R. S. B., Matsumoto, K. and Mulyadi, K. (1997). The yellow butterfly species of the genus *Eurema* Hubner, causing severe defoliation in the forestry plantations of *Albizzia* and *Paraserianthes falcataria* (L.) Nielson in the western part of Indonesia. *JIRCAS Journal*, **4**, 41–9.

Jactel, H., Brockeroff, E. and Duelli, P. (2005). A test of the biodiversity–stability theory: meta-analysis of tree species diversity effects on insect pest infestations, and re-examination of responsible factors. In *Forest Diversity and Function: Temperate and Boreal Systems*, ed. M. Schererer-Lorenzen, C. Korner and E. D. Schulze. Berlin: Springer-Verlag, pp. 235–62.

Jain, A., Roychoudhury, N., Sharma, S., Bhargava, A. and Pant, N. C. (1998). Host plant resistance to insect pests in teak (*Tectona grandis* L.f.) with reference to biochemical parameters. *Indian Journal of Forestry*, **21**, 285–9.

Jain, A., Roychoudhury, N. and Bhargava, A. (2000). Role of foliar protein and polyphenol and their relationship to clonal resistance in teak against the leaf skeletoniser, *Paliga machoeralis* Walker (Lepidoptera: Pyralidae). *Journal of Tropical Forest Science*, **12**, 221–6.

Jain, A., Singh, A. K., Banerjee, S. K. and Shukla, P. K. (2002). Chemical screening of different clones of *Tectona grandis* in relation to resistance against their key defoliators. *Indian Journal of Forestry*, **25**, 254–73.

Jain, N. C. (1996). Growing pest problems in *Prosopis cineraria* in arid zones of India. In *Impact of Diseases and Insect Pests in Tropical Forests*, ed. K. S. S. Nair, J. K. Sharma and R. V. Varma. Peechi, India: Kerala Forest Research Institute, pp 507–10.

Jamal, A. (1994). Major insect pests of gum arabic trees *Acacia senegal* Willd. and *Acacia seyal* L. in Western Sudan. *Journal of Applied Entomology*, **117**, 10–20.

Janzen, D. H. (1974). Tropical backwater rivers, animals and mast fruiting by the Dipterocarpaceae. *Biotropica*, **6**, 69–103.

Jones, W. E., Grace, J. K. and Tamashiro, M. (1996). Virulence of seven isolates of *Beauveria bassiana* and *Metarhizium anisopliae* to *Coptotermes formosanus* (Isoptera: Rhinotermitidae). *Environmental Enomology*, **25**, 481–7.

Joseph, K. J. (1998). Biology and breeding behaviour of the elephant dung beetle, *Heliocopris dominus* Bates (Coprinae: Scarabaeidae). *Entomon*, **23**, 325–9.

Joseph, K.J. and Venkitesan, T.S. (1996). Studies on the psyllid *Heteropsylla cubana* on subabul plantations in Kerala, India. In *Impact of Diseases and Insect Pests in Tropical Forests*, ed. K.S.S. Nair, J.K. Sharma and R.V. Varma. Peechi, India: Kerala Forest Research Institute, pp. 416–22.

Kalia, S. and Lall, R.B. (2000). Efficacy of three varietal toxins of *Bacillus thuringiensis* tested against some important forest insect pests of multipurpose forest tree species. *Indian Forester*, **126**, 62–6.

Kalshoven, L.G.E. (1950–1). *De Plagen van de Cultuurgewassen in Indonesie*, Van Hoeve W. Gravenhage/Bandoeng, Vol. **1** and **2**. Revised and translated by Van der Laan P.A. (1981) as *Pests of Crops in Indonesia*. Jakarta: P.T. Ichtiar Baru–Van Hoeve.

Kalshoven, L.G.E. (1953). Important outbreaks of insect pests in the forests of Indonesia. *Transactions of the 9th International Congress of Entomology*, **2**, 229–34.

Kamath, M.K., Senivasa, E. and Bola, I. (1996). Impact of termites on mahogany (*Swietenia macrophylla*) plantations in Fiji. In *Impact of Diseases and Insect Pests in Tropical Forests*, ed. K.S.S. Nair, J.K. Sharma and R.V. Varma. Peechi, India: Kerala Forest Research Institute, pp. 298–303.

Kan, W.H. (1958). Investigations on the injury and its control of silver spotted longicorn beetle to beef-wood. *Bulletin of Taiwan Forest Research Institute*, **58**, 1–24.

Kasno and Husaeni, E.A. (2002). An integrated control of sengon stem borer in Java. In *Pest Management in Tropical Forest Plantations*, ed. C. Hutacharern, B. Napompeth, G. Allard and F.R. Wylie. Bangkok: FAO Forestry Research Support Programme for Asia and the Pacific, pp. 127–31.

Kaur, S., Varshney, V.K. and Dayal, R. (2003). GC-MS Analysis of essential oil of *Shorea robusta* bast. *Journal of Asian Natural Products Research*, **5**, 231–4.

Kaushik, R.C. (1956). Teli variety of *Tectona grandis*. In *Proceedings of 9th Silvicultural Conference, Part 2*. Dehra Dun, India: Forest Research Institute and Colleges, pp. 92–4.

Kellman, M. and Tackaberry, R. (1997). *Tropical Environments: The Functioning and Management of Tropical Ecosystems*. London: Routledge.

Kerr, J. (1782). Natural History of the insect which produces the gum lacca. *Philosophical Transactions of the Royal Society of London*, **71**, 374–82.

KFRI (Kerala Forest Research Institute). (1981). *Termite Control in Eucalypt Plantations, KFRI Information Bulletin 3*. Peechi, India: Kerala Forest Research Institute.

Khan, A.H. and Chatterjee, P.N. (1944). Undergrowth in teak plantations as a factor in reducing defoliation. *Indian Forester*, **70**, 365–9.

Khan, H.R., Kumar, S. and Prasad, L. (1988a). Studies on seasonal activity of some agro-forestry insect pests by light trap. *Indian Forester*, **114**, 215–29.

Khan, H.R., Bhandari, R.S., Prasad, L. and Kumar, S. (1988b). Population dynamics of *Hyblaea puera* Cram. (Lepidoptera: Hyblaeidae) and *Eutectona machaeralis* Walk. (Lepidoptera: Pyralidae) in teak forest of Madhya Pradesh (India). *Indian Forester*, **114**, 803–13.

Khan, M.A. and Ahmad, D. (1991). Comparative toxicity of some chlorinated hydrocarbon insecticides to poplar leaf beetle, *Chrysomela populi* L. (Coleoptera, Chrysomelidae). *Indian Journal of Forestry*, **14**, 42–5.

Kigomo, B.N. (1990). Bamboo resource in East African region. In *Bamboo: Current Research*, ed. I.V.R. Rao, R. Gananaharan and C.B. Sastry. Peechi, India: Kerala Forest Research Institute and International Development Research Centre, Canada, pp. 22–8.

Kim, S.S. and Sohn, J.C. (2003). A new record of the family Hyblaeidae (Lepidoptera) from Korea. *Korean Journal of Entomology*, **33**, 127–31.

Kirton, L.G., Brown, V.K. and Azmi, M. (1999). Do forest-floor wood residues in plantations increase the incidence of termite attack: testing current theory. *Journal of Tropical Forest Science*, **11**, 218–39.

Koenig, J.C. (1779). *Beschaftigungen der Berlinischen Gesellschaft Naturforchender Freunde.* (English translation: Fletcher, T.B. (1921).) *Proceedings of the 4th Entomology Meet.* Pusa, India, pp. 312–33.

Koricheva, J., Larsson, S. and Haukioja, E. (1998). Insect performance on experimentally stressed woody plants: a meta-analysis. *Annual Review of Entomology*, **43**, 195–216.

Kulhavy, D. L., Smith, L.A. and Ross, W.G. (2001). Impact of the Texas leaf-cutting ant (*Atta texana* (Buckley)) (Hymenoptera: Formicidae) on a forested landscape. In *Proceedings: Integrated Management and Dynamics of Forest Defoliating Insects*, ed. A.M. Liebhold, M.L. McManus, I.S. Otvos and S.L.C. Fosbroke. Philadelphia: USDA Forest Service, pp. 85–90.

Kumar, P., Dabgar, V.M., Santosh, K. and Swaminath, M.H. (1999). A new pest to *Terminalia tomentosa* Wight. *Myforest*, **35**, 259–61.

Lapis, E.B. (1987). The pine shoot moths, *Dioryctria rubella Hamps.* (Lepidoptera: Pyralidae) and *Petrova cristata* Wals. (Lepidoptera: Tortricidae), in the Philippines. In *Forest Pests and Diseases in Southeast Asia, BIOTROP Special Publication No. 26,* ed. E.D. de Guzman and S.T. Nuhamara. Bogor, Indonesia: Southeast Asian Regional Center for Tropical Biology, pp. 97–103.

Laranjeiro, A.J. (1994). Integrated pest management at Aracruz Celulose. *Forest Ecology and Management*, **65**, 45–52.

Larrea, D.M. (1999). Effect of growth stage, and nitrogen and carbon concentration, on *Swietenia macrophylla* (Meliaceae) seedling folivory by insects. *Ecologia en Bolivia*, **32**, 29–35.

Lavelle, P., Dangerfield, M., Fragoso, C. *et al.* (1994). The relationship between soil microfauna and tropical soil fertility. In *The Biological Management of Soil Fertility*, ed. P.L. Woomer and M.J. Swift. Chichester, UK: John Wiley & Sons, pp. 137–69.

Lawson, S.A., Wylie, F.R., Wylie, R.L. and Ryan, P. (2002). Longicorn beetles (*Phoracantha* spp.) and giant wood moths (*Endoxyla* spp.): emerging threats in tropical and subtropical eucalypt plantations in Queensland, Australia. In *Pest Management in Tropical Forest Plantations*, ed. C. Hutacharern, B. Napompeth, G. Allard and F.R. Wylie. Bangkok: FAO Forestry Research Support Programme for Asia and the Pacific, pp. 33–45.

Leather, S. R. (2000). Herbivory, phenology, morphology and the expression of sex in trees: who is in the driver's seat? *Oikos*, **90**, 194–6.

Lefroy, H. M. (1909). *Indian Insect Life: A Manual of the Insects of the Plains (Tropical India)*. Calcutta: Thacker, Spink & Co.

Lima, dos P. P. S. and Filho, E. B. (1985). The leaf cutting ants: serious pests of forests in Brazil. In *Noxious Insects to Pine and Eucalypt Plantations in the Tropics*, ed. J. H. Pedrosa-Macedo. Curitiba, Brazil: Federal University of Parana, Brazil and International Union of Forestry Research Organizations, pp. 147–8.

Logan, J. W. M., Cowie, R. H. and Wood, T. G. (1990). Termite (Isoptera) control in agriculture and forestry by non-chemical methods: a review. *Bulletin of Entomological Research*, **80**, 309–30.

Loganathan, J. and David, P. M. M. (1999). Rainfall: a major factor leading to outbreak of teak (*Tectona grandis*) defoliator, *Hyblaea puera* Cramer (Lepidoptera: Hyblaeidae) in commercial teak plantation. *Indian Forester*, **125**, 316–20.

Loganathan, J. and David, P. M. M. (2000). Laboratory and field evaluation of *Bacillus thuringiensis* Berliner products against the teak defoliator, *Hyblaea puera* Cramer. *Insect Science and its Application*, 61–5.

Loganathan, J., Soman, P. and Maragatham, S. (2001). Monitoring of two major pests of teak in intensively managed plantation through light trap study. *Indian Forester*, **127**, 1047–52.

Long, Z. T., Mohler, C. L. and Carson, W. P. (2003). Extending the resource concentration hypothesis to plant communities: effects of litter and herbivores. *Ecology*, **84**, 652–65.

López-Bellido, L. and Fuentes, M. (1997). *Leucaena leucocephala* (Lamk) de Wit. In *Plant Resources of South-East Asia. Vol. 11, Auxiliary Plants*, ed. I. F. Hanum and L. J. B. van der Masen. Bogor, Indonesia: PROSEA Foundation, pp. 175–84.

Lowe, R. G. (1961). Control of termite attack on *Eucalyptus citriodora* Hook. *Empire Forestry Review*, **40**, 73–8.

Lucero, R. M. (1987). Pine shoot borer infestation in Abra Province of the Philippines. In *Forest Pests and Diseases in Southeast Asia, BIOTROP Special Publication No. 26*, ed. E. D. de Guzman and S. T. Nuhamara. Bogor, Indonesia: Southeast Asian Regional Center for Tropical Biology, pp. 157–62.

Lugo, A. E., Farnworth, E. G., Pool, D., Jerez, P. and Kaufman, G. (1973). The impact of the leaf-cutter ant *Atta colombica* on the energy flow of a tropical wet forest. *Ecology*, **54**, 1292–301.

Mackenzie, J. M. D. (1921). Some notes on forest insect pests in Burma. *Indian Forester*, **47**, 309–17.

Mailu, A. M., Khamala, C. P. M. and Rose, D. J. W. (1980). Population dynamics of pine woolly aphid, *Pineus pini* (Gmelin) (Hemiptera: Adelgidae), in Kenya. *Bulletin of Entomological Research*, **70**, 483–90.

Maloney, P. E. and Rizzo, D. M. (2002). Pathogens and insects in a pristine forest ecosystem: the Sierra San Pedro Martir, Baja, Mexico. *Canadian Journal of Forestry Research*, **32**, 448–57.

Mangoendihardjo, S., Mahrub, E. and Warrow, J. (1990). Endemic natural enemies of the leucaena psyllid in Indonesia. In *Leucaena Psyllid: Problems and Management*, ed. B. Napompeth and K. G. MacDicken. Bangkok: Winrock International/IDRC/NFTA/F-FRED, pp. 159–61.

Mary, M. V. and Sankaran, K. V. (1991). *Ex-situ decomposition of leaf litters of* Tectona grandis, Eucalyptus tereticornis *and* Albizia falcataria, *KFRI Research Report 71*. Peechi, India: Kerala Forest Research Institute.

Mathew, G. (1982). *A survey of beetles damaging commercially important stored timber in Kerala, KFRI Research Report 10*. Peechi, India: Kerala Forest Research Institute.

Mathew, G. (1986). Insects associated with forest plantations of *Gmelina arborea* Roxb. in Kerala, India. *Indian Journal of Forestry*, **94**, 308–11.

Mathew, G. (1990). *Biology and Ecology of the Teak Trunk Borer, Cossus cadambae Moore and its Possible Control, KFRI Research Report No. 68*. Peechi, India: Kerala Forest Research Institute.

Mathew, G. (1991). Biology, seasonal population trends and impact of the teak carpenterworm, *Alcterogystia cadambae* (Moore) (Lepidoptera: Cossidae). *Annals of Entomology*, **9**, 39–46.

Mathew, G. (1993). Nursery pests of indigenous tree species and their management in Kerala, India. In *Diseases and Insects in Forest Nurseries, proceedings of IUFRO Conference in Dijon, France*, ed. R. Perrin and J. R. Sutherland. Paris: INRA, pp. 145–51.

Mathew, G. (1995). Field performance of some indigenous tree species with respect to pest incidence in Kerala (India). *Indian Journal of Forestry*, **18**, 133–40.

Mathew, G. (2002). Distribution, host range and possible management of *Indarbela quadrinotata* attacking forest plantations of *Paraserianthes falcataria* in Kerala, India. In *Pest Management in Tropical Forest Plantations*, ed. C. Hutacharern, B. Napompeth, G. Allard and F. R. Wylie. Bangkok: FAO Forestry Research Support Programme for Asia and the Pacific, pp. 133–8.

Mathew, G. and Nair, K. S. S. (1985). Insects associated with forest plantations of *Paraserianthes falcataria* in Kerala, India. *Malaysian Forester*, **49**, 200–5.

Mathew, G. and Nair, K. S. S. (1986). Bagworms (Lepidoptera, Psychidae) of Kerala: their potential as pests of tree crops. In *Proceedings of the 3rd Oriental Entomology Symposium, vol. 2*, ed. G. Koshy *et al.* Trivandrum, India: University of Kerala, pp. 163–7.

Mathew, G. and Rahamathulla, V. K. (1995). Biodiversity in the Western Ghats: a study with reference to moths (Lepidoptera: Heterocera) in the Silent Valley National Park, India. *Entomon*, **20**, 25–33.

Mathew, G. and Rugmini, P. (1996). Impact of the borer *Alcterogystia cadambae* (Moore) (Lepidoptera: Cossidae) in forest plantations of teak in Kerala, India. In *Impact of Diseases and Insect Pests in Tropical Forests*, ed. K. S. S. Nair, J. K. Sharma and R. V. Varma. Peechi, India: Kerala Forest Research Institute, pp. 304–10.

Mathew, G. and Sudheendrakumar, V. V. (1992). Outbreak of *Udongo montana* (Heteroptera: Pentatomidae) on bamboo in natural forests and adjoining plantation in Wynad. *Bamboo Information Centre India Bulletin*, **2**, 17–18.

Mathew, G., Rugmini, P. and Jayaraman, K. (1989). Studies on spatial distribution in the teak carpenterworm *Cossus cadambae* Moore (Lepidoptera, Cossidae). *Journal of Research on the Lepidoptera*, **28**, 88–96.

Mathur, R.N. (1942). On the biology of the parasites of the shisham defoliators in the Punjab plantations. *Indian Forest Records (Entomology) (New Series)*, **7**, 11–65.

Mathur, R.N. (1943). Bamboo defoliators. *Indian Journal of Entomology*, **5**, 117–28.

Mathur, R.N. (1960). Pests of teak and their control. *Indian Forest Records (New Series), Entomology*, **10**, 43–66.

Mathur, R.N. (1979). Biology of *Tingis (Caenotingis) beesoni* Drake (Heteroptera: Tingitidae). *Indian Forest Bulletin*, **276**, 1–6.

Mathur, R.N. and Singh, B. (1960). A list of insect pests of forest plants in India and the adjacent countries, Part 5: List of insect pests of plant genera 'D' to 'F' (*Dactyloclenium* to *Funtumia*). *Indian Forest Bulletin*, **171**(4), 1–165.

Mathur, R.N. and Singh, B. (1960–1). A list of insect pests of forest plants in India and the adjacent countries, Parts 4–10. Indian Forest Bulletin (New Series), **171**(3–9), 1–45, 1–165, 1–91, 1–148, 1–130, 1–88, 1–116.

Mathur, R.N. and Singh, B. (1961). A list of insect pests of forest plants in India and the adjacent countries, Part 10. List of insect pests of plant genera 'T' to 'Z' (*Tabernaemontana* to *Zizyphus*). *Indian Forest Bulletin (New Series)*, **171**(9), 1–116.

Mathur, R.N., Chatterjee, P.N. and Sen-Sarma, P.K. (1970). Biology, ecology and control of *Ailanthus* defoliator *Atteva fabriciella* Swed. (Lepidoptera: Yponomeutidae) in Madhya Pradesh. *Indian Forester*, **96**, 538–52.

Matsumoto, K. (1994). *Studies on the Ecological Characteristics and Methods of Control of Insect Pests of Trees in Reforested Areas in Indonesia*. Final Report submitted to Agency for Forestry Research and Development. Jakarta: AFRD, Ministry of Forestry.

Matsumoto, K. and Irianto, R.S.B. (1998). Adult biology of the albizia borer, *Xystrocera festiva* Thomson (Coleoptera: Cerambycidae), based on laboratory breeding, with particular reference to its oviposition schedule. *Journal of Tropical Forest Science*, **10**, 367–78.

Matsumoto, K. and Kotulai, J.R. (2000). Insect pests of mahoganies with particular attention to shoot borers. In *Proceedings of the Seminar on High Value Timber Species for Plantation Establishment: Teak and Mahoganies, JIRCAS Working Report No. 16*. Sabah, Malaysia: Japan International Cooperation Agency, pp. 67–79.

Matsumoto, K. and Kotulai, J.R. (2002). Field tests on the effectiveness of *Azadirachta* companion planting as a shoot borer repellent to protect mahogany. *JIRCAS Journal*, **10**, 1–8.

Matsumoto, K., Mulyadi, K. and Irianto, R.S.B. (1997). A promising method to protect mahogany plantations from attack by the shoot borer, *Hypsipyla robusta* Moore (Lepidoptera: Pyralidae). *JIRCAS Journal*, **5**, 23–9.

Mattson, W.J. and Addy, N.D. (1975). Phytophagous insects as regulators of forest primary production. *Science*, **190**, 515–22.

Mauldin, J., Jones, S. and Beal, R. (1987). Viewing termiticides. *Pest Control*, **55**, 46–59.

May, R.M. (1994). Past efforts and future prospects towards understanding how many species there are. In *Biodiversity and Global Change*, ed. O.T. Solbrig, H.M. van Emden and P.G.W.J. van Oordt. Wallingford, Oxon: CAB International, pp. 71–84.

Mayhew, J.E. and Newton, A.C. (1998). *The Silviculture of Mahogany*. Wallingford, Oxon: CAB International.

McClay, A.S. (1990). Distribution of leucaena psyllid and its natural enemies in Mexico: implications for biological control. In *Leucaena Psyllid: Problems and Management*, ed. B. Napompeth and K.G. MacDicken. Bangkok: Winrock International/IDRC/NFTA/F-FRED, pp. 139–43.

McGaughery, W.H. (1994). Problems of insect resistance to *Bacillus thuringiensis*. *Agriculture, Ecosystems and Environment*, **49**, 95–102.

Menendez, J.M., Echevarria, E., Garcia, A. *et al.* (1986). Use of *Bacillus thuringiensis* in the control of *Rhyacionia frustrana*. *Revista Forestal Baracoa*, **16**, 101–11.

Mesa, A. de (1939). Molave (*Vitex parviflora* Juss.) and its carpenter moth enemy (*Xyleutes ceramicus* Walk.). *Philippine Journal of Forestry*, **2**, 45–53.

Meshram, P.B. and Garg, V.V. (2000). A new report of *Parasa lepida* Cramer (Lepidoptera: Limacodidae) and *Trypanophora semihyalina* Kollar (Lepidoptera: Zygaenidae) as pests of *Gmelina arborea*. *Indian Forester*, **126**, 690–1.

Meshram, P.B. and Tiwari, C.K. (2003). Lace bug *Tingis beesoni* (Drake) causing top dying of *Gmelina arborea* (Linn.) in Hi Tech plantations and evaluation of certain pesticides against the bug. *Journal of Applied Zoological Researches*, **14**, 200–3.

Meshram, P.B., Pathak, S.C. and Jamaluddin, N.S. (1990). Population dynamics and seasonal abundance of some forest insect pests (nursery stage) through light trap. *Indian Forester*, **116**, 494–503.

Meshram, P.B., Husen, M. and Joshi, K.C. (1993). A new report of ambrosia beetle, *Xylosandrus compactus* Eichoff. (Coleoptera: Scolytidae) as a pest of African mahogany, *Khaya* sp. *Indian Forester*, **119**, 75–7.

Meshram, P.B., Joshi, K.C. and Sarkar, A.K. (1994). Relative resistance of certain clones of *Tectona grandis* to teak leaf skeletonizer, *Eutectona machaeralis* Walk. (Lepidoptera: Pyralidae). *Indian Forester*, **120**, 58–61.

Meshram, P.B., Pande, P.K. and Banerjee, S.K. (2001). Impact of pest problems in *Gmelina arborea* Linn. plantations in Western Maharashtra. *Indian Forester*, **127**, 1377–86.

Messer, A.D., Wanta, N.N. and Sunjaya. (1992). Biological and ecological studies of *Calliteara cerigoides* (Lepidoptera: Lymantriidae), a polyphagous defoliator of Southeast Asian Dipterocarpaceae. *Japanese Journal of Entomology*, **60**, 191–202.

Miller, J.R. and Strickler, K.L. (1984). Finding and accepting host plants. In *Chemical Ecology of Insects*, ed. V.J. Bell and R.T. Carde. London: Chapman & Hall Ltd., pp. 127–57.

Milner, R.J. (2000). Current status of *Metarhizium* as a mycoinsecticide in Australia. *Biocontrol News and Information*, **21**, 47N–57N.

Milner, R.J., Staples, J.A. and Lutton, G.G. (1998). The selection of isolates of the hyphomycete fungus, *Metarhizium anisopliae*, for control of termites in Australia. *Biological Control*, **11**, 240–7.

Mishra, S.C. (1992). Comparative natural resistance of different clones of *Tectona grandis* L.f. to teak skeletoniser *Eutectona machaeralis* Wlk. (Lepidoptera: Pyralidae). *Indian Forester*, **118**, 274–8.

Mishra, S.C., Veer, V. and Chandra, A. (1985). *Aristobia horridula* Hope (Coleoptera: Lamiidae) a new pest of shisham (*Dalbergia sissoo* Roxb.) in West Bengal. *Indian Forester*, **111**, 738–41.

Misra, R.M. (1993). *Beauveria bassiana* (Balsamo) Vuillemin, a fungal pathogen of *Hypsipyla robusta* Moore (Lepidoptera: Pyralidae). *Indian Journal of Forestry*, **16**, 236–8.

Misra, R.M. and Singh, P. (1993). Bioassay of Thuricide: a microbial insecticide against important forest pests, II. *Indian Forester*, **199**, 33–5.

Misra, R.M., Mishra, R.K. and Bhandari, R.S. (2001). Biological agents of *Heteropsylla cubana* on *Leucaena leucocephala* in Kalsi, Dehra Dun, Uttaranchal. *Indian Journal of Forestry*, **24**, 403–8.

Mitchell, M.R. (1989). Comparison of non-persistent insecticides in controlled release granules with a persistent organochlorine insecticide for the control of termites in young *Eucalyptus* plantations in Zimbabwe. *Commonwealth Forestry Review*, **68**, 281–94.

Mohanadas, K. (2000). *Management of the Shoot Borer* Hypsipyla robusta *(Lepidoptera: Phycitidae) in Mahogany Plantations, KFRI Research Report No. 184*. Peechi, India: Kerala Forest Research Institute.

Mohanadas, K. and Varma, R.V. (1984). A new host record for *Atteva fabriciella*, a pest of *Ailanthus*. *Journal of Tree Sciences*, **3**, 128.

Montagnini, F., Gonzalez, E., Porras, C. and Rheingans, R. (1995). Mixed and pure forest plantations in the humid neotropics: a comparison of early growth, pest damage and establishment costs. *Commonwealth Forestry Review*, **74**, 306–14.

Morgan, F.D. and Suratmo, F.G. (1976). Host preferences of *Hypsipyla robusta* (Moore) (Lepidoptera: Pyralidae) in West Java. *Australian Forestry*, **39**, 103–12.

Murphy S.T. (1996). Status and impact of invasive conifer aphid pests in Africa. In *Impact of Diseases and Insect Pests in Tropical Forests*, ed. K.S.S. Nair, J.K. Sharma and R.V. Varma. Peechi, India: Kerala Forest Research Institute, pp. 289–7.

Murphy, S.T., Abraham, Y.J. and Cross, A.E. (1991). Ecology and economic importance of the aphid pests, *Pineus* sp. and *Eulachnus rileyi* on exotic pine plantations in southern and eastern Africa. In *Exotic Aphid Pests of Conifers: a Crisis in African Forestry*, ed. FAO. Rome: FAO, pp. 48–53.

Nair, K.S.S. (1982). *Seasonal Incidence, Host Range and Control of the Teak Sapling Borer,* Sahyadrassus malabaricus, *KFRI Research Report No. 16*. Peechi, India: Kerala Forest Research Institute.

Nair, K. S. S. (1986a). The problem of insect defoliation of teak: to spray or not to spray. In *Proceedings of 2nd Forestry Conference, vol. 2,* ed. N.K. Mathur. Dehra Dun, India: Forestry Research Institude and Colleges, pp. 876–9.

Nair, K. S. S. (1986b). Important insect pest problems of forest plantations in tropical India. In *Proceedings of the 18th IUFRO World Congress, vol. 1.* Ljubljana, Yugoslavia: International Union of Forestry Research Organizations, pp. 134–45.

Nair, K. S. S. (1987a). Migration, a mechanism of parasite evasion. In *Advances in Biological Control Research in India*, ed. K.J. Joseph and U.C. Abdurahiman. Calicut, India: University of Calicut, pp. 84–6.

Nair, K. S. S. (1987b). Life history, ecology and pest status of the sapling borer, *Sahyadrassus malabaricus* (Lepidoptera: Hepialidae). *Entomon*, **12**, 167–73.

Nair, K. S. S. (1987c). Control of the sapling borer, *Sahyadrassus malabaricus* (Lepidoptera: Hepialidae) in forest plantations. *Entomon*, **12**, 137–9.

Nair, K. S. S. (1988). The teak defoliator in Kerala, India. In *Dynamics of Forest Insect Populations*, ed. A. A. Berryman. New York: Plenum Press, pp. 267–89.

Nair, K. S. S. (1991). Social, economic and policy aspects of integrated pest management of forest defoliators in India. *Forest Ecology and Management*, **39**, 283–8.

Nair, K. S. S. (1994). Forestry: pesticide application practices. In *The Pesticides Industry*, ed. B.V. David. Madras, India: HC Kothari Group Publications Division, pp. 86–101.

Nair, K. S. S. (1998). KFRI's tryst with the teak defoliator. *Evergreen* (KFRI, Peechi), **40**, 1–7.

Nair, K. S. S. (ed.) (2000). *Insect Pests and Diseases in Indonesian Forests: An Assessment of the Major Threats, Research Efforts and Literature*. Bogor, Indonesia: Center for International Forestry Research.

Nair, K. S. S. (2001a). *Pest Outbreaks in Tropical Forest Plantations: Is There a Greater Risk for Exotic Tree Species*? Bogor, Indonesia: Center for International Forestry Research.

Nair, K. S. S. (2001b). Management of the teak defoliator, *Hyblaea puera*: current status. In *Protection of World Forests from Insect Pests: Advances in Research, IUFRO World Series Vol. 11*, ed. R. Alfaro, K. Day, S. Salom *et al.* Malaysia: International Union of Forestry Research Organizations, pp. 231–8.

Nair, K. S. S. and Mathew, G. (1988). *Biology and Control of Insect Pests of Fast-Growing Hardwood Species, KFRI Research Report 51*. Peechi, India: Kerala Forest Research Institute.

Nair, K. S. S. and Mathew, G. (1992). Biology, infestation characteristics and impact of the bagworm, *Pteroma plagiophelps* Hamps. in forest plantations of *Paraserianthes falcataria*. *Entomon*, **17**, 1–13.

Nair, K. S. S. and Mohanadas, K. (1996). Early events in the outbreak of the teak caterpillar, *Hyblaea puera*. *International Journal of Ecology and Environmental Sciences*, **22**, 271–9.

Nair, K. S. S. and Sudheendrakumar, V.V. (1986). The teak defoliator, *Hyblaea puera*: defoliation dynamics and evidences for short-range migration of moths. *Proceedings of the Indian Academy of Sciences (Animal Sciences)*, **95**, 7–21.

Nair, K. S. S. and Sudheendrakumar, V. V. (1992). Economics of pest control in forest plantations. In *Socio-Economic Research in Forestry*, ed. S. Chand Basha, P. K. Muraleedharan, K. K. Seethalakshmi, K. V. Sankaran and K. K. N. Nair. Peechi, India: Kerala Forest Research Institute and the Ford Foundation, pp. 94–100.

Nair, K. S. S. and Sumardi. (2000). Insect pests and diseases of major plantation species. In *Insect Pests and Diseases in Indonesian Forests: An Assessment of the Major Threats, Research Efforts and Literature*, ed. K. S. S. Nair. Bogor, Indonesia: Center for International Forestry Research, pp. 15–38.

Nair, K. S. S. and Varma, R. V. (1981). *Termite Control in Eucalypt Plantations, KFRI Research Report No. 6*. Peechi, India: Kerala Forest Research Institute.

Nair, K. S. S. and Varma, R. V. (1985). Some ecological aspects of the termite problem in young eucalypt plantations in Kerala, India. *Forest Ecology and Management*, **12**, 287–303.

Nair, K. S. S., Sudheendrakumar, V. V., Varma, R. V. and Chacko, K. C. (1985). *Studies on the Seasonal Incidence of Defoliators and the Effect of Defoliation on Volume Increment of Teak, KFRI Research Report No. 30*. Peechi, India: Kerala Forest Research Institute.

Nair, K. S. S., Mathew, G., Mohanadas, K. and Menon, A. R. R. (1986a). *A Study of Insect Pest Incidence in Natural Forests, KFRI Research Report No. 44*. Peechi, India: Kerala Forest Research Institute.

Nair, K. S. S., Mathew, G., Varma, R. V. and Sudheendrakumar, V. V. (1986b). Insect pests of eucalypts in India. In *Eucalyptus in India: Past, Present and Future*, ed. J. K. Sharma, C. T. S. Nair, S. Kedharnath and S. Kondas. Peechi, India: Kerala Forest Research Institute, pp. 325–35.

Nair, K. S. S., Mohanadas, K. and Sudheendrakumar, V. V. (1995). Biological control of the teak defoliator, *Hyblaea puera* Cramer (Lepidoptera, Hyblaeidae) using insect parasitoids: problems and prospects. In *Biological Control of Social Forest and Plantation Crops Insects*, ed. T. N. Ananthakrishnan. New Delhi: Oxford and IBH Publishing Co. Pvt. Ltd., pp. 77–95.

Nair, K. S. S., Sudheendrakumar, V. V., Varma, R. V., Chacko, K. C. and Jayaraman, K. (1996a). Effect of defoliation by *Hyblaea puera* and *Eutectona machaeralis* (Lepidoptera) on volume increment of teak. In *Impact of Diseases and Insect Pests in Tropical Forests*, ed. K. S. S. Nair, J. K. Sharma and R. V. Varma. Peechi, India: Kerala Forest Research Institute, pp. 257–73.

Nair, K. S. S., Babjan, B., Sajeev, T. V. *et al.* (1996b). Field efficacy of nuclear polyhedrosis virus for protection of teak against the defoliator, *Hyblaea puera* Cramer (Lepidoptera: Hyblaeidae). *Journal of Biological Control*, **10**, 79–85.

Nair, K. S. S., Varma, R. V., Mathew, G., Sudheendrakumar, V. V. and Mohanadas, K. (1996c). Perceived impact of insect pest problems in forest plantations in Kerala. In *Impact of Diseases and Insect Pests in Tropical Forests*, ed. K. S. S. Nair, J. K. Sharma and R. V. Varma. Peechi, India: Kerala Forest Research Institute, pp. 317–23.

Nair, K. S. S., Sudheendrakumar, V. V., Mohanadas, K. *et al.* (1997). Search for teak trees resistant to the defoliator, *Hyblaea puera* Cramer (Lepidoptera, Hyblaeidae). In *Ecology and Evolution of Plant-feeding Insects in Natural and Man-made*

Environments, ed. A. Raman. New Delhi: International Scientific Publications, pp. 109–22.

Nair, K. S. S., Sudheendrakumar, V. V., Varma, R. V. and Mohanadas, K. (1998a). *Tracing the Epicentres of Teak Defoliator Outbreaks in Kerala, KFRI Research Report No. 147*. Peechi, India: Kerala Forest Research Institute.

Nair, K. S. S., Varma, R. V., Sudheendrakumar, V. V., Mohanadas, K. and Ali, M. I. M. (1998b). *Management of the Teak Defoliator (*Hyblaea puera*) Using Nuclear Polyherosis Virus (NPV), KFRI Research Report No. 151*. Peechi, India: Kerala Forest Research Institute.

Nair, K. S. S., Kallarackal, J. and Chacko, K. C. (1999). Environmental impact of plantations. In *Plantation Crops of Kerala: Challenges and Options*. Trivandrum, India: State Committee on Science, Technology and Environment, pp. 51–69.

Nair, M. R. G. K. (1989). *A Monograph on Crop Pests of Kerala and their Control*, 2nd edn. Thrissur, India: Kerala Agricultural University.

Nakamuta, K., Matsumoto, K. and Kotulai, J. R. (2002a). Field trapping of the mahogany shoot borer, *Hypsipyla robusta* (Lepidoptera: Pyralidae), with synthetic sex pheromones in Sabah, Malaysia. *JIRCAS Journal*, **10**, 9–12.

Nakamuta, K., Nakashima, T., Ikeda, T., Eungwijarnpanya, S., Yincharoen, S. and Hutacharern, C. (2002b). Sex pheromone of the teak beehole borer, *Xyleutes ceramicus*. In *Pest Management in Tropical Forest Plantations*, ed. C. Hutacharern, B. Napompeth, G. Allard and F. R. Wylie. Bangkok: FAO Forestry Research Support Programme for Asia and the Pacific, pp. 93–8.

Nambiar, E. K. S., Gintings, A. N., Ruhiat, D., Natadiwirya, M., Harwood, C. E. and Booth, T. H. (eds.) (1998). *Sustained Productivity of Short and Medium Rotation Plantation Forests for Commercial and Community Benefit in Indonesia: an Analysis of Research Priorities*. Canberra: CSIRO Forestry and Forest Products, pp. 13–16.

Napompeth, B. (1990a). Leucaena psyllid problems in Asia and the Pacific. In *Leucaena Psyllid: Problems and Management*, ed. B. Napompeth and K. G. MacDicken. Bangkok: Winrock International/IDRC/NFTA/F-FRED, pp. 1–11.

Napompeth, B. (1990b). Utilization of natural enemies for biological control of the leucaena psyllid in Thailand. In *Leucaena Psyllid: Problems and Management*, ed. B. Napompeth and K. G. MacDicken. Bangkok: Winrock International/IDRC/NFTA/F-FRED, pp. 175–80.

Napompeth, B. (1994). *Leucaena Psyllid in the Asia-Pacific Region: Implications for its Management in Africa*. Bangkok: FAO Regional Office for Asia and the Pacific.

Napompeth, B. and MacDicken, K. G. (eds.) (1990). *Leucaena Psyllid: Problems and Management*. Bangkok: Winrock International/IDRC/NFTA/F-FRED.

Napompeth, B. and Maneeratana, T. (1990). Biological and partial ecological life tables of *Heteropsylla cubana* Crawford and its predator, *Curinus coerulus* Mulsant, in Thailand. In *Leucaena Psyllid: Problems and Management*, ed. B. Napompeth and K. G. MacDicken. Bangkok: Winrock International/IDRC/NFTA/F-FRED, pp. 130–8.

Narendran, T. C. (2001). Taxonomic entomology: research and education in India. *Current Science*, **81**, 445–7.

Narendran, T.C. and Kovac, D. (1995). Insecta (Hymenoptera: Eurytomidae) associated with bamboo in West Malaysia. *The Raffles Bulletin of Zoology*, **43**, 329–36.

Nascimento, M.T. and Proctor, J. (1994). Insect defoliation of a monodominant Amazonian rainforest. *Journal of Tropical Ecology*, **10**, 633–6.

Natawiria, D. (1990). Insect pests in plantation forests in Indonesia. In *Pests and Diseases of Forest Plantations*, ed. C. Hutacharern, K.G. MacDicken, M.H. Ivory and K.S.S. Nair. Bangkok: FAO Regional Office for Asia and the Pacific, pp. 56–61.

Nautiyal, J.C. (1988). *Forest Economics: Principles and Applications*. Dehra Dun, India: Natraj Publishers.

Newton, A.C., Baker, P., Ramnarine, S., Mesen, J.F. and Leakey, R.R.B. (1993). The mahogany shoot borer: prospects for control. *Forest Ecology and Management*, **57**, 301–28.

Ngatiman and Tangketasik, J. (1987). Some insect pests on trial plantation of PT ITCI, Balikapan, East Kalimantan, Indonesia. *Tropical Forest Research Journal Samarinda*, **2**, 41–53.

Nkunika, P.O.Y. (1980). A survey of the termite species associated with *Eucalyptus* plantations in Zambia. *Zambia Journal of Science and Technology*, **5**, 33–8.

Nogueira, S.B. and De-Souza, A.J. (1987). Heartwood termite, *Coptotermes testaceus* (Isoptera: Rhinotermitidae), a serious pest of eucalypts in cerrado areas. *Brasil Florestal*, **15**, 27–9.

Notoatmodjo, S.S. (1963). Cara-cara mencegah serengan masal dan boktor *Xystrocera festiva* Pasco pada tegakan *Albizia falcataria* [Prevention of massive attacks of *Xystrocera festiva* on *Albizia falcataria*]. *Aporan LPH*, **92**, 1–12.

Nottingale, P.J. and Schultz, T.C. (1987). What is a forest pest? In *Insect Outbreaks*, ed. P. Barbosa and J.C. Schultz. London: Academic Press, pp. 59–80.

Odiyo, P.O. (1979). Forecasting infestations of a migrant pest: the African armyworm *Spodoptera exempta* (Walk.). *Philosophical Transactions of the Royal Society of London*, B 287, 403–13.

Ogol, C.K.P.O. and Spence, J.R. (1997). Abundance, population dynamics and impact of the leucaena psyllid *Heteropsylla cubana* Crawford in a maize-leucaena agroforestry system in Kenya. *Insect Science and its Application*, **17**, 183–92.

Oka, A.G. and Vaishampayan, S.M. (1981). Whitegrub menace in teak nurseries in Maharashtra. In *Progress in Soil Biology and Ecology of India*, ed. G.K. Veeresh. Bangalore, India: University of Agricultural Sciences, pp. 161–4.

Oka, I.N. (1990). Progress and future activities of the leucaena psyllid research programmes in Indonesia. In *Leucaena Psyllid: Problems and Management*, ed. B. Napompeth and K.G. MacDicken. Bangkok: Winrock International/IDRC/NFTA/F-FRED, pp. 25–7.

Pandey, D. (1995). *Forest Resources Assessment 1990: Tropical Forest Plantation Resources, FAO Forestry paper 128*. Rome: FAO.

Pandey, D. (1997). *Hardwood Plantations in the Tropics and Subtropics: Tropical Forest Plantation Areas 1995, Project GCP/INT/628/UK*. Rome: FAO.

Parry, M. S. (1959). Control of termites in *Eucalyptus* plantations. *Empire Forest Review*, **38**, 288–92.

Patel, M. and Sahu, A. K. (1995). Control of termite attack on some high pulp yielding trees. *Indian Forester*, **121**, 295–9.

Patil, B. V. and Thontadarya, T. S. (1981). Record of *Beauveria bassiana* (Balsamo) Vuillemin on teak skeletonizer, *Pyrausta machaeralis* (Walker). *Indian Forester*, **107**, 698–9.

Patil, B. V. and Thontadarya, T. S. (1983a). Seasonal incidence of teak skeletonizer *Pyrausta machaeralis* Walker (Lepidoptera: Pyralidae) in Prabhunagar Forest. *Indian Journal of Ecology*, **10**, 204–9.

Patil, B. V. and Thontadarya, T. S. (1983b). Natural enemy complex of the teak skeletonizer, *Pyrausta machaeralis* Walker (Lepidoptera: Pyralidae) in Karnataka. *Entomon*, **8**, 249–55.

Patil, B. V. and Thontadarya, T. S. (1983c). Studies on the acceptance and biology of different *Trichogramma* spp. on the teak skeletonizer, *Pyrausta machaeralis* Walker. *Indian Forester*, **109**, 292–7.

Patil, B. V. and Thontadarya, T. S. (1984). Efficacy of egg parasite, *Trichogramma* spp. in parasitising the eggs of the teak skeletonizer, *Pyrausta machaeralis* Walker (Lepidoptera: Pyralidae). *Indian Forester*, **110**, 413–18.

Patil, B. V. and Thontadarya, T. S. (1986). Diapause in teak skeletonizer, *Pyrausta machaeralis* and its parasite, *Cremnops atricornis*. *Mysore Journal of Agricultural Sciences*, **20**, 38–42.

Patil, B. V. and Thontadarya, T. S. (1987a). Biology of the teak skeletonizer, *Pyrausta machaeralis* Walker (Lepidoptera: Pyralidae). *Mysore Journal of Agricultural Sciences*, **21**, 32–9.

Patil, B. V. and Thontadarya, T. S. (1987b). Studies on the induction and termination of hibernation of the teak skeletonizer, *Eutectona machaeralis* Walker (Lepidoptera: Pyralidae). *Entomon*, **12**, 179–82.

Pedgley, D. E. (1990). Concentration of flying insects by the wind. *Philosophical Transactions of the Royal Society of London*, 328B, 631–53.

Pedrosa-Macedo, J. H. (ed.) (1985). *Noxious Insects to Pine and Eucalypt Plantations in the Tropics*. Brazil: Federal University of Parana and International Union of Forestry Research Organizations.

Perum Perhutani. (1995). *A Glance at Perum Perhutani (Forest State Corporation), Indonesia*. Jakarta: Perum Perhutani.

Pillai, S. R. M. and Gopi, K. C. (1990a). The bagworm *Pteroma plagiophleps* Hamp. (Lepidoptera, Psychidae) attack on *Acacia nilotica* (Linn.) Wild. ex del. *Indian Forester*, **116**, 581–3.

Pillai, S. R. M. and Gopi, K. C. (1990b). Further records of insect pests on *Acacia nilotica* (Linn.) Will. Ex Del. in nurseries and young plantations and the need for control measures. *Indian Journal of Forestry*, **13**, 8–13.

Pillai, S. R. M. and Gopi, K. C. (1990c). Epidemic outbreak of mealy bug *Ferrisiana virgata* (Cockerell) (Pseudococcidae: Homoptera) in su-babul (*Leucaena leucocephala* (Lam.) de Wit) plantations. *Indian Forester*, **116**, 822–4.

Pillai, S.R.M. and Gopi, K.C. (1991). Insects associated with *Albizia falcataria* (L.) Fosberg in South India. *Myforest*, **27**, 91–106.

Pillai, S.R.M. and Thakur, M.L. (1990). On some aspects of biology of *Ithome lassula* Hodges (Cosmopterygidae): a microlepidopteran pest introduced to India with the host *Leucaena leucocephala* (Lam.) de Wit. *Indian Forester*, **116**, 643–7.

Piyakarnchana, T. (1981). Severe defoliation of *Avicennia alba* BL. by larvae of *Cleora injectaria* Walker. *Journal of the Science Society of Thailand*, **7**, 33–6.

Pogue, M.G. and Aiello, A. (1999). Description of the immature stages of three species of *Eulepidotis* Guenée (Lepidoptera: Noctuidae) with notes on their natural history. *Proceedings of the Entomological Society of Washington*, 101, 300–11.

Price, P.W. (1997). *Insect Ecology*, 3rd edn. New York: John Wiley & Sons, Inc.

Quiniones, S.S. and Zamora, R.A. (1987). Forest pests and diseases in the Philippines. In *Forest Pests and Diseases in Southeast Asia, BIOTROP Special Publication No. 26*, ed. E.D. de Guzman and S.T. Nuhamara. Bogor, Indonesia: Southeast Asian Regional Center for Tropical Biology, pp. 43–65.

Rabindra, R.J., Varma, R.V., Ali, M.I.M. and Sudheendrakumar, V.V. (1997). Tests for cross-infectivity of the nuclear polyhedrosis virus of the teak defoliator, *Hyblaea puera* to some lepidopterous insects. *Pest Management in Horticultural Ecosystems*, **3**, 109–11.

Raetano, C.G., Wilcken, C.F. and Crocomo, W.B. (1997). Termite control in eucalypt forests with fipronil (Regent 20G) applied to the soil surface. *Revista Arvore*, **21**, 289–93.

Rahayu, S., Subyanto and Kuswanto. (1998). The occurrence of pest and disease of *Shorea* spp.: A preliminary study in Wanariset and Bukit Soeharto Forest Area in East Kalimantan, Indonesia. In *Proceedings of the Seminar on Ecological Approach for Productivity and Sustainability of Dipterocarp Forests*, ed. H.M.S. Sabarnudin, S. Suhardi and Y. Okimori. Yogyakarta: Gajah Mada University and Kansai Environmental Engineering Center, pp. 53–7.

Rajagopal, D. (1982). Relative incidence of termites on exotic species of *Eucalyptus* in Karnataka. *Myforest*, **18**, 9–13.

Rajak, R.C., Agarwal, G.P., Khan, A.R. and Sandhu, S.S. (1993). Susceptibility of teak defoliator (*Hyblaea puera* Cramer) and teak skeletonizer (*Eutectona machaeralis* Walker) to *Beauveria bassiana* (Bals.) Vuill. *Indian Journal of Experimental Biology*, **31**, 80–2.

Ralph, R. (1985). Occurrence of *Celosterna scabrator* var. *spinator* on *Eucalyptus*. *Journal of the Bombay Natural History Society*, **82**, 682–3.

Ralph, R. (1990). Infestation by the borer *Celosterna scabrator* variety *scabrator* on *Acacia nilotica* in north Karnataka. *Commonwealth Forestry Review*, **61**, 91–4.

Rauf, A., Hidayat, P., Maryana, N. and Winasa, I.W. (1990). Biology and demography of *Heteropsylla cubana* Crawford (Homoptera: Psyllidae). In *Leucaena Psyllid: Problems and Management*, ed. B. Napompeth and K.G. MacDicken. Bangkok: Winrock International/IDRC/NFTA/F-FRED, pp. 114–18.

Rees, C.J.C. (1983). Microclimate and the flying Hemiptera fauna of a primary lowland rain forest in Sulawesi. In *Tropical Rain Forest: Ecology and Management*, ed. S.L. Sutton, T.S. Whitmore and A.C. Chadwick. Oxford: Blackwell Scientific Publications, pp. 121–36.

Remadevi, O.K. and Muthukrishnan, R. (1998). Farmer trials on the control of the defoliators, *Eutectona machaeralis* Walker and *Hyblaea puera* Cramer on teak saplings. In *Advances in IPM for Horticultural Crops*, ed. P.P. Reddy, T.I. Kumar and A Verghese. Bangalore, India: Indian Institute of Horticultural Research, pp. 179–82.

Resende, V.F., Zanuncio, J.C., Guedes, R.N.C. and Nogueira, P.B. (1995). Comparative effects of carbosulfan and aldrin in the protection of eucalyptus seedlings against soil termites. *Anais da Sociedade Entomologica do Brasil*, **24**, 645–8.

Richards, O.W. and Davies, R.G. (1977). Imms' General Textbook of Entomology. 10th edn., vols 1 and 2. London: Chapman & Hall.

Riley, J.R., Reynolds, D.R. and Farmery, M.J. (1983). Observations on the flight behaviour of the armyworm moth, *Spodoptera exempta*, at an emergence site using radar and infra-red optical techniques. *Ecological Entomology*, **8**, 395–418.

Risch, S.J. (1981). Insect herbivore abundance in tropical monocultures and polycultures: an experimental test of two hypotheses. *Ecology*, **62**, 1325–40.

Risch, S.J., Andow, D. and Altieri, M.A. (1983). Agroecosystem diversity and pest control: data, tentative conclusions, and new research directions. *Environmental Entomology*, **12**, 625–9.

Roberts, H. (1978). When ambrosia beetles attack mahogany trees in Fiji. *Unasylva*, **29**, 25–8.

Roonwal, M.L. (1978). The biology, ecology and control of the sal heartwood borer, *Hoplocerambyx spinicornis*: a review of recent work. *Indian Journal of Entomology*, **1**, 107–20.

Roonwal, M.L. (1979). *Termite Life and Termite Control in Tropical South Asia*. Jodhpur, India: Scientific Publishers.

Roonwal, M.L. (1990). Field-ecological observations on the *Ailanthus excelsa* defoliator, *Eligma narcissus indica* (Lepidoptera: Noctuidae), in Peninsular India. *Indian Journal of Forestry*, **13**, 81–4.

Roonwal, M.L. and Rathore, N.S. (1984). New termite pests of eucalyptus in India and their control. *Zeitschrift fur Angewandte Entomologie*, **98**, 225–30.

Root, R.B. (1973). Organisation of a plant-arthropod association in simple and diverse habitats: the fauna of collards (*Brassica oleracea*). *Ecological Monographs*, **43**, 94–125.

Rose, D.J.W., Page, W.W., Dewhurst, C.F. *et al.* (1985). Downwind migration of the African army worm moth, *Spodoptera exempta*, studied by mark-and-capture and by radar. *Ecological Entomology*, **10**, 299–313.

Ross, H. (1962). *How to Collect and Preserve Insects. Circular 39*. Illinois: Illinois Natual History Survey, p. 30.

Roychoudhury, N. and Joshi, K.C. (1996). Search for natural resistance in teak clones against *Eutectona machaeralis* Walker (Lepidoptera: Pyralidae). *Indian Journal of Forestry*, **19**, 205–13.

Roychoudhury, N., Joshi, K.C., Sambath, S. and Pandey, D.K. (1994). Effectiveness of three varietal toxins of *Bacillus thuringiensis* against teak leaf skeletonizer, *Eutectona machaeralis* Walker (Pyralidae). *Myforest*, **30**, 57–60.

Roychoudhury, N., Jain, A. and Joshi, K.C. (1995a). Resistance in teak clones against leaf skeletonizer, *Eutectona machaeralis* Walker (Lepidoptera: Pyralidae). In *Advances in Forestry Research in India,* vol. 13, ed. R. Prakash. Dehra Dun, India: International Book Distributors, pp. 140–58.

Roychoudhury, N., Jain, A. and Joshi, K.C. (1995b). Alteration of growth and development in leaf skeletonizer, *Eutectona machaeralis* Walker, due to variations in teak leaves of different maturity. *Indian Journal of Experimental Biology*, **33**, 227–9.

Roychoudhury, N., Joshi, K.C. and Pal, R. (1997a). Larval feeding preference of leaf skeletonizer, *Eutectona machaeralis* Walker, on teak leaves of different maturity. In *Advances in Forestry Research in India, vol. 16*, ed. R. Prakash. Dehra Dun, India: International Book Distributors, pp. 172–9.

Roychoudhury, N., Jain, A., Joshi, K.C. and Lal, R.B. (1997b). Natural resistance in teak clones to leaf skeletonizer, *Eutectona machaeralis* Walker: an appraisal. *The Indian Forester*, **123**, 1027–35.

Sajap, A.S. and Jan, P.T. (1990). Pathogenicity of two entomogenous fungi *Beauveria bassiana* and *Metarhizium anisopliae* in the termite *Coptotermes curvignathus*. In *Pests and Diseases of Forest Plantations*, ed. C. Hutacharern, K.G. MacDicken, M.H. Ivory and K.S.S. Nair. Bangkok: FAO Regional Office for Asia and the Pacific, pp. 266–71.

Sajap, A.S., Yaacob, A.W. and Aidah, M. (1997). Biology of *Spirama retorta* (Lepidoptera: Noctuidae), a new pest of *Acacia mangium* in Peninsular Malaysia. *Journal of Tropical Forest Science*, **10**, 167–75.

Sajeev, T.V. (1999). *Spatial dynamics of teak defoliator* (Hyblaea puera *Cramer*) *outbreaks: patterns and causes*. Unpublished Ph.D. thesis, Cochin University of Science and Technology, Kerala, India.

Sakchoowong, W. (2002). Effects of entomopathogenic fungi (*Beauveria bassiana*, and *Metarhizium anisopliae*) on teak defoliator (*Hyblaea puera* Cramer, Lepidoptera: Hybleidae) in laboratory. In *Pest Management in Tropical Forest Plantations*, ed. C. Hutacharern, B. Napompeth, G. Allard and F.R. Wylie. Bangkok: FAO Forestry Research Support Programme for Asia and the Pacific, pp. 105–10.

Salazar, R. (1984). Preliminary notes on the Nantucket pine tip moth, *Rhyacionia frustrana* (Lepidoptera: Tortricidae) in Costa Rica. *Turrialba*, **34**, 250–2.

Sands, W.A. (1962). The evaluation of insecticides as soil and mound poisons against termites in agriculture and forestry in West Africa. *Bulletin of Entomological Research*, **53**, 179–92.

Sankaran, K.V., Mohanadas, K. and Ali, M.I.M. (1989). *Beauveria bassiana* (Bals.) Vuill., a possible biocontrol agent against *Myllocerus viridanus* Fabr. and *Calopepla leayana* Latreille in south India. *Current Science*, **58**, 467–9.

Sankaran, T. (1970). The oil palm bagworms of Sabah and the possibilities of their biological control. *PANS*, **16**, 43–55.

Santhakumaran, L.N., Remadevi, O.K. and Sivaramakrishnan, V.R. (1995). A new record of the insect defoliator, *Pteroma plagiophleps* Hamp. (Lepidoptera: Psychidae) from mangroves along the Goa coast (India). *Indian Forester*, **121**, 153–5.

Santhosh, K. and Kumar, P. (2003). New incidence of *Alcterogystia cadambae* Moore (Lepidoptera: Cossidae) on *Butea monosperma* Taub. *Myforest*, **39**, 89–91.

Santoso, E. and Hardi, T. (1991). Pest and disease control of timber estates in East Kalimantan. *Jurnal Penelitian dan Pengembangan Kehutanan*, **7**, 14–17.

Sargent, C. and Bass, S. (eds.) (1992). *Plantation Politics: Forest Plantations in Development.* London: Earthscan Publications Ltd.

Sasidharan, N. (1997). *Forest Trees of Kerala: A Checklist Including Exotics, KFRI Handbook No. 2.* Peechi, India: Kerala Forest Research Institute.

Saur, E., Imbert, D., Etienne, J., Mian, D. and Dodd, R.S. (1999). Insect herbivory on mangrove leaves in Guadeloupe: effects on biomass and mineral content. *Hydrobiologia*, **413**, 89–93.

Schabel, H.G. and Madoffe, S. (2001). Status and management of small coleopterous borers in living timbers in tropical Africa. In *Protection of World Forests from Insect Pests: Advances in Research, IUFRO World Series Vol. 11*, ed. R. Alfaro, K. Day, S. Salom *et al.* Malaysia: International Union of Forestry Research Organizations, pp. 223–30.

Schabel, H.G., Hilje, L., Nair, K.S.S. and Varma, R.V. (1999). Economic entomology in tropical forest plantations: an update. *Journal of Tropical Forest Science*, **11**, 303–15.

Schowalter, T.D., Hargrove, W.W. and Crossley D.A. Jr. (1986). Herbivory in forested ecosystems. *Annual Review of Entomology*, **31**, 177–96.

Selander, J. and Bubala, M. (1983). *A survey of pest insects in forest plantations in Zambia, Research Note No. 33.* Zambia: Division of Forest Research, Forest Department.

Selander, J. and Nkunika, P.O.Y. (1981). *Termites as pests of* Eucalyptus *in Zambia: discussion paper 17th IUFRO World Congress.* Kyoto, Japan: International Union of Forestry Research Organizations.

Selander, J., Loyttyniemi, K. and Chendauka, B.S. (1989). *Control of termites in eucalypt plantations: testing protection methods and evaluating carbosulfan as an alternative of organochlorines, Research Note No. 46.* Kitwe: Division of Forest Research, Forest Department.

Senguttuvan, T., Chinniah, C., Varma, R.V. and Nair, K.S.S. (2000). Knock-down toxicity of insecticides and Bt formulations on larvae of teak defoliator, *Hyblaea puera*. *Indian Journal of Forestry*, **23**, 160–3.

Sen-Sarma, P. K. (1996). Impact of termites in the forest ecosystem. In *Impact of Diseases and Insect Pests in Tropical Forests*, ed. K. S. S. Nair, J. K. Sharma and R. V. Varma. Peechi, India: Kerala Forest Research Institute, pp. 343–8.

Sen-Sarma, P. K. and Thakur, M. L. (1983). Insect pests of *Eucalyptus* and their control. *Indian Forester*, **109**, 864–81.

Sharma, G., Dutta, B. K. and Seth, R. (2001). Potential of *Bacillus thuringiensis* subsp. *kurstaki* in protection of *Gmelina arborea* Roxb. *Journal of Entomological Research*, **25**, 283–8.

Sharma, N. and Joshi, K. C. (2004). Entomopathogenic fungi attacking sal heartwood borer, *Hoplocerambyx spinicornis* Newman. *Indian Journal of Forestry*, **27**, 133–40.

Sharma, Y. M. L. (1987). Inventory and resource of bamboos. In *Recent Research on Bamboos*, ed. A. N. Rao, G. Dhanarajan and C. B. Sastry. Hangzhou: Chinese Academy of Forestry and International Development Research Centre, Canada, pp. 4–17.

Shivayogeswara, B., Mallikarjunaiah, H., Krishna Prasad, N. K. and Das, K. G. S. (1988). Incidence of babool borer *Celosterna scabrator* Fabricius (Lamiinae, Cerambycidae, Coleoptera) on *Eucalyptus* in Malanad tracts of Shimoga. *Indian Forester*, **114**, 410–13.

Shujauddin and Khan, H. R. (1999). New record of damage to poplar due to *Leucoptera sphenograpta* Meyrick (Lepidoptera: Lyonetiidae), a leaf miner on poplar in Uttar Pradesh. *Indian Forester*, **125**, 1255–60.

Simon and Schuster. (1999). *Simon & Schuster New Millenium Encyclopedia and Home Reference Library*, Deluxe Edition (Electronic Version). New York: Viacom Inc.

Singh, A. P. and Singh, K. P. (1995). Damage evaluation of *Nodostoma waterhousie* Jacoby (Coleoptera: Chrysomelidae) on different clones, provenances and species of poplar in Himachal Pradesh. *Indian Journal of Forestry*, **18**, 242–4.

Singh, P. (1985). Aerial spraying of chemicals to control teak defoliators. In *Proceedings of 2nd Forestry Conference, vol. 2*, ed. N. K. Mathur. Dehra Dun, India: Forest Research Institute and Colleges, pp. 901–7.

Singh, P. (1990). Current status of pests of Bamboos in India. In *Bamboos: Current Research*, ed. I. V. R. Rao, R. Gananaharan and C. B. Sastry. Peechi, India: Kerala Forest Research Institute and International Development Research Centre, Canada, pp. 190–4.

Singh, P. and Bhandari, R. S. (1988). Insect pests of bamboos and their control. *Indian Forester*, **114**, 670–83.

Singh, P. and Gupta, B. K. (1978). Laboratory evaluations of insecticides as contact sprays against forest pests. I. Teak skeletoniser: *Pyrausta machaeralis* Walker (Lepidoptera: Pyralidae). *Indian Forester*, **104**, 359–66.

Singh, P. and Mishra, R. M. (1978). Bioassay of thuricide, a microbial insecticide against important forest pests. *Indian Forester*, **104**, 838–42.

Singh, P. and Misra, R. M. (1987). New record of *Beauveria tenella* (Delacroix) Siemaszko on teak skeletonizer *Eutectona machaeralis* Walker (Lepidoptera: Pyralidae). *Indian Forester*, **113**, 476–8.

Singh, P. and Thapa, R. S. (1988). Defoliation epidemic of *Ascotis selenaria imparata* Walk. (Lepidoptera: Geometridae) in sal forest of Asaroro Range, West Dehra Dun Division. *Indian Forester*, **114**, 269–74.

Singh, P., Prasad, G. and Johri, A. B. (1978). Aerial spraying of insecticides to control teak defoliators. *Indian Forester*, **104**, 781–2.

Singh, P., Fasih, M. and Prasad, G. (1982). Insect pests of exotic pines in India. *Indian Forester*, **108**, 93–107.

Singh, P., Joshi, K. C. and Gurung, D. (1988). Efficacy of some insecticides against the larvae of khasi pine shoot borer, *Dioryctria castanea* Bradely (Lepidoptera: Pyralidae). *Indian Forester*, **114**, 29–34.

Singh, R., Goraya, G. S., Singh, C., Kumar, H. and Kumar, S. (2001). Mortality of chir pine trees by insect borers in Morni Hills, Haryana: A case study. *Indian Forester*, **127**, 1279–86.

Sivaramakrishnan, V. R. (1986). Notes on recent outbreak of *Celosterna scabrator* Fabr. (Coleoptera: Cerambycidae) on *Eucalyptus* in Karnataka. *Myforest*, **22**, 103–5.

Sivaramakrishnan, V. R. and Remadevi, O. K. (1996). Insect pests in forest nurseries in Karnataka, India, with notes on insecticidal control of a psyllid on *Albizzia lebbek*. In *Impact of Diseases and Insect Pests in Tropical Forests*, ed. K. S. S. Nair, J. K. Sharma and R. V. Varma. Peechi, India: Kerala Forest Research Institute, pp. 460–3.

Skilling, D. D. and Batzer, H. O. (1995). *World Directory of Forest Pathologists and Entomologists*. Minnesota: US Department of Agriculture.

Snook, L. K. and Negreros-Castillo, P. (2004). Regenerating mahogany (*Swietenia macrophylla* King) on clearings in Mexico's Maya forest: the effects of clearing method and cleaning on seedling survival and growth. *Forest Ecology and Management*, **189**, 143–60.

Soepangkat, S. (1998). Protection programme for pests in IPF-trans PT Hutan Kusuma. In *Proceedings Workshop Permasalahan dan Strategi Pengalolaan Hama di Areal Hutan Tanaman*, ed. G. Suratmo, S. Hadi, E. A. Husaeni *et al*. Bogor. Indonesia: Fakultas Kehutanan dan Departemen Kehutanan, pp. 49–54.

Soria, F. J., Villagran, M., Tio, R. del and Ocete, M. E. (1995). Incidence of *Curculio elephas* Gyll. (Coleoptera, Curculionidae) in cork and holm oak forests in the natural park Sierra Norte de Sevilla. *Boletin de Sanidad Vegetal Plagas*, **21**, 195–201.

Southwood, T. R. E. (1977). The relevance of population dynamic theory to pest status. In *Origins of Pest, Parasite, Disease and Weed Problems*, ed. J. M. Cherrett and G. R. Sagar. Oxford: Blackwell Scientific, pp. 35–54.

Speechly, H. (1978). *Insects attacking trees in the Bai Bang area of Vietnam*. Paper presented in Meeting of IUFRO Working Groups S 2.06.12 and 2.07.07 on Pests and Diseases of Tropical Pines. Bogota, Colombia: International Union of Forestry Research Organizations.

Speight, M. R. (1996). The relationship between host tree stresses and insect attack in tropical forest plantations and its relevance to pest management. In *Impact of*

Diseases and Insect Pests in Tropical Forests, ed. K. S. S. Nair, J. K. Sharma and R.V. Varma. Peechi, India: Kerala Forest Research Institute, pp. 363–72.

Speight, M. R. and Speechly, H. T. (1982a). Pine shoot moths in S. E. Asia I. Distribution, biology and impact. *Commonwealth Forestry Review*, **61**, 121–34.

Speight, M. R. and Speechly, H. T. (1982b). Pine shoot moths in S. E. Asia II. Potential control measures. *Commonwealth Forestry Review*, **61**, 203–11.

Speight, M. R. and Wylie, F. R. (2001). *Insect Pests in Tropical Forestry*. Wallingford, UK: CABI International.

Speight, M. R., Hunter, M. D. and Watt, A. D. (1999). *Ecology of Insects: Concepts and Applications*. Oxford: Blackwell Science.

Srivastava, A. K., Kapoor, N. and Negi, B. K. (1984). *Araeothrips longisetis* Ananthakrishnan (Tubulifera: Thysanoptera: Insecta) causing damage to the leaves of *Shorea robusta* Roxb. (Dipterocarpaceae) in Dehra Dun. *Indian Journal of Forestry*, **7**, 155.

Stalev, Z. (1989). Species composition of Geometridae in forests with a predominance of *Fagus orientalis*. *Gorsko Stopanstvo*, **45**, 15–16.

Staples, J. A. and Milner, R. J. (2000). A laboratory evaluation of the repellency of *Metarhizium anisopliae* conidia to *Coptotermes lacteus* (Isoptera: Rhinotermitidae). *Sociobiology*, **36**, 1–16.

Stebbing, E. P. (1899). *Injurious Insects of Indian Forests*. New Delhi: Government of India.

Stebbing, E. P. (1903). *Departmental Notes on Insects That Affect Forestry*. Calcutta: Government of India Central Printing Office.

Stebbing, E. P. (1906). *A note on the life history of* Hoplocerambyx spinicornis *(the Singbhum sal borer), Forest Bulletin No. 8*. Calcutta: Government of India.

Stebbing, E. P. (1914). *Indian Forest Insects of Economic Importance: Coleoptera*. London: Eyre and Spottiswoode Ltd.

Stork, N. E. (1987). Guild structure of arthropods from Bornean rainforest trees. *Ecological Entomology*, **12**, 69–80.

Strauss, S. H., Howe, G. T. and Goldfarb, B. (1991). Prospects for genetic engineering of insect resistance in forest trees. *Forest Ecology and Management*, **43**, 181–209.

Strong, D. R., Lawton, J. M. and Southwood, T. R. E. (1984). *Insects on Plants: Community Patterns and Mechanisms*. Oxford: Blackwell Scientific Publications.

Su, X. and Yu, X. D. (1979). A preliminary study of cotton locust: an insect pest of beef-wood on Dong-Hai Island, Kwangtung (Guangdong) Province (China). *Scientia Silvae Sinicae*, **15**, 171–7.

Subrmonian, S. and Krishnamurthy, S. V. (2002). Outbreak of hairy caterpillar *Euproctis lunata* Walker on acacia trees. *Insect Environment*, **8**, 112.

Subyanto, Sultoni, A. and Musyafa. (1992). Inger-inger (*Neotermes tectonae* Damm.): one of the problems on teak management in Central Java. In *One Century of Sustainable Forest Management with Special Reference to Teak in Java*, ed. H. Simon. Yogyakarta, Indonesia: Perum Perhutani and Gajah Mada University, pp. 209–12.

Sudheendrakumar, V.V. (1986). *Studies on the Natural Enemies of the Teak Pests,* Hyblaea puera *and* Eutectona machaeralis, *KFRI Research Report No. 38*. Peechi, India: Kerala Forest Research Institute.

Sudheendrakumar, V.V. (2003). Reproductive behaviour of *Hyblaea puera* Cramer (Lepidoptera: Hyblaeidae). *Entomon*, **28**, 77–84.

Sudheendrakumar, V.V. and Bharathan, V. (2002). Feasibility of using indigenous parasitoids for biological control of the teak defoliator, *Hyblaea puera* (Lepidoptera: Hyblaeidae). In *Pest Management in Tropical Forest Plantations*, ed. C. Hutacharern, B. Napompeth, G. Allard and F.R. Wylie. Bangkok: FAO Forestry Research Support Programme for Asia and the Pacific, pp. 111–21.

Sudheendrakumar, V.V., Ali, M.I.A. and Varma, R.V. (1988). Nuclear polyhedrosis virus of the teak defoliator, *Hyblaea puera*. *Journal of Invertebrate Pathology*, **51**, 307–8.

Sudheendrakumar, V.V., Nair, K.S.S. and Chacko, K.C. (1993). Phenology and seasonal growth trend of teak at Nilambur (Kerala), India. *Annals of Forestry*, **1**, 42–6.

Sudheendrakumar, V.V., Evans, H.F., Varma, R.V. *et al.* (2001). Management of the teak defoliator, *Hyblaea puera* using baculovirus within a control window concept. In *Tropical Forestry Research: Challenges in the New Millennium*, ed. R.V. Varma, K.V. Bhat, E.M. Muralidharan and J.K. Sharma. Peechi, India: Kerala Forest Research Institute, pp. 106–14.

Suharti, M., Irianto, R.S.B. and Santosa, S. (1994). Behaviour of the stem borer *Xystrocera festiva* Pascoe on *Paraserianthes falcataria* and integrated control. *Buletin Penelitian Hutan*, **558**, 39–53.

Suharti, M., Asmaliyah, H. and Hawiati, W.P. (1995). Neem (*Azadirachta indica*) trees as a natural insecticide resource to control forest pests. *Bulletin Penelitian Hutan*, **589**, 1–26.

Suharti, M., Sitepu, I.R. and Anggraeni, I. (2000). Behaviour, intensity and effect of the attack of stem borer, *Indarbela acutistriata* on sengon in KPH Kediri. *Buletin Penelitian Hutan*, **623**, 37–50.

Supriana, N. and Natawiria, D. (1987). Current potential pine forest pests in Indonesia: a particular case in North Sumatra. In *Forest Pests and Diseases in Southeast Asia, BIOTROP Special Publication No. 26,* ed. E.D. de Guzman and S.T. Nuhamara. Bogor, Indonesia: Southeast Asian Regional Center for Tropical Biology, pp. 139–45.

Su-Qiong, MacLean, D.A. and Needham, T.D. (1996). The influence of hardwood content on balsam fir defoliation by spruce budworm. *Canadian Journal of Forest Research*, **26**, 1620–8.

Suratmo, F.G. (1977). Infestation of the leading shoots of mahogany (*Swietenia macrophylla* King) by *Hypsipyla robusta* (Moore) in West Java, Indonesia. In *Proceedings of the Symposium on Forest Pests and Diseases in Southeast Asia 1976*. Bogor, Indonesia: BIOTROP, pp. 121–32.

Suratmo, F.G. (1987). Current potentially dangerous forest pests in Indonesia. In *Forest Pests and Diseases in Southeast Asia, BIOTROP Special Publication No. 26,* ed. E.D. de Guzman and S.T. Nuhamara. Bogor, Indonesia: Southeast Asian Regional Center for Tropical Biology, pp. 91–6.

Suratmo, F.G. (1996). Emerging insect pest problems in tropical plantation forest in Indonesia. In *Impact of Diseases and Insect Pests in Tropical Forests*, ed. K.S.S. Nair, J.K. Sharma and R.V. Varma. Peechi, India: Kerala Forest Research Institute, pp. 502–6.

Sutton, S.L. (1983). The spatial distribution of flying insects in tropical rain forests. In *Tropical Rain Forest: Ecology and Management*, ed. S.L. Sutton, T.S. Whitmore and A.C. Chadwick. Oxford: Blackwell Scientific Publications, pp. 77–91.

Tabashnik, B.E., Carriere, Y., Dennehe, T.J. *et al.* (2003). Insect resistance to transgenic Bt crops: lessons from the laboratory and field. *Journal of Economic Entomology*, **96**, 1031–8.

Tarrago, M.F.S. and Costa, E.C. (1990). Occurrence of *Thyrinteina arnobia* (Stoll, 1782) (Lepidoptera: Geometridae) on black acacia in Rio Grande do Sul. *Revista do Centro de Ciencias Rurais*, **20**, 219–21.

Teaknet. (1995). Major teak resources of the Asia-Pacific region. *Teaknet Newsletter*, 1, 7.

Thakur, M.L. (2000). *Forest Entomology: Ecology and Management*. Dehra Dun, India: Sai Publishers.

Thakur, M.L., Kumar, S., Negi, A. and Rawat, D.S. (1989). Chemical control of termites in *Eucalyptus* hybrid. *Indian Forester*, **115**, 733–43.

Thapa, R.S. (1970). Bionomics and control of laran defoliator, *Margaronia hilaralis* Walk. (Lepidoptera, Pyralidae). *The Malaysian Forester*, **33**, 55–62.

Thapa, R.S. and Bhandari, R.S. (1976). Biology, ecology and control of Kadam defoliator *Arthroschista hilaralis* Walk. (Lepidoptera: Pyralidae) in plantations in West Bengal. *Indian Forester*, **102**, 388–401.

Tho, Y.P. (1987). Forest pests in Peninsular Malaysia. In *Forest Pests and Diseases in Southeast Asia, BIOTROP Special Publication No. 26*, ed. E.D. de Guzman and S.T. Nuhamara. Bogor, Indonesia: Southeast Asian Regional Center for Tropical Biology, pp. 11–20.

Thompson, R.C. (1897). Insect ravages among teak and *Anogeissus latifolia*. *Indian Forester*, **23**, 325.

Tilakaratna, D. (1991). Parasites of the teak defoliator, *Hyblaea puera*. *Sri Lanka Forester*, **20**, 23–5.

Tin, N.T. (1990). Biology of the sawfly *Shizocera* sp. (Hymenoptera: Argidae), in North Vietnam. In *Pests and Diseases of Forest Plantations*, ed. C. Hutacharern, K.G. MacDicken, M.H. Ivory and K.S.S. Nair. Bangkok: FAO Regional Office for Asia and the Pacific, pp. 194–7.

Tobi, D.R., Grehan, J.R. and Parker, B.L. (1993). Review of the ecological and economic significance of forest Hepialidae (Insecta: Lepidoptera). *Forest Ecology and Management*, **56**, 1–12.

Tsay, J., Lee, M.J. and Chen, R.S. (2001). Evaluation of *Beauveria bassiana* for controlling casuarina tussock moth (*Lymantria xylina* Swinhoe) in casuarina plantations. *Taiwan Journal of Forest Science*, **16**, 201–7.

Turchin, P. (1995). Population regulation: old arguments and a new synthesis. In *Population Dynamics: New Approaches and Synthesis*, ed. N. Cappuccino and P.W. Price. California: Academic Press, pp. 19–40.

Turnbull, J.W., Midgley, S.J. and Cossalter, C. (1998). Tropical acacias planted in Asia: an overview. In *Recent Developments in Acacia Planting*, ed. J.W. Turnbull, H.R. Crompton and K. Pinyopusarerk. Canberra: Australian Center for International Agricultural Research, pp. 14–28.

USDA. (1980). *Guidelines for the Control of Insect and Mite Pests of Foods, Fibers, Feeds, Ornamentals, Livestock, Households, Forest and Forest Products, US Department of Agriculture, Agriculture Handbook No. 571*. Washington, DC: US Government Printing Office.

Vaishampayan, S.M., Verma, R. and Bhowmik, A.K. (1987). Possible migration of teak defoliator, *Hyblaea puera* Cramer (Lepidoptera: Hyblaeidae) in relation to the movement of the south-west monsoon as indicated by light trap catches. *Indian Journal of Agricultural Sciences*, **57**, 41–6.

Valenciaga, N., Barrientos, A. and Mora, C. (1999). Performance of the beneficial entomofauna in *Leucaena leucocephala* (Lam) de Wit areas. *Cuban Journal of Agricultural Science*, **33**, 319–24.

Van Den Beldt, R.J. and Napompeth, B. (1992). Leucaena psyllid comes to Africa. *Agroforestry Today*, **4**, 11–12.

Varma, R.V. (1986). *Seasonal Incidence and Possible Control of Important Insect Pests in Plantations of Ailanthus triphysa, KFRI Research Report No. 39*. Peechi, India: Kerala Forest Research Institute.

Varma, R.V. (1991). *Spatial and Temporal Distribution of* Ailanthus *Pests,* Eligma narcissus *and* Atteva fabriciella, *KFRI Research Report No. 78*. Peechi, India: Kerala Forest Research Institute.

Varma, R.V. (1996). Impact of *Atteva fabriciella* on growth of *Ailanthus triphysa*. In *Impact of Diseases and Insect Pests in Tropical Forests*, ed. K.S.S. Nair, J.K. Sharma and R.V. Varma. Peechi, India: Kerala Forest Research Institute, pp. 311–16.

Varma, R.V. (2001). *Termite Control in Clonally Propagated Root Trainer Raised Planting Stock, KFRI Research Report No. 215*. Peechi, India: Kerala Forest Research Institute.

Varma, R.V. and Ali, M.I.M. (1986). *Bacillus firmus* as a new insect pathogen on a lepidopteran pest of *Ailanthus triphysa*. *Journal of Invertebrate Pathology*, **47**, 379–80.

Varma, R.V. and Nair, K.S.S. (1997). *Evaluation of Newer Termiticides Including Plant Products for Forest Plantation Establishment, KFRI Research Report No. 127*. Peechi, India: Kerala Forest Research Institute.

Varma, R.V., Mohanadas, K., Mathew, G. and Nair, K.S.S. (1989). Laboratory evaluation of some insecticides against *Pteroma plagiophleps* Hampson, a bagworm pest of *Albizia falcataria*. *Indian Journal of Plant Protection*, **17**, 89–90.

Varma, R.V., Balagopalan, M. and Nair, K.S.S. (1995). Field evaluation of neem cake as a possible protectant against subterranean termites attacking *Eucalyptus* seedlings. *Annals of Forestry*, **3**, 178–80.

Vasudev, V., Nagaraj, H. J., Nagabhushana and Hegde, S. N. (2001). Altitudinal and seasonal distribution of *Drosophila* species in the North Canara region of Western Ghats. *Entomon*, **26** (Special Issue), 326–31.

Vasudevan, K. G. (1971). *Working Plan for Nilambur Forest Division, 1967–8 to 1976–7.* Trivandrum, India: Government of Kerala.

Vazquez, A. M. C. (1980). Forest protection at the Forest Experiment Station 'El Tormento'. *Cienia Forestal*, **5**, 49–58.

Veenakumari, K. and Mohanraj, P. (1996). Folivorous insects damaging teak, *Tectona grandis* L. in the Andaman Islands, Bay of Bengal, Indian Ocean. *Journal of Entomological Research*, **20**, 177–8.

Velkov, V. V., Medvinsky, A. B., Sokolov, M. S. and Marchenko, A. I. (2005). Will transgenic plants adversely affect the environment? *Journal of Biosciences*, **30**, 515–48.

Verma, T. D. and Gaur, R. K. (1994). Bioecology of *Dioryctria abietella* Denis and Schiff: a pest of conifers in the north-western Himalaya. *Biotropica*, **7**, 12–17.

Villacarlos, L. and Wilding, N. (1994). Four new species of Entomophthorales infecting the leucaena psyllid *Heteropsylla cubana* in the Philippines. *Mycological Research*, **98**, 153–64.

Villacarlos, L. T., Paglinawan, R. W. and Robin, R. P. (1990). Factors affecting Leucaena psyllid populations in Leyte, Philippines. In *Leucaena Psyllid: Problems and Management*, ed. B. Napompeth and K. G. MacDicken. Bangkok: Winrock International/IDRC/NFTA/F-FRED, pp. 122–9.

Vir, S. and Parihar, D. R. (1993). Insect pests of *Acacia senegal* (L.) Willd. and *Acacia tortilis* (Forsk.) Hyne in the Thar Desert of India. *Myforest*, **29**, 105–11.

Waage, J. (1990). Exploration for biological control agents of leucaena psyllid in tropical America. In *Leucaena Psyllid: Problems and Management*, ed. B. Napompeth and K. G. MacDicken. Bangkok: Winrock International/IDRC/NFTA/F-FRED, pp. 144–52.

Wagner, M. R., Atuahene, S. K. N. and Cobbinah, J. R. (1991). *Forest Entomology in West Tropical Africa: Forest Insects of Ghana*. Dordrecht, The Netherlands: Kluwer Academic Publishers.

Wardell, D. A. (1987). Control of termites in nurseries and young plantations in Africa: established practices and alternative courses of action. *Commonwealth Forestry Review*, **66**, 77–89.

Wazihullah, A. K. M., Islam, S. S., Rahman, F. and Das, S. (1996). Reduced attack of keora (*Sonneratia apetala*) by stem borer in mixed-species plantations in coastal Bangladesh. *Journal of Tropical Forest Science*, **8**, 476–80.

Weisser, W. W. and Siemann, E. (eds.) (2004). *Insects and Ecosystem Function*. Berlin: Springer-Verlag.

Wheeler, M. R. (1986). Additions to the catalogue of the world's Drosophilidae. In *The Genetics and Biology of Drososphila, vol. 3e*, ed. M. Ashburner, H. L. Carson and J. N. Thompson. London: Academic Press, pp. 395–409.

Whitmore, T. C. (1984). *Tropical Rain Forests of the Far East*, 2nd edn. Oxford: Oxford University Press.

Whitmore, T. C. (1998). *An Introduction to Tropical Rain Forests*. Oxford: Clarendon Press.

Whitten, A. J. and Damanik, S. J. (1986). Mass defoliation of mangroves in Sumatra, Indonesia. *Biotropica*, **18**, 176.

Wilkinson, W. (1962). The principles of termite control in forestry. In *Proceedings of the 8th British Commonwealth Forestry Conference*. East Africa: Comonwealth Forestry Association, pp. 1–8.

Wilson, E. O. (1971). *The Insect Societies*. Cambridge: the Belknap Press of Harvard University Press.

Wint, G. R. W. (1983). Leaf damage in tropical rain forest canopies. In *Tropical Rain Forest: Ecology and Management*, ed. S. L. Sutton, T. S. Whitmore and A. C. Chadwick. Oxford: Blackwell Scientific Publications, pp. 229–39.

Wolda, H. (1979). Abundance and diversity of Homoptera in the canopy of a tropical forest. *Ecological Entomology*, **4**, 181–90.

Wolda, H. and Foster, R. (1978). *Zunacetha annulata* (Lepidoptera: Dioptidae), an outbreak insect in a neotropical forest. *Geo Eco Trop*, **2**, 443–54.

Wong, M., Wright, S. J., Hubbell, S. P. and Foster, R. B. (1990). The spatial pattern and reproductive consequences of outbreak defoliation in *Quararibea asterolepis*, a tropical tree. *Journal of Ecology*, **78**, 579–88.

World Health Organization. (1973). *The Use of Viruses for the Control of Insect Pests and Disease Vectors: report of a joint FAO/WHO meeting on insect viruses. WHO Technical Report Series No. 531*. Geneva: WHO.

Wu, S. H., Chen, C. C. and Wang, T. H. (1979). A preliminary study on the teak defoliator *Pyrausta machaeralis* Walker. *Acta Entomologica Sinica*, **22**, 156–63.

Wylie, F. R. (1992). *A comparison of insect pest problems in eucalypt plantations in Australia and in southern China*. Paper presented at 19th International Congress of Entomology. Beijing: International Congress of Entomology.

Wylie, F. R. and Floyd, R. B. (2002). The insect threat to eucalypt plantations in tropical areas of Australia and Asia. In *Pest Management in Tropical Forest Plantations*, ed. C. Hutacharern, B. Napompeth, G. Allard and F. R. Wylie. Bangkok: FAO Forestry Research Support Programme for Asia and the Pacific, pp. 11–17.

Wylie, F. R. and Peters, B. C. (1993). Insect pest problems of eucalypt plantations in Australia: 1. Queensland. *Australian Forestry*, **56**, 358–62.

Wylie, R., Floyd, R., Elliott, H. *et al.* (1998). Insect pests of tropical acacias: a new project in Southeast Asia and Northern Australia. In *Recent Developments in Acacia Planting*, ed. J. W. Turnbull, H. R. Crompton and K. Pinyopusarerk. Canberra: Australian Center for International Agricultural Research, pp. 234–9.

Yamazaki, S., Taketani, A., Fujita, K., Vasques, C. P. and Ikeda, T. (1990). Ecology of *Hypsipyla grandella* and its seasonal changes in population density in Peruvian Amazon forest. *Japan Agricultural Research Quarterly*, **24**, 149–55.

Yemane, A. (1990). Further considerations on pest management of grassland reforestation in the Philippines. In *Pests and Diseases of Forest Plantations*, ed. C. Hutacharern, K. G. MacDicken, M. H. Ivory and K. S. S. Nair. Bangkok: FAO Regional Office for Asia and the Pacific, pp. 214–18.

Yonghi, P. and Kuiguang, C. (1996). Forest insect pests in Yunnan Province, China and methods for their management. In *Impact of Diseases and Insect Pests in Tropical Forests*, ed. K. S. S. Nair, J. K. Sharma and R. V. Varma. Peechi, India: Kerala Forest Research Institute, pp. 384–8.

Zakaria, M. (1990). Diseases of forest plantation species in Peninsular Malaysia. In *Pests and Diseases of Forest Plantations*, ed. C. Hutacharern, K. G. MacDicken, M. H. Ivory and K. S. S. Nair. Bangkok: FAO Regional Office for Asia and the Pacific, pp. 94–9.

Zanuncio, J. C., Guedes, R. N. C., Cruz, A. P. D. and Moreira, A. M. (1992). Efficiency of aerial application of *Bacillus thuringiensis* and deltamethrin in the control of *Theirinteina arnobia* Stoll (Lepidoptera: Geometridae) in a eucalyptus grove in the Para region. *Acta Amazonica*, **22**, 485–92.

Zhu, J. H. (2002). Study on application of entomopathogenic nematodes to control *Odontotermes formosanus* Shiraki on *Eucalyptus*. *Journal of Fujian College of Forestry*, **22**, 366–70.

Zobel, B. J., Wyk, G., van and Stahl, P. (1987). *Growing Exotic Forests*. New York: Wiley.

Zulfiyah, A. (1998). Pest problems and threat in industrial plantation forests at PT Musi Hutan Persada, South Sumatra. In *Proceedings Workshop Permasalahan dan Strategi Pengelolaan Hama di Areal Hutan Tanaman*, ed. G. Suratmo, S. Hadi, E. A. Husaeni *et al.* Bogor, Indonesia: Fakultas Kehutanan dan Departemen Kehutanan, pp. 153–6.

Index

Abbreviations: Fig = Figure; sp = species; spp = several species; Tab = Table